PLANT WEALTH OF THE LOWER GANGA DELTA

(An Eco-Taxonomical Approach)

Kumudranjan Naskar

Volume I

1993
DAYA PUBLISHING HOUSE
DELHI - 110035

ISBN 81-7035-117-0 (Set)
ISBN 81-7035-118-9 (Vol. I)

Published by:
Anil Mittal for **Daya Publishing House,**
1123/74, Deva Ram Park, Tri Nagar,
Delhi - 110035. Ph: 7231826.

Laser Typeset by: ShiKi Associates, Ph: 5505295.

Printed at: **Raj Offset Printers,** Delhi.

Bound at: Book Wrap, Delhi-35.

PRINTED IN INDIA

Dedicated to the Sacred Memory of

My Late Parents

Ram Sadhan Naskar

and

Tulsi Rani Naskar

ACKNOWLEDGEMENTS

It is a great pleasure for me to acknowledge all the yeomen services rendered by the followings for completion of the work.. Words are too inadequate to express my feelings:

To Dr. D.N. Guha Bakshi, my respected teacher and Ex-Scientist, Botanical Survey of India, Howrah, for his keen interest and directing guidelines for my research works;

To Dr. Gour Gopal Maity, Lecturer, Kalyani University, Nadia for his sincere help both in and off the field studies, especially for solving some critical nomenclatural notes;

To my parents, my Boudi (Smt. Biva Naskar) and my loving wife Smt. Sipra Naskar for their constant help and encouragements;

To Dr. A.K. Mahapatra, Sri Rabindranath Mandal, Sri Ram Chandra Mandal of Minakhan, Sri Dinataran Sarkar of Masjidbati, Sri Jatish Das and Sri Bimal Kumar Halder of BSI for helping me during field collections;

To Sri Mahadev Naskar, Sri Anil Naskar, Sri Mantu Biswas of Botanical Survey of India, Howrah, Dr. Prantosh Roy, Lecturer, Agartala Women's College, Tripura, Sri A.K. Barui and Sri. D. Saha of CICFRI, Salt Lake for helping me in finding out proper identity of the specimens and other ways;

To Sri P.K. Ghosh, Sri P.K. Das Gupta, Sri Anup Biswas and Sri Rathindranath Mandal for preparing the sketches and photographs;

I am also grateful to the Directors, Central Inland Capture Fisheries Research Institute, Barrackpore and Dr. Ajoy Kumar Ghosh, Principal Scientist, CICFRI, Salt Lake, Calcutta, for their valuable suggestions and encouragements for this work.

Last but not least, I am grateful to Mr. Anil Mittal and others for Daya Publishing House, Delhi for quick publication of the book.

KUMUDRANJAN NASKAR
(Author)

FOREWORD

The State of West Bengal can be divided into two natural geographical divisions, the Himalayan and the Plain Divisions. The latter is a part of the vast land of Lower Ganga Plain, extending from the District-West Dinajpur in the north to the Sundarbans in the South. The flora of this portion is the interaction of intense bio-edaphic influences. Floristic study of this area started from the establishment of the Botanic Garden at Shibpur in 1787. Celebrated botanists and Phyto-geographers like Roxburgh, Buchanan Hamilton, Hooker, Kurz, Clarke and Prain *et al.* worked on the floristic composition and substaintially contributed much to our knowledge of botany of this region. Prain (1903a) observed, "The time for the preparation of a complete flora of the lower Provinces has not yet come; much special work is still called in many of the more outlaying districts". In this context, the taxonomic researches of the districts of Howrah, Hooghly, Nadia, Murshidabad, Malda, Midnapore and 24-Parganas are worth mentioning.

The first botanist to do significant botanical contribution on lower Ganga Plain was William Roxburgh. The famous *Hortus Bengalensis* and *Flora Indica* bear elequent testimony of the pioneer works. Two others contemporary botanists—William Carey and Buchanan Hamilton helped Roxburgh to collect plants from Lower Ganga delta including Sundarbans. The *Hortus Suburbanus Calcuttensis* by Voigt (1845) and J.W. Masters' *Calcutta Flora* (1839-1840) are other publications which deserve special attention. J.D. Hooker came to Calcutta in 1848 and collected extensively plants from Sundarbans and vicinity.

Kurz also collected plants from Lower Bengal. Clarke's works mainly centred round weed flora and he presented a beautiful description of plants of the Sundarbans after Heinig.

In the trail of publication of the *Flora of British India* (Ed. by J.D. Hooker, 1872-1897), Prain published *Bengal Plants* (1903a), *Flora of Sundribuns* (1903b) and the *Vegetation of the Districts of Hugli, Howrah and 24 Pergunnahs* (1905). The taxonomic researches were on the wane during the two world wars and the interest was revived after independence of India.

Apart from the critical evaluation of Sundarbans by Prain, few geographical, taxonomical, ecological studies on these regions were made by Chanda, Mukerjee, Mukherjee & Mukherjee, Naskar & Guha Bakshi. Naskar and Guha Bakshi's *'Mangrove Swamps of the Sundarbans: An Ecological Perspectives'* (1987) may be reckoned with as a very useful publication in this respect.

In view of the recent changes in the taxonomic categories and owing to the realignment of political boundaries, the State of West Bengal presently require an uptodate flora (Basak & Guha Bakshi, 1977).

The need for publication of the district flora in a concerted way from horticultural, botanical and genetical aspects was also stressed by the Committee for Survey, Conservation and Utilization of Resources of Indian Science Congress (Sethna, 1977).

It is evident that the period of exploration has practically come to an end to Lower Ganga plain. The present goal of floristic is to prepare accurate keys and descriptions of plants of this region and to provide explanations for the development and distribution of the flora and to suggest the strategies for their survival. This fact will ultimately form the basis for monographic works and other aspects of vegetational history (Basak & Guha Bakshi, l.c.).

The Plant Wealth of the Lower Ganga Delta—An Eco- Taxonomical Approach by Dr. Kumudranjan Naskar, ARS is a right endeavour in the right direction.

An ardent Plant Taxonomist, Dr. Naskar is a son of the soil; born and brought up in the District 24-Parganas and in the threshold of the Sundarbans, he developed a special aptitude about the plants of the region since his student days and is well conversant with each and every plant of the locality. In course of his service career also, he follows his quest for plant collections in a missionary zest and zeal, processed them and published a good number of publications—Mangrove Swamps of Sundarbans: An Ecological Perspectives (1987), Aquatic and Semi- Aquatic Plants of the Lower Ganga Delta (1990) and about a quarter century published papers on this locality.

The District 24-Parganas, one of the biggest districts of the pre-independent India (next to Vishakhapatnam of Andhra-Pradesh and Mymensingh of undivided Bengal) allure the Plant Taxonomists, Plant Geographers and others because of the Sundarbans. Situated in close proximity of Calcutta, the District enjoys an unenviable position from different aspects also. Naturally Dr. Naskar's choice of the District for his

Ph.D. dissertation was acclaimed by the botanists and he did his job admirably and ardously.

The District 24-Parganas (very recently partitioned as South and North 24-Parganas) is the lowermost portion of the West Bengal and Lower Gangetic Plain with an area of about 14,136 km^2 (about 15.70% of the total area of West Bengal of which Sundarbans is about 4267 km^2: about 30% of the total area of the District with a population of about 1,75,00,000).

In this work, Dr. Naskar dealt with 1175 Angiosperm species under 680 genera of 154 families. He has presented the physiography, history, physiognomy, vegetation and other relevant data meticulously and in a very precise manner.

It is my great pleasure in forwarding the book to the scholastic community as well as to common men, which I believe, will be very helpful. The Plant Taxonomists, Ecologists, Foresters, Botanists, Researchers, Pharmacognosits, Plant hunters, in, other words, all walks of life will be immensely benefitted by this gigantic and monumental treatise.

DWIJENDRA NARAYAN GUHA BAKSHI
Scientist- 'SE' (Retired)
Botanical Survey of India
Indian Botanic Garden, Howrah.
INDIA

66 Prince Baktiar Shah Road,
Calcutta-700033.
The 30th May, 1992.

PREFACE

Late Lamented pioneer botanist, Rev. Fr. Santapau (*J. Indian bot. Soc. 35:* 1-6.1956) pointed out that in India the paramount importance at the present moment is of intensive survey rather than extensive studies and strongly recommended for intensive search on the complete flora of our country. Many foreign elements have become part and parcel of our flora and have become completely naturalised.

The Botanical Survey of India in the *'Flora of India Project'* has evolved a comprehensive scheme for the study of the present day flora of the country. After the monumental work of J.D. Hooker and his associates (1872-1897) and Sir David Prain (1903a, 1903b and 1905) on Bengal plants, a reassessment of the Flora of Bengal has become an imperative necessity of our home state, West Bengal.

India gained independence in 1947 and in that year as well, Bengal was partitioned as West Bengal and East Pakistan (now Bangladesh). After that there was a tremendous exodus of population from the eastern wing of Bengal. As a result, there was a great socio-economic changes and the impact was so fierce that a ruthless denudation of the natural vegetation, especially in West Bengal. The virginity of the forests was also lost and there was a great change of upheaval in the vegetation.

The floristic studies of almost all the Districts of West Bengal have been completed barring a few; mention may be made of the following:

S.S.R. Bennet	1968	-	Howrah District Flora
S.R. Das	1968	-	Nadia District Flora
S.Sen	1972	-	Hooghly District Flora
M.N. Sanyal	1973	-	Bankura District Flora
D.N. Guha Bakshi	1974	-	Murshidabad District Flora
R.K. Basak	1976	-	Birbhum District Flora
B.P. Guha	1976	-	Birbhum District Flora
P.K. Bhattacharya	1976	-	Burdwan District Flora
K.C. Malick	1978	-	Purulia District Flora
J.K. Sikdar	1980	-	Jalpaiguri District Flora
S.K. Majhi	1980	-	Midnapur District Flora
K.R. Naskar	1981	-	24 Parganas District Flora
B.C. Mandal	1992	-	Maldah District Flora

The work of the above-mentioned districts have been completed.

A few other workers are also working for the remaining districts of West Bengal. A good number of publications have come out in respect of floristic works of the District of 24-Parganas and vicinities after independence of India, which are as follows:

Culshaw (1950), Majumdar, R.B. (1956,1962), Anonymous (1957), Chatterjee, D.(1958), Ghosh, R.B. (1960), Mukerjee, S.K. (1960,1961), Patil, R.P. (1961), Dutta & Mitra (1961), Choudhuri, Mallick & Sen (1962), Banerjee, A.K. (1964), Ganguly, D.K. (1965), Majumdar, N.C. (1965), Datta & Majumdar (1966), Dutta & Ganguly (1967), Datta & Bhattacharya (1969), Chanda & Mukherjee (1969), Bose & Mukherjee (1969), Guha Bakshi & Sen (1971), Mukherjee, B.B. (1972), Mitra, R.L. (1973), Dasgupta, R. (1973), Caratini, Chanda & Thanikaimoni (1974), Mukherjee, A.K. (1975), Rai Choudhury & Tribedi (1976), Blasco, F. (1975), Chanda, S. (1977), Mahapatra & Chakraverty (1977), Mahapatra, A.K. (1978), Mukherjee & Mukherjee (1978), Paria & Sahoo (1978,1979), Naskar, Guha Bakshi and Maity (1979), Mitra & Banerjee (1979), Maity & Guha Bakshi (1981) Naskar & Guha Bakshi (1987), Naskar (1990) and few others.

An X-rayed investigation will reveal that these works are sporadic and not systematic and no useful purpose would serve in respect of floristic elements of the District 24 Parganas, especially where the world famous mangrove forest of the Sundarbans are laying with its virginity still now untrodden in many remote corners.

In order to wipe out the lacuna and an urge from inner instinct to contribute something towards the national developments, this Lower Ganga Delta has been chosen by me for my studies floristically including the Sundarbans, both extensive and intensive means.

A flashback:

From the very dawn of floristic studies of this 'Lower Gangetic Delta', especially the Sundarbans, a galaxy of worldfamed botanists and taxonomists contributed their mites for giving a true pen-picture of the vegetation of these regions. Quite a good number of elites participated in the studies from time to time. The Celebrates are:

Dr. William Roxburgh, the father of Indian Botany (1776), William Jones, founder of the Asiatic Society of Bengal (1784), Col. Robert Kyd, founder of the then Royal Botanic Garden

(1787), now Indian Botanic Garden, Sibpore, Howrah; John Flemming, Dr. Wallich, (1820,1824), Buchanan Hamilton (1825), William Carey (1832), J.O. Voight (1845), W. Griffith (1836), Thomas Thomson (1856), Inc Clelland, J.D. Hooker (1907), T. Anderson (1865), C.B. Clarke (1871), J.W. Masters (1840), W. Schlich, Ellis, R.L. Heining (1893), G.A. Gammie, Lace, Sir David Prain (1903a, 1903b,1903c,1905) and a few other names to conjure with.

About 90 years have elapsed and the situation changed enormously. India got independence and in the trail of partition of Bengal, along with great change of socio-economic life of the state, the impact was perceptibly felt in the geo-climatic as well as eco-floristic pattern of these regions. About 2/3 parts of the Sundarbans are now under the jurisdiction of the Bangladesh (formerly of East Pakisthan). Naturally an overhaul study and reassessment have become a natural corollary. The data will be useful from agricultural point of view and plant based industries. Many foreign elements have been naturalised to the present day situations and a large number of monographic works were published, which resulted better understanding of the taxa and have evolvéd a series of nomenclatural changes.

Along with the description of the taxa the field data, *i.e.*, physiognomy, habitat, distribution, phenology, frequency or availability, economic importances and the vernacular names and other ecological data are very much helpful and of paramount importance for their utilization and the future studies.

In course of this study, I have described in detail for 1175 Angiosperm plant species under 154 Families along with their economic aspects. Some new species from the district as well as from West Bengal were also added. The mangrove flora of the Sundarbans are also described critically.

All these information and searches contained in the treatise are authentic. This Flora is the first complete work of its kind and hence it is a valuable contribution to the knowledge of the Indian Botany.

Dated: Baruipur, W.B.
The 30th June, 1992.

Kumudranjan Naskar
(Author)

CONTENTS

LEGENDS TO THE MAPS, HISTOGRAMS AND GRAPHS

LEGENDS TO THE TABLES AND LISTS

LEGENDS TO THE SKETCHES OF THE PLANTS

LEGENDS OF PHOTOGRAPHS

Plate-1. ***Rhizophora mucronata*** Lamk. of the Sundarbans, India.

a. Long view of stilt roots on the river bank

b. Close view of stilt roots and wood cutter in the river flats

c. Close view of bow like stilt roots

Plate-2. ***Rhizophora mucronata*** Lamk. of the Sundarbans, India.

a. Viviparous germinated fruits and hypocotyles

b. Hypocotyles and different stages of viviparous germinations

c. Bunch of viviparous germinated fruits and hypocotyles

Plate-3. ***Rhizophora mucronata*** Lamk. and ***Kandelia candel*** (L.) Druce

a. Flowering twigs of *Kandelia candel* (L.) Druce (Right hand) and *Rhizophora mucronata* Lamk. (Left hand)

b. Viviparous germinated fruits of *Rhizophora mucronata* Lamk. (Right hand) and *Kandelia* candel (L.) Druce (Left hand)

Plate-4. ***Rhizophora apiculata*** Blume

a. Flowering twigs

b. Pendulous stilt roots in hands

c. Viviparous germinated seedlings

Plate-5. ***Bruguiera gymnorrhiza*** (L.) Lamk.

a. Single tree on the river bank

b. Twig with viviparous germinated hypocotyle

c. Flowering twig

d. Flowering twig (close view) and viviparous germinated hypocotyle

Plate-6. Three common species of ***Bruguiera*** Lamk. (Rhizophoraceae) from the Sundarbans, India

a. *Bruguiera gymnorrhiza* (L.) Lamk. & *Bruguiera sexangula* (Lour.) Poir-twigs with mature flower and persistent calyx

b. & c. *Bruguiera cylindrica* (L.) Blume—twigs with viviparous germinated hypocotyles

d. *Bruguiera sexangula* (Lour.) Poir.—Single mature flower with persistent calyx on a twig

Plate-7. ***Kandelia candel*** (L.) Druce

a. Flowering twig

b. Branches with viviparous germinated hypocotyle/seedling

Plate-8. ***Ceriops decandra*** (Griff.) Ding Hou

a. Plants with exposed base roots/flanges or broom like
b. Tidal inundated plants
c. Twigs with viviparous germinated fruits and hypocotyles

Plate-9. ***Three mangrove flora***

a. *Ceriops tagal* (Perr.) Robinson—branches with viviparous germinated hypocotyles
b. *Lumnitzera racemosa* Willd.— plant in the inter- tidal zone
c. *Heritiera littoralis* Dryand.— well developed buttress

Plate-10. ***Heritiera fomes*** Buch. Ham.

a. Flowering twig (close view)
b. Flowering twig (long view)
c. Twig with fruits
d. Seed coat in hand and potted seedling

Plate-11. *Mangrove Flora of the Sundarbans, India*

a. Blooming twig of *Bruguiera gymnorrhiza* (L.) Lamk.
b. Herbarium of *Heritiera fomes* Buch. Ham.
c. Germinated seeds of *Heritiera littoralis* Dryand (bigger size in left side) and seeds & potted seedling of *Heritiera fomes* Buch. Ham. (smaller size in right side)

Plate-12. ***Avicennia alba*** Blume of the Sundarbans, India.

a. Fruiting branch with leaves
b. Fruiting bunch
c. Tidal inundated *Avicennia* forest

Plate-13. Different Species of ***Avicennia*** L.

a. Different stages of seedling development of *Avicennia officinalis* L.
b. Flowering twigs of *Avicennia marina* (Forssk.) Vierh.
c. Flowering twigs of *Avicennia officinalis* L.
d. Pneumatophores and stilt roots of *Avicennia officinalis* L.

Plate-14. ***Avicennia officinalis*** L.

a. River bank dense forest tree
b. Plant with dense pneumatophores (aerial roots)

Plate-15. River flat, Succession of ***Porteresia coarctata*** (Roxb.) Takeoka (front grass) and ***Avicennia*** spp. (back trees).

Plate-16. ***Sonneratia caseolaris*** (L.) Engler of the Sundarbans, India.

a. Close view of flowers and fruits in hands
b. Long view of the tree and fruiting twig

Plate-17. ***Sonneratia apetala*** Buch. Ham.

a. Full grown tree with drooping leaves—back of *Porteresia* sp.
b. Bunch of fruits and leaves in hands
c. Tree trunk and aerial roots

Plate-18. ***Pneumatophores of the true mangrove species of the Sundarbans, India.***

a. *Xylocarpus mekongensis* Pierre – woody pneumatophores on the muddy soil
b. *Sonneratia apetala* Buch. Ham. – exposed horizontal spreading roots and creet aerial pneumatophores on the river banks

Plate-19. ***Aegiceras corniculatum*** (L.) Blanco of the Sundarbans, India.

a. Twig with umbel inflorescence, flower buds and mature fruits
b. Twig with imature fruits
c. Flowering branches (long view)
d. Flowering and fruiting branches (close view)

Plate-20. ***Aegialitis rotundifolia*** Roxb.—a true mangroves flora of the Sundarbans, India.

a. Twig with mature fruits
b. Herbarium specimen of *Aegialitis rotundifolia* Roxb.
c. Different stages of seeds germination and seedling development

Plate-21. ***Excoecaria agallocha*** L. act as soil binder in the mangrove forests of the Sundarbans, India.

a. Exposed roots acted as the soil binder
b. Flowering (male) twig—with axillary catkin inflorescence
c. Gall formation on trunk base

Plate-22. ***Nypa fruticans*** (Thunb.) Wurmb. -an endangered palm of the Sundarbans, India.

a. Long view of palm (*Nypa* sp.) thicket

b. Close view of two plants

Plate-23. ***Phoenix paludosa*** Roxb. -dense thickets in the mangrove forests of the Sundarbans.

a. Dense thickets on the river bank on consolidated soil

b. Tidal inundated thicket with fruit bunches

Plate-24. Two common and most economic palm flora (Arecaceae) in the mangrove forests of the Sundarbans, India.

a. ***Phoenix paludosa*** Roxb. -fruiting heads

b. ***Nypa fruticans*** (Thunb.) Wurb. -Planted seedling in nursery at Bagna, Sundarbans, India

Plate-25. Different views of the mangrove forests of the Lower Ganga Delta, Sundarbans of India.

a. Watch Tower at Jhilla Block inside the thickets of *Phoenix paludosa* Roxb.

b. ***Excoecaria agallocha*** L. -trees, on the coastal area and sea waves at Henry's Island, Sundarbans

c. Narrow tidal creek during low-tide, exposed bank at Pirkhali Block with mangrove flora

Plate-26. ***Xylocarpus*** Species.

a. Full grown tree of *Xylocarpus mekongensis* Pierre with woody pneumatophores

b. *Xylocarpus granatum* Koeing with pendulous fruits

c. Single pendulous fruit of *Xylocarpus granatum* Koeing

d. Flowering twigs of *Xylocarpus granatum* Pierre

e. *Xylocarpus granatum* Pierre - twig with inflorescence

f. Single fruit and single seed of *Xylocarpus granatum* Pierre

Plate-27. ***Xylocarpus mekongensis*** Pierre - fruiting twigs, with mature fruit.

Plate-28. ***Acanthus ilicifolius*** L. - a halophytic back mangal of the Sundarbans, India.

a. Dense thickets on the river bank

b. Single flowering twig

Plate-29. Mangrove Associates of the Sundarbans, India.

a. *Manilkara hexandra* (Roxb.) Arn.- a non-mangrove of the Sundarbans

b. *Macrosolen cochinchinensis* Van. - epiphytic on mangrove tree

Plate-30. Different types of plants

a. ***Clerodendrone inerme*** Gaertn. and ***Crinum defixum*** Ker-Gawl-river bank mangrove associates

b. ***Psidium guajava*** L. - an introduced fruit plant with fruits

c. ***Derris trifoliata*** Lour. - flowering twig, a back mangal climber

Plate-31. Both indigenous weed and exotic fruit plant.

a. ***Scirpus articulatus*** L. - sedge of the marshy places

b. ***Spondius pinnata*** (Linn. f.) Kury. - fruiting branch

Plate-32. Flora of the Lower Ganga Delta.

a. ***Eichhornia crassipes*** (Mart.) Solms. - an aquatic weed on the freshwater bhery

b. ***Brownlowia lanceolata*** Benth. - a true mangrove shrub of the Sundarbans

Plate-33. Flora of the Flower Ganga Delta.

a. ***Hydrilla verticillata*** Royle - a freshwater aquatic weed, harvested for carp feeding

b. ***Aegialilis rotundifolia*** Roxb. - a true mangrove shrub, natural regenerated seedling

Plate-34. Beach flora, on the Sandy soil of Sagar Island - Sundarbans, India.

a. ***Ipomoea pes-caprae*** (L.) Sweet - with flowers - act as sand binder

b. ***Pandanus tectorius*** Soland ex Parkinson - with fruits

Plate-35. ***Sarcolobus carinatus*** Wall. - a mangrove associated climber

a. Close view with leaves and flowers

b. Twining twig with inflorescence

1

INTRODUCTION

Since the publication of Roxburgh's *Hortus Bengalensis* (1814), Voigt's *Hortus Suburbanus Calcuttensis* (1845), Prain's *Bengal Plants* (1903a), *Flora of Sundribuns* (1903b) and the *Vegetation of the District of Hugli-Houra and 24-Pergunahas* (1905), no systematic attempts have been made until very recently to study the floristic composition of West Bengal.

Culshaw (1950) and a few subsequent workers listed some sporadic collections of plants from different parts of Bengal, but these are inadequate to form a true and representative collections of West Bengal.

Bennet (1980) and Guha Bakshi (1984) have published the Howrah and Murshidabad districts flora, respectively. Naskar & Guha Bakshi (1987) and Naskar (1990) have also published the Mangrove Flora of the Sundarbans and Aquatic & Semi-Aquatic Plants of the Lower Ganga Delta, respectively; while the other districts Flora have not yet published. There is no detailed or thorough floristic accounts of the Lower Ganga Delta, particularly the Flora of the 24-Parganas District in West Bengal; though Botanists were interested mostly about the Flora of the Mangrove Vegetation of the Sundarbans, since the year 1776. Majumdar (1956, 1962) worked out extensively about the *'Grasses'* and *'Weed Flora'* of the District 24-Parganas. Majumdar (1965), Datta & Majumdar (1966) and Dutta & Ganguly (1967) published the *'Flora of Calcutta and its neighbourhood.'*

Besides them, Patil (1961), Ganguly (1965), Mukherjee (1975), Naskar & Guha Bakshi (1979, 1980, 1983a, 1983b, 1983c, 1984, 1985a, 1985b, 1986, 1987a, 1987b, 1988, 1989, 1990) have also contributed some valuable publications partly on the Flora of the Lower Ganga Delta.

The District 24-Parganas (S. & N.) is situated in the southern most part of the Lower Ganga delta and the world renowned mangrove forests of the Sundarbans have spread over 4266.6 km^2. This mangrove vegetation covers all the southern portion of the District 24-Parganas and in its back the vast wetlands are situated. The Aquatic and Semi-aquatic flora of the Lower Ganga wetlands and the Mangrove flora were identified and studied by Naskar (1990) and Naskar & Guha Bakshi (1987); but the other floras were not published so far.

In the recent time, extensive and intensive explorations were undertaken for the utilization of natural resources; which supply information on the distributions, physiognomy, phenology and the degree of abundance etc. This practice is still going on in search of these natural resources, particularly for the plant resources; there are monumental works for the perfect utilization in the daily life of human beings.

In this way, the new enumeration, new search and revision of the Flora are going on for particular uses of medicinal, timber yielding and other economic plants and to get a detailed vegetational map. For these proper uses, it is most essential to have full knowledge for their occurrence, frequency and phenology.

We have to drive the full benefit from our plants by studying their useful properties; we need precise information on their exact distributions and frequency. Detailed floristic studies will greatly help in the economic development of an over populated country, like ours.

The present work is an attempt to study the Angiosperm Flora of the Lower Ganga Delta (24-Parganas) in West Bengal with special reference to the Mangrove Flora of the world renowned Sundarbans.

Extensive collections were undertaken covering all the seasons and from all parts of the Lower Ganga Delta in 24-Parganas. All these collected flora were identified and studied mainly in the Central National Herbarium (CAL), Indian Botanic Garden, Shibpore, Howrah.

Identifications of plants, latest nomenclatures with basionyms and synonyms, all relevent citations, popular and vernacular names, descriptions of the families, genera and exomorphic characters of the species with their physiognomy, field notes, usefulness etc. have studied in detail, together with artificial keys in all these categories.

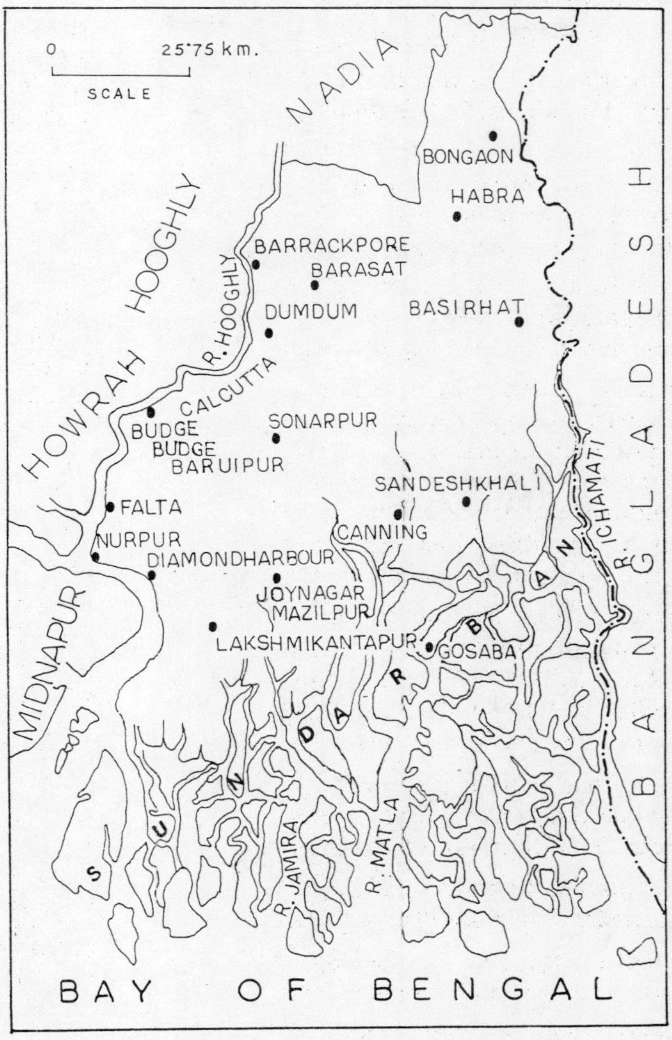

Map 1: 24 Parganas (Undivided) District of West Bengal, India

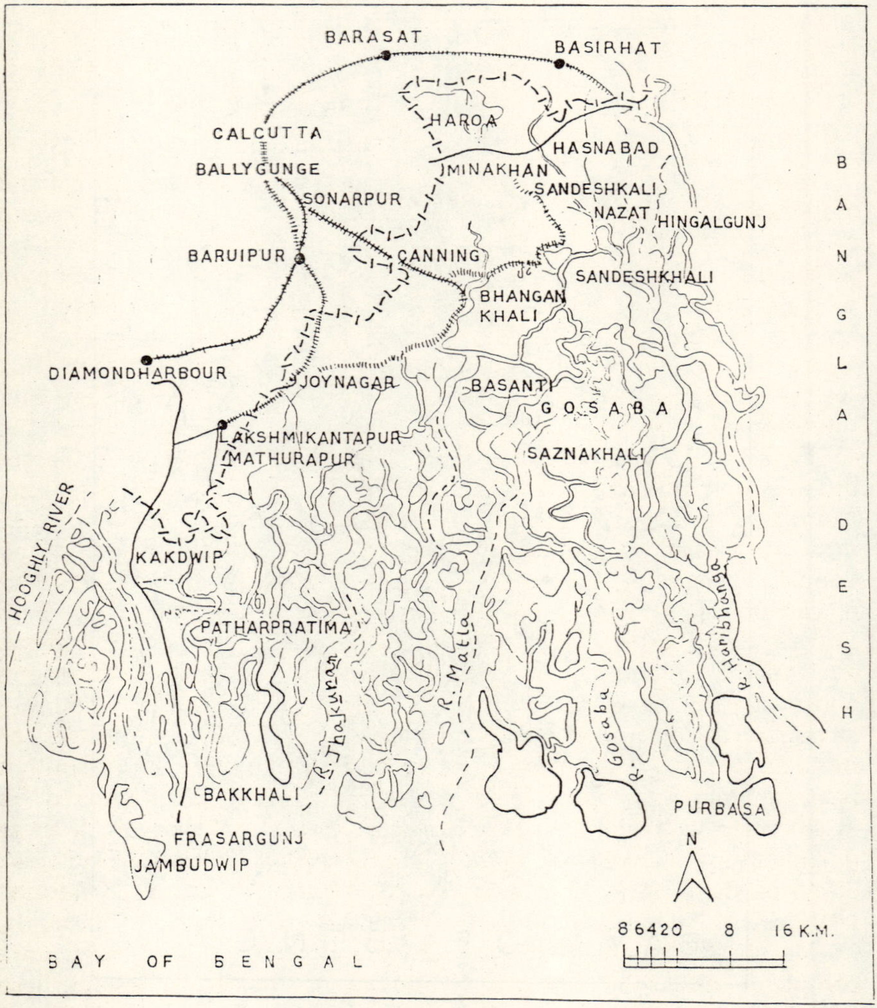

Map 2: The Sundarbans, 24 Parganas of West Bengal, India

2

DISTRICT 24-PARGANAS, WEST BENGAL

The District 24 Parganas is the lowermost portion of West Bengal and along with Nadia, Hooghly, Howrah and Midnapore, is popularly known as Lower Bengal or Lower Gangetic Plain. This District 24 Parganas lies within the limits of the Gangetic Delta—its physical nature is very common with the deltaic land, as it is flat, little raised (6m-9m above the mean sea level) and bordering by several river, channels and embankments; the soil here is mostly alluvial. The north-western half of the District is somewhat well raised, as it is the older formation in comparison to south-eastern half; the latter is low laying and younger formation towards the seaboard of the south. Prain (1905) described the entire District 24 Parganas into two halves by an imaginary straight line (Dampier & Hodges Line, 1831), running from the Mud-point (adjacent to Sagar-Islands) towards the north-east, upto Basirhat and designated the upper half as 'Rice- swamp of Central Bengal'. The striking feature of these two halves is that the north-west half is carrying few rivers or canals, *viz.*, Ichhamati, a narrow flow of Bidhyadhari; besides this, Tally's Nullah, Adiganga or Buriganga, Surjyapur Canal, Piyali canal, Beliaghata canal, Nonakhal, Harua Gang and vast area of low marshy lands of Salt-Lake area; all these lands are sluggish. The south-east half is completely distinct from the north-west, as it is a net-work of tidal channels, rivers, creeks and in the extreme south it is a conglomeration of islands (about 56 in numbers), separated by interconnected tidal estuaries, tributaries, canals and creeks. All these exhibit a typical environment–the 'Hooghly-Matla-Raimangal Estuaries'. Some of these areas, are Sagar Islands, Khasimara Islands, Ghorama Islands, Lohachara Islands, Bed Ford Mark, Jammu Dwip, Henry's Islands, Luthian Islands; and several other larger and smaller islands, which are completely under the mangrove forests of the Sundarbans. Besides these islands (dwip), there are several other landmasses, fully encircled by broad and dangerous rivers; such types of areas are Kakdwip, Namkhana, Frazergunj, Bakkhali, Satjelia, Mollakhali, Sandeshkhali, Kachuberia, Ranigarah (Nafargunj), Basanti, Golabari, Pathankhali, Gosaba, Bhangan

Khali, Rangabelia, Mollarchak, Surjyaberia, Heroabhanga, Beltali, Jharkhali, Kaikhali, Birajmani, Mahishmari, Rendakhali, Gurguria-lat, Selfeli, Goranbose, Marigunj, Dongajora, Madhabpur, Daharani etc. Almost all these areas are locally famous as *'lat'*. Some of these *lat* areas are also known by plot numbers, *viz.*, A - Plot, K-Plot, J-Plot etc. These deltaic areas were cleared or reclaimed and inhabitated only 150 to 200 years back.

All these land masses are broken up by swamps, intersected by hundreds of river channels; these rivers are mostly carrying saline sea water and mangrove vegetation are the most characteristic features of these areas. The world famous Sundarbans is situated in the south-east portion of the district, covering an area about 9630 kms^2; and all these *'lat'* areas are recently reclaimed and cultivated, which were under the dense and dangerous forests before 150-200 years. At present, only a limited areas of about 4266.60 sq.kms. (Naskar, 1981) are under the reserve forest of the Sundarbans (Map-2).

24-Parganas District Boundaries

The District 24 Parganas (undivided) lies within 21° 31′ N to 23° 15′ N (North latitude) and 88°10′E-89°51′E (East longitude). It was the largest district of West Bengal (15.70% of the total area of West Bengal) and occupied an area of 14,136.0 kms^2 (of which Sundarbans forest area is about 4266.6 $kms.^2$, about 30% of the total district area) with a population of 10726751 and the population density was 758/kms^2 (1981 Census).

The district looks like an irregular parallelogram in shape and is bounded by the followings (Map-1).

(a) Nadia District of the Indian territory and Jessore of Bangladesh on the north;

(b) Khulna of Bangladesh on the east; the river Ichhamati demarcates the eastern part and the international border of the District 24 Parganas from Bangladesh, for a long distance from Bongon to Hasnabad and the river Harinbari demarcates rest of the eastern border;

(c) Bay of Bengal on the south; and

(d) The River Hooghly on the west; this river Hooghly separates the District 24 Parganas from the Districts Hooghly, Howrah and Midnapur of West Bengal.

List-1 (Census Report, 1981)

		In West Bengal	*In 24 Parganas*
1.	Population	54580647	10726751
2.	Population density (/Kms^2.)	615	758
3.	Rural area Sq.Kms. Total population	– –	13311 5479162
4.	Urban area Sq. Kms. Total population	– –	484.9 2970320
5.	Decennial population growth during 1961 to 1971 (%)	26.87	34.53
6.	Literacy rate (%)	33.20	33.45
7.	Workers (%)	30.2	25.87
8.	Cultivators (%)	29.76	24.33
9.	Agricultural labourers (%)	25.23	27.65
10.	Total Number of Villages	41392	3875
11.	Total Number of Towns	–	233

Calcutta has been segregated out during 1903 as a separate district from 24 Parganas, but its area is only 104 kms^2 and is fully encircled by the District 24 Parganas. Even now, it is very difficult to draw a clear demarcation line of Calcutta as it is so intimately connected with the District 24 Parganas both South and North.

The District 24 Parganas derived its name from the number of *'parganas'* or fiscal-divisions, *viz,:* (i) Amirpur, (ii) Akbarpur, (iii) Azimabad, (iv) Baila, (v) Baridhati, (vi) Basandhari, (vii) Calcutta, (viii) Dokhin Sagar, (ix) Garh, (x) Hathiagarh, (xi) Ikhtiarpur, (xii) Kharijuri, (xiii) Khaspur, (xiv) Maidanmal (or Mednimall), (xv) Magura, (xvi) Manur, (xvii) Mayda, (xviii) Munragacha, (xix) Paikan, (xx) Pechakull, (xxi) Satal (xxii) Shahnagar, (xxiii) Shahpur, and (xxiv) Uttar Pargana; which were ceded to the East India Company in the year 1757 by Mir Jafar, the then Nawab Nazim of Bengal.

The District headquarter is in Alipore, Calcutta-700027, a Southern-Suburb of Calcutta.

There were six sub-divisions under the District 24-Parganas; *e.g.*, (i) Sadar Sub-Division at Alipur, (ii) Southern Sub-Division at Diamond Harbour, (iii) Eastern Sub-Division at Basirhat, (iv) North-Eastern Sub-Division at Bangaon, (v) Northern Sub-Division at Barrackpore and the (vi) Central Sub- Division at Barasat.

It was also proposed to divide the District into two parts, *viz.*, i) the District 24 Parganas for the northern half and the District Sundarbans for the southern part; it was supposed to serve better administration and better care for the Sundarbans as an individual district. But during 1988 the District 24 Parganas has been divided into two separate Districts, *i.e.*, 24 Parganas South and 24 Parganas North; Alipore is the District Head Quarter of the 24 Parganas South and in this part the two Sub-Divisions, *viz.*, Alipore and Diamond Harbour are included. The District Head Quarter of the 24 Parganas North is at Barasat and the remaining four Sub-Divisions, *viz.*, Barasat, Basirhat, Barrackpore and Bongaon remain within 24 Parganas North District.

Numerous towns, with busy jute and cotton mills stretch along the whole length of the river Hooghly from Garden Reach northwards; but away from its bank, the population is almost entirely rural and are devoted to agriculture. The 24 Parganas is the most distinctively urban district in West Bengal. The population clusters most thickly in the Suburbs of Calcutta, where the density is larger than the average population.

The net sown area is 46% of the total area; 75% of the gross cropped area is under paddy cultivation and the relative yield index value of rice crop is 102 (Rep. National Agr. Com. 1976). The most striking thing is that the rice crop in 24 Parganas is unirrigated but its yield is close to all India average.

Three principal crops of rice are, (a) *Sown time*:- February to May, *Boro*; (b) *Sown time*: -June to September, *Aman* and (c) *Sown time*:- October to January, *Aus*. But of the 3 crops, *Aman* is by far the largest and popular crop. Irrigated high yielding paddy cultivation is now becoming popular in the L. Ganga Delta region.

Jute is the second crop in the District and it is cultivated mainly in the low laying paddy fields, during the period June to September.

Cotton, wheat, maize, soyabean, kusum, different types of pulses, fodder crops and many others are the subsidiary crops of the District.

List-2
The Principal Areas of the District Under Crop Production

Crop		Area	Crop		Area
1. Paddy	–	637000 ha	6. Mesta	–	6000 ha
2. Jute	–	47000 ha	7. Gram	–	120000 ha
3. Wheat	–	9000 ha	8. Other pulses	–	53000 ha
4. Maize	–	4000 ha	9. Sugar cane	–	2000 ha
5. Oil seeds	–	9000 ha	10. Misc.	–	60000 ha

Modified after (Rep. National Agr. Com. 1976).

History of the District 24 Parganas

The Raghuvansa, the Mahabharata and some of the Puranas referred to this portion of the Ganga delta.

The great Chinese Pilgrim, Hiuen Tsiang, in the seventh century described this District as low laying country, bordering on the sea and rich in crops, flowers and fruits, as a part of Sumatra; the climate of this tract was soft and the inhabitants were of small stature, black complexion and hardy.

Bipradas (1495) described the voyage of the Great Merchant, Chand Saudagar from Burdwan to sea and mentioned several river side villages from Bhatpara to Baruipur. He proceeded downward from Adiganga.

Calcutta (Kalikata) was mentioned in the 'Ain-i-Akbari' (1582) and there was some hints for 24 Parganas and Nadia Districts under the revenue division of Satgaon.

At that time, the actual rulers of the Sundarbans were the Great Hindu Chief Pratapaditya and the other Barah Bhuiyas; they were really the Vassals of the Emperor, but enjoyed full independence in the south and south-east of the Ganga Delta.

Fernandez, a Priest (1598) gave an account and mentioned a route layed through the Sundarbans.

Babu Radha Kumud Mookerjee wrote in his "History of Indian Shipping" that *'Chandecon'* (the Sagar Islands), the Capital of Pratapaditya, was one of the main centre of the Hindu naval activities during Man Singh's viceroyalty; number of men- of-war were always found to be ready for battle and in a seaworthy condition at that naval station. It is said that the Portuguese admiral of Pratapaditya, named Rodda, defeated the Mughal forces in a battle fought at the confluence of the Adiganga and Bidyadhari; at this spot there were a group of temples, with an old image of Kali (Babu Pratap Chandra Ghosh, Bengal Past & Present, April 1908 vol.2.29).

In the latter half of the sixteenth century, the Portuguese shared trade near the town Hooghly. The English first came for trade and tried to establish their products in India and collected the raw materials from India. There had for sometime been friction between the English and the Mughals, owing to the extensions of the latter and their interference with the Company's trade and hedges and Company's Agents in Bengal.

Charnock gradually established his trade and violated the order of Nawab. In September, 1687, Charnock came to Sutanuti and made a permanent settlement. He opened negotiations to build a factory, and meanwhile hutted the remnant of his troops on the high eastern bank of the Hooghly. In February, 1690 the Emperor granted the English a new licence for trade and the viceroy of Bengal issued a permit authorizing them to return to Bengal, while, in return for the payment of Rs.3,000/-; Charnock received a guarantee that their trade should be free and exempt from local exactions.

After receiving His Highness of guarantee on Sunday, 24th August, 1690 at the noon time, the weather-beaten band anchored for the third time in the long pool of Calcutta. With a poor guarantee of only 30 soldiers which was thence forward without a break to grow into the British Capital in India. Charnock tried heart and soul and faced several miserable incidents and was able to built Calcutta. The name Calcutta was identified with *'golgotha'*, the place of skulls. Within a decade this lonely river bank became a busy mart with 1200 English inhabitants; during this period about 460 were buried and the miseries were only for feverstriken. Sutanuti was chosen for commercial and strategic reasons. The Hooghly river tapped the trade of the Ganga-valley and Sutanuti was situated at the highest point at which the river was navigable for sea-going vessels.

The Portuguese also occupied 'Tardaha' on Bidyadhari at the spot where Tolly's Nullah now joins the river Hoogh'y. The Dutch had established a factory for salting pork at Barrackpore, before the end of the seventeenth century and latter maintained a station at Falta for sea going vessels.

In the early part of the eighteenth century, the ill-fated, short lived Ostend Company appeared among the ranks of the competitors for the trade of Bengal. Before sailing for Europe, the Captain applied to the Nawab Murshid Kuli Khan for some lands for erecting a factory there, in case a company was incorporated. His request was readily granted by the Nawab as he could introduce the rivals to the English. He, therefore, assigned the Flemish the village of Bankibazar, on the bank of the Hooghly, about 5 km north of Barrackpore.

The people were exposed not to the sudden fury of cyclones, but to the constant raids by piratical free booters and slave dealers. In a cyclone on the 30th September, 1737, Calcutta looked like a place that had been bombarded by an enemy. Moreover, the raids of the Portuguese and the Magh pirates had long been the terror of the people in this riverain tracts.

According to the East Indian Chronicle (1758) in February, 1717 the Maghs carried off from this part of Bengal more than 1800 adult men, women and the children; they were taken to Arakan, where the king chosed the artisans, about one fourth of the numbers, to be his slaves; and the rest were sold into slavery at prices varying from Rs.20/--70/- only. Slavery was also common in Calcutta, as we may realize from the remarks of Sir William Jones (1785), Chief Judge of the Supreme Court.

On the 20th June, 1756 Calcutta was taken by Nawab Siraj-ul-Daula and the tragedy of the black hole followed. At Falta, the English remained for six months, until a strong force had been collected, and they felt themselves fit for the battle. The English received a reinforcement of 230 men, mostly Europeans, under major Kilpatrick, who had been sent from Madras before the news of the fall of Calcutta and waited for arrival of vessals from England, Bombay and elsewhere, that were bound for Calcutta. On the 2nd January, 1757 Calcutta was recovered by the English and by the end of this month the Nawab advanced to retake Calcutta with an army of more than 40,000 men.

The English had only 1350 Europeans and 800 sepoys, and with this small force Clive attacked the Nawab, who had taken up a position between the Salt Lake and the Maratha ditch. The action took place in one foggy morning near the present Circular Road and the English lost their way and Clive saw a wild buffalo and imagined how close the jungle was to the infant city. The Salt Lake occupied a much larger area and came up to a few km of Calcutta. The battle was not, therefore, as decisive as it would otherwise have been, but the reverse was sufficient for Siraj-ul-Daula, who became alarmed for his own safety and the communication of his army. Accordingly, he signed a treaty, by which he restored to the English the goods and villages that he had seized and promised compensation for what had been damaged or destroyed; recognized all their former privileges and permitted them to establish a mint and erect fortification.

On the 20th December, 1757 Mir Jafar, the new Nawab of Bengal, made an assignment to the East India Company of the Zamindari or land-holder's right over a tract of the country, known as the Zamindari of Calcutta or as the 24 Parganas Zamindari, from the number of parganas included in it

The East India Company received only the Zamindari rights, the right to collect rents from the cultivators.

The exigencies of space forbid any but a brief mention of the work of reclamation and development carried out in the Sundarbans towards the close of the eighteenth century by Tilman Henckell, who was the Judge and Magistrate of Jessore in 1781 and had the jurisdiction over the Sundarbans tract, to the south. In the Sundarbans, Henckell inaugurated a system of reclamation which after many vicissitude had converted large areas of forest into fertile rice field. His object was to introduce a body of peasant proprietors, holding directly under the Government, and with this purpose he granted about 150 leases in 1785. At the same time,he established three stations in the heart of the Sundarbans, in order to assist in their development by providing markets for the sale of produce and the supply of the boatmen plying along the waterways. One of these markets was situated at Henckellganj (now pronounced as Hingalgunj) at the junction of the Jamuna and the Kalindi river.

The next most notable event in the history of the district was Sepoys Mutiny, which broke out among the sepoyes at Barrackpore during the Burmese war in 1857.

The subsequent history for this District was not peaceful though uneventful development and progress were in industries and in agricultures. Its communications have been improved and the population were also increased fast. There the *'Jungle'* (forests) has been driven back with the reclamation of new agricultural colony.

Originally, this District was constituted with 444 estates, but in 1816 some estates were transferred from Burdwan and the number increased to 564; but these estates were retransferred to Burdwan in 1862. For a considerable time, the District was divided into two parts, *viz.*, the Alipore and the Barasat divisions; each of which formed a separate magistrate District. In the year 1861, the District was divided into the following eight sub-divisions, namely (i) Diamond Harbour, (ii) Baruipur, (iii) Alipore, (iv) Dum Dum, (v) Barrackpore, (vi) Barasat, (vii) Basirhat, and (viii) Satkhira. The Satkhira sub-division was detached from 24 Parganas in 1882 and made part of the Khulna District of Bangladesh; and the Baruipur Sub-division was abolished in 1883, Dum Dum and Barrackpore in 1893. The Barrackpore Sub-division was reconstitued in the year 1904 from the portions of the Alipur (Sadar) Sub-division and Barasat Sub-division separated out from East Pakistan (Bangladesh) and added with the District 24 Parganas. Moreover, the suburb Calcutta was also designated as a separate District from 24 Parganas for better management and control but it is intimately mixed with the District 24 Parganas.

In 1905, the Sundarbans Act (Bengal Act I of 1905) was passed by which Regulation IX of 1816 repealed the Office of the Commissioner in the Sundarbans was abolished; therefore the functions were transferred to the Collectors.

Geology of the 24 Parganas District

The soil of the District is mainly composed of deltaic alluvium clay and is formed from the finest clay or silt, which are carried down by the great river Ganga, Bhagirathi and Hooghly in the West and the Padma and even the Brahmaputra on the east. So, it is also known as the tributaries of the 'Ganga-Padma river system'. Sand is found only in the upper surface, where water current or the tidal amplitude ceased the easy settlement of the clay particles. Further, this lower delta region was formed from the clay particles of Chhotanagpur of Bihar; which were carried down by the rivers Damodar, Rupnarayan and Kanshai; not only the clay and sand were carried by these rivers but also the stones were carried in the south-west part of the District and ultimately helped to build the landmass in the present forms. During boring of tubewells for drinking water in the Sagar Islands, it is clearly estimated that, even in the depth of 300 to 400m, the presence of the finest clay and yellow sand of the Chhotanagpur types

(24 Parganas Dist. Gaz. 1913). A boring at the Calcutta reached a depth of 150 m without any signs of either a rocky bottom or marine beds, the fragments of freshwater shells were also found as low as 125m below the surface. The upper strata containing sand passing into stiff blue clay and numerous shells were also common. Calcutta Review (1859) mentioned the presence of *'Sundri'* (*Heritiera* sp.) tree in Port Canning, which is now restricted only in the dense forests of the Sundarbans. In an excavation from a depth of 3m at Canning Town, a forest of the large *'Sundri'* *(Heritiera fomes)* trees were found *in-situ* and no portions were noticed above the ground level. From this, it can easily be assumed that the past Sundarbans are now under the water or surface, and there are many other documentary evidences in support of the theory. Ghosh (1941) drew some more interesting facts about the geology of Calcutta from the botanical point of view. Chanda and Mukherjee (1969), Mukherjee (1972) and Vishnu Mittre and Gupta (1972) claimed on the basis of palynological studies and established that the area where the present metropolis of Calcutta is situated, was the mangrove forests or swamp about 3000 years back; and these deltaic region was estimated to form or came up from the sea only 7,000 to 8,000 years back.

Excavation from all the corners of the district, clearly demonstrate the glorious past and threw some light on its origin and age, as well as its nature of formation. It appears that these deltaic tracts stretching from Khulna to Calcutta must at sometime have undergone a subsidence.

On 'The Gangetic Delta' Calcutta Review (1857) quoted, "If we consider the unsubstaintial nature of the foundation of the Sundarbans, which, at a depth of only 40m from the surface, consists of a bed of semi-fluid mud, 10-15m in thickness and then imagine the terrific convulsions that have at different periods shaken the delta to its deepest foundations. We must not be surprised to find that the liquid mass, unable to support and superincumbent weight, has repeatedly bulged seawater, reducing the level of the delta submerging whole forest together with their fauna and flora".

Colonel Gastrell considered that it was caused suddenly by great earthquake—the enormous wave rolled in the form of Bay over the Sundarbans and destroying all its path.

Dr. Thomas Oldham (1870) wrote in an article published in the 'Proceedings of the Asiatic Society of Bengal', "I suppose no one will hesitate to acknowledge that the whole of the country, including Sundarbans proper, laying between the Hooghly on the west and the Meghna on the east is only the delta caused by the deposition of the debris carried down by the river Ganges and Brahmaputra and their tributaries. It is also equally well known that in such flats the streams are constantly altering their courses, eating away on one bank and depositiong on the other bank until the channel in which they formerly followed becomes chocked up, and the water is compelled to seek another course. It is also certain that, in this peculiar delta, the general course of the main waters of the Ganges has gradually tracked from the west towards the east, until, of late years, the larger body of waters of the Ganges has united with those of the Brahmapurta, and have together proceeded to the sea as the Meghna."

The Soil

The soils of the District 24 Parganas are of new alluvium, known as 'khader' and are of four types 1) *Matial* or clay soil, 2) *Dorasa* or loam soil, 3) *Balia* or sandy soil and 4) *Nona* or saline soil. The colour of the soil is generally grey or light brown. The matial is further sub-divided into three types namely, (i) *Kala matial*, (ii) *Ranga matial* and (iii) *Jhajhara matial*. *Kala matial* is a stiff black clay of great natural fertiiity, on which all kinds of crops can be grown; *ranga matial* is of reddish colour—it cracks in the dry season and sinks into holes in the rains. There is no marked differentiation into various horizons and the profile is often devoid of stratification. It is well suited for winter rice and on higher levels can be used for the cultivation of jute and other bhadoi crops; *Jhajhara matial*, which is inferior to the other two types, blackish in colour but is easy to plough even when dry. The *dorasa* soil is a mixture of clay and sand and it is used for bhadoi and rabi crops, also suitable for sugarcane. The dihi lands or the elevated lands surrounding the village sites come under this category. While *baila* is a common name for all soils in which the proportion of sand exceeds that of the clay. But *nona* is a wet saline soil, which in ordinary years does not dry up enough to permit cultivation. The soil in the Sundarbans tracts may be divided into the following four classes, namely *matial, dorasa, dhops* or *chura* and *dhal*. The alluvium soil are generally deficient in nitrogen (N_2) and humus, and occasionally in phosphorus; potash is usually adequate. The lime content shows a great variation. The rainfall is heavy and the rivers are sluggish, the soil along the river banks contain an abundance of fine silt, and those away from river are rich in clay.

River Systems of the District 24-Parganas

The rivers of the District are mainly spread in the southern and south-eastern portion. The slope of the District is negligible and is from north to south. For that, the water current is not so meagre and their channels show a tendency of choking by gradual silting, along with the heavy growth of aquatic vegetation. The river-bed of Piali rises up by silting and now it is converted into paddy field and some narrow canal (nala) only remained as the evidence of its past existence. According to the Barendra Research Society's Atlas (1924-1933), the Adi-Ganga was passing from the side of Kalighat, Baruipur, Joynagar, Mazilpur, Bishnupur, Kashinagar, Kakdwip, Sikarpur, Chemagari Canal, Marichkhali canal, Rudranagar, Bishnupur, Golokpur, Narendrapur, Narayniabad, Magra Canal and Gangasagar canal. In that period, the river Baratala has not any flow. The Sagar Island was separated afterwards from the main landmass of Kakdwip by the Baratala river; and the main stream of the Adi-Ganga was vanished from Sagar Island.

Major Tally (1785), an English Merchant, connected the Adi-Ganga with Bidhyadhari for the betterment of his business of tiles (tally). The name Tallygunge was derived after his name; the Tally nullah is also flowing today but the main stream of Adi-Ganga has choked and it is now transformed into rice-fields, ponds or jhills. On the side way of the Kulpi Road from Garia to Joynagar, this flow is existed which is known as Adi-Ganga. The vegetation of this tract is dense, as the land is highly fertile. From this river bed at Boral, Rajpur, Harinavi, Baruipur, Joynagar the evidence of several ancient civilization of 2000 years back has been dug out. The river Bhagirathi has also changed its flow at that time and it has flowed from the Adi-Ganga bed to the present position through Sibpur, Uluberia, Budge Budge etc.

The rivers of the District mainly carry the saline water; and the mangrove species are the main characteristic flora on the banks of those rivers. These rivers are of two types i) One group is originated from the river Hooghly and Mathabhanga, to the north of the district; and ii) others are originated as the tributaries of the Bay of Bengal. These rivers cannot spread over the land; these are only serving to receive local drainage and carry it off into the tidal water. The salinity of the western region is more than that of the eastern part and forest types also vary a wide range. The Bay of Bengal is the chief source of saline water and the river Saptamukhi, Thakuran and others are carrying brackishwater; only the river Hooghly is carrying fresh water from upstream. The fresh water of the eastern part is discharged through Bidya, Gosaba and the Harinbhanga river. The water level reaches upto 4-5 m above mean sea level during high tides and a considerable area of the land mass become exposed during low tides.

Some of the principal rivers of the district are Hooghly, Bidyadhari, Matla, Ichhamati, Karatal, Gosaba, Piali and many others of the southern tributaries of the Sundarbans.

Hooghly River:- Hooghly, the lowermost portion of the Holy Ganga, is the western boundary of the District 24 Parganas. From the left bank of this river the Bhangarkhal is connected and flows south-east from the Salt Lake (Calcutta), and ultimately joins with Sakha Bidhyadhari, near Kulti (Ghusighata). Tally nullah is another side canal of the eastern bank of the Hoogly river and passes south-ward of Calcutta. The tidal water passes through this canal, but the nala is blocked within a limited distance. The river Hooghly changes its position to west and then south and south- east; ultimately it bifurcates on the northern limit of the Sagar Islands. The eastern channel is called the Baratala river or Muri-Ganga and western one ultimately meet with Haldi River and Rasulpur river of Midnapur. At the junction of the Hooghly and Haldi River the Mud-point or the Goramara, Khasimara and Lohachora Islands are situated. From Nurpur downwards the width of the river Hooghly attend about 5 km or more and looks as the gulf of the sea. Some of the other small streams of the Hooghly are the Diamond Harbour canal, Kholakhali Creek, Falta-khal, Nila-khal and Harua khal.

During the rainy season, the Hoogly carries enormous volume of fresh water; the Hooghly is also a tidal river, and the tides running up very strongly; it is navigable by medium to small liners and vessels and is the gateway of the foreign trade of Bengal.

Bidhyadhari: The Bidyadhari is a tidal river with a very circuitous course. It begins in the Sundarbans near Kalinagar and passes through Malancha-Minakhan-Ghusighata, Ranigachi and Haroa, where it is known as Haroa Gang; then bends westwards and ultimately joins the Nona Khal. Previously, it has joined Tally's Nullah and Beliaghata Canal; but this nullah or the canal has silted up and there remains only a little sign of the previous flow. The stream of Shyambazar-khal passes through Bhangar khal and meets the Bidyadhari near Haroa and one storm-water and sewage connection of Calcutta, started from Tapsia passes through Baman Ghata of Salt Lake and also meet at Ghusighata with the river Sakha Bidyadhari.

Piali: It leaved Bidyadhari about 14 kms north to the Bamanghata and flowed southward then southwest and one stream joined with Matla river. But this stream is heavily silted up and now transformed into the low cultivated lands and a narrow canal is only existed to a little extent near Uttar Bhag Pumping station then Dhosha-Amtala and Herobhanga; and ultimately discharge to the Matla river near Rendho khali. The Surjyapur khal also meets with the Piali near Dhosha. Now this river is converted to storage water by a cross barrage at Kella.

Jamuna or Ichhamati: The Jamuna enters in the District from Nadia and flows south-east to Tibi, where it is joined by the river Ichhamati. From there, it flows in south-east direction through Bongaon, Baduria, Basirhat, Taki, Hasnadad and from Debhata to Basantapur. This part is usually known as Ichhamati. From Basantapur, it flows through Sundarbans to Khulna of Bangladesh and ultimately meets Raimangal estuary.

Kalindi: Kalindi branches off from the Jamuna at Basantapur and forms the eastern boundary of the District, down to the sea, where it merges in the Raimangal estuary. Kalagachia river is formed by the combination of several rivers of the Sundarbans or water channels in the Sandeshkhali thana near Chuchkhali and flows in south-east direction. It is connected with the Matla river and with the river Atharabanka.

Matla River: Previously, the river was connected with Piali and Bidyadhari at Canning, but recently this flow is completely suspended, as a result Matla river gradually silted up. So that navigation was heavily hampered during low tide. Once the Matla river was so strong that navigation was very troublesome. At present, main stream of the Matla river passes south-west and is divided into two arms near Purandar. The main stream passes through Kultali-Garanbose and then the reserve forests of Sundarbans and ultimately discharges to the Bay with several branches. The other stream passes through Basanti, Pathankhali, Surjyaberia, Masjidbati and meets with Bidhya river and ultimately join the Bay in Sundarbans proper. From Pathankhali, one stream passes to the north-west direction and meets with Kalagachia river of Raimangal estuary. This stream is locally known as Basanti Khal or Kalagachia River.

From the Matla river, there are several branches passing in both the sides, one of which is Kuria Bhanga river and others are less familiar. People of Sundarbans use this river Matla as a transport media to Canning town using country boat, motor launch or small steamer.

Gosaba River: It passes north from Gosaba and ultimately meets with Raimangal near Chuchkhali of Sandeshkhali thana, and this is the connector of a number of rivers of the eastern part of the District.

Other rivers: The southern portion of the District is traversed by a large number of rivers and canals, both of these are tidal and originated from the Bay of Bengal. It forms a complete net-work system by these rivers of both huge and narrow streams. All of them gradually flow south-ward and meet the Bay of Bengal. And this landmasses with the networks of tidal forests fall under the world famous Sundarbans mangrove vegetation, the site for Royal Bengal Tiger, deers, wild pigs, marsh crocodiles etc.

Some of the important streams are (i) Baratala River or Channel Creek - an offshoot of the Hoogly river, (ii) Saptamukhi river, (iii) Thakuran and Jamir river (iv) Kaikalmari - an offshoot of Matla river, (v) Gosaba river, (vi) Hariabhanga river etc.

Khals or Nullas: A number of 'Khals' or 'Nulla' and canals are spread througout the district, these are both man-made and natural sources, between the innumerable streams of the larger rivers and estuaries. All those channels serve to draw the water from one block to another and ultimately the irrigation system largely depends on those channels. The rainwater remain or collected in these canals and is drained off to the rivers and estuaries which ultimately save the District from overflowing and also help in normal and natural agricultural system.

There are some sluice gates constructed between the junction of the canals and rivers or esturaries, which can regulate the water from back flow during high tide; some of these important sluice gates are Uttarbhag, Dosha, Diamond Harbour, Ghusighata etc.

Marshes: There are smaller and larger marshes spread over throughout the district. Their presence are the natural results of the configutation of the country. The river banks have silted up till the highest levels, from which the ground gradually slopes downwards forming a depressed tract between each set of two rivers. The depressed portion so constituted are natural basins, in which water stand and from which it has no exit. These are:- (i) *Dhapa or salt water lake* was a tract like this and lately it is slowly filled up with silt of the canals of Bidyadhari and also the waste deposit and rubbish sewage sludge of Calcutta Corporation. But the Salt Lake is now raised up with the silt from the Ganga and is going to be a beautiful planned city in the name, Bidhan Nagar. The South-East portion of this Salt Lake is still remain in its past form and is largely used for fish culture. A bird sanctury was proposed in this area and (2) *Kulgachi Bils* - situated between the Bidyadhari and Katakhal in the Deganga, Malatipur and Haroa area. This area is now converted to agricultural fields and fisheries and some portion remain fallow as wetlands and reed jungles.

Other such tract is at *Bariti-Bil*, between Balinda and Basirhat. Bayara-Bil is situated on the east of Jamuna and is covered with reed jungles. Garia math was the example of such a bil, which was covered with reed jungle and recently some portion is filled up with some dugout ponds and canals and reclaimed by the refugees or for the inhabitance of the city. Brace Bridge and the surrounding areas were under this low laying tract; but several industries are gradually settled up here after reclaming and silting up by rubbish dust. There were marshy lands in the Dum Dum area, which are now reclaimed by refugees but some smallest tracts are now also exist here and there throughout the District as marshes or low lands, with heavy growth of reeds, *Eichhornia* sp. and some other aquatic weeds.

The main Estuaries

The chief estuaries or arms of the Bay of Bengal from west to east are the Hooghly, Baratala or Channel Creek, Saptamukhi, Jamira or Thakuran, Matla, Goasuba, Raimangal or Hariabhanga Harinbari or Herobhanga.

Some Important Islands

These numerous estuaries separated the main land masses into several smaller and larger Islands of which the main are (i) Sagar Islands with Khasimara, Ghoramara, Loha-Chora and Bedford marks, situated in between the Hooghly and Baratala rivers and again cut up into many smaller Islands by criss-cross streams, (2) Frazergunj-west of the Saptamukhi, on this Island Bakkhali is situated; which is sea-beach and recently attracts the interest of the tourists. Henry's creek separates the Henry's Islands from Bakkhali, (3) Lothian Island, is in the month of the Saptamukhi, (4) Bulcherry (Balchari) Island is situated in between the Jamira and Matla, (5) Halliday Island is in the Matla, (6) Dalhousie Island, between Matla and Goasuba, and (7) Bangaduni Island at the month of Goasuba. Recently another Island, Jammu Dwip is formed on the South-

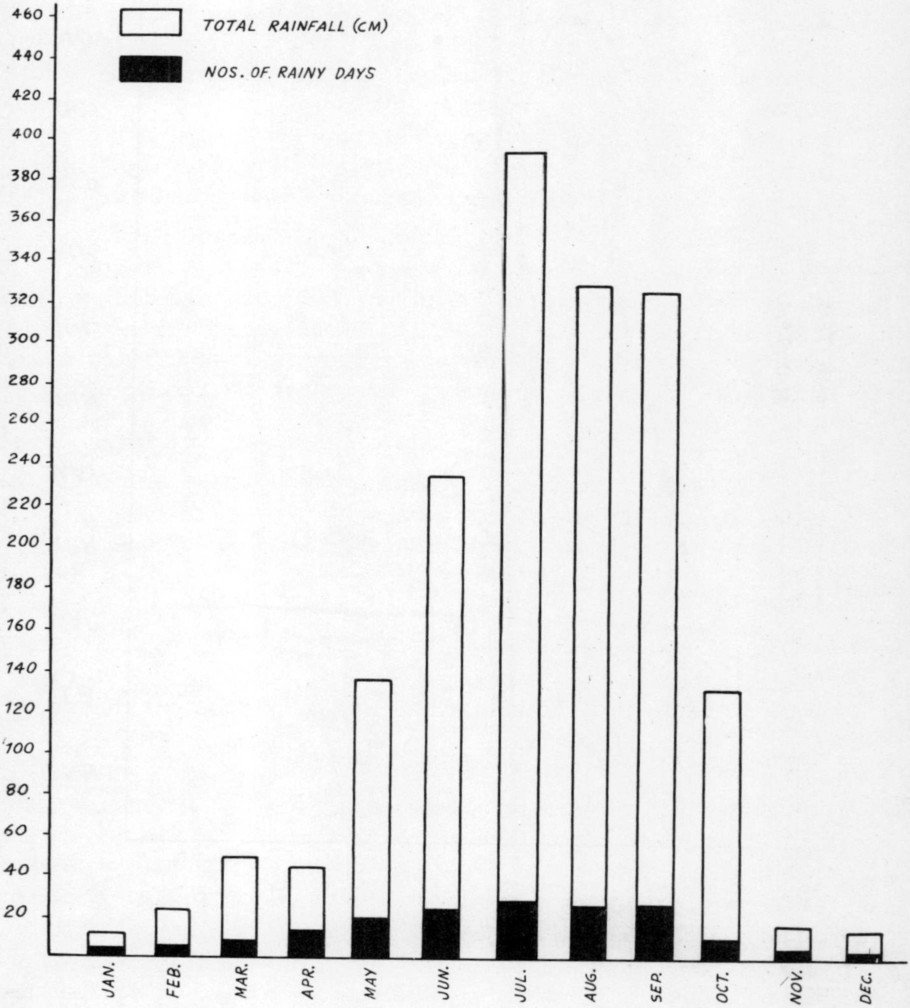

Hist. 1: Rainfall and Rainydays Historgram (24 Parganas, Alipore)

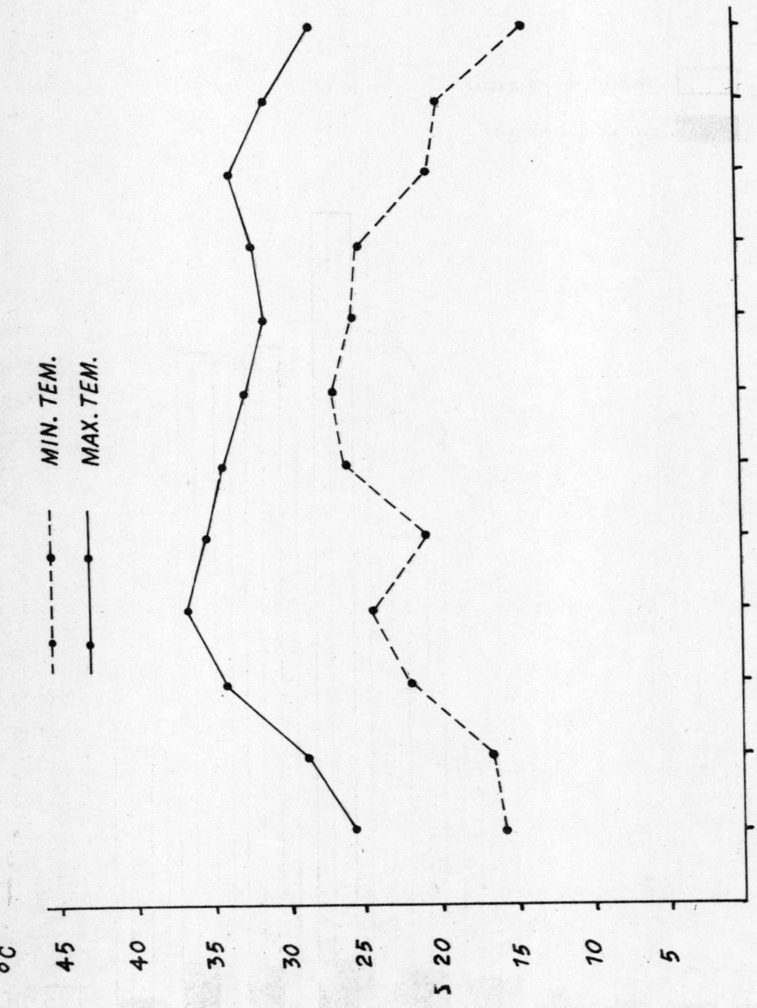

Graph-1. Maximum and Minimum Temperature (24 Parganas, Alipore)

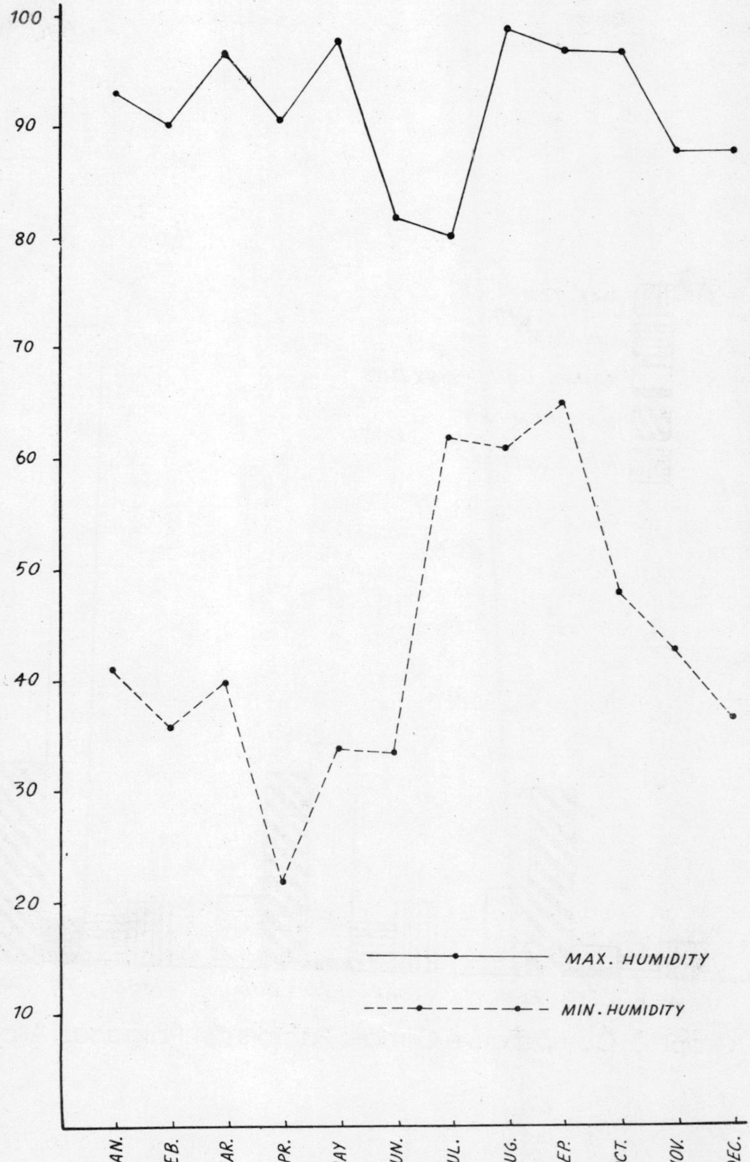

Graph-2. Relative Humidity (24-Parganas, Alipore)

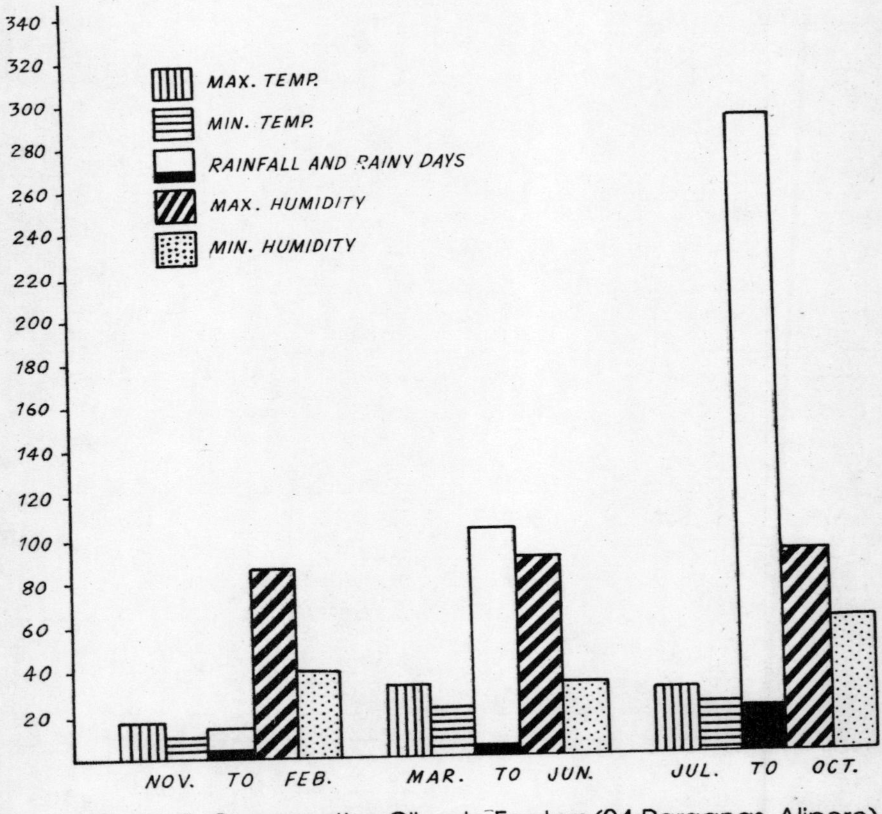

Graph-3. Comparative Climate Factors (24-Parganas, Alipore)

East corner of Sagar Islands. In the year, 1979 a new Island Purbasha was noticed on the exterme south-east of the district.

All these islands are traversed by number of rivers and as those are sea facing, land are full of sand and salt; the characteristic vegetation are the mangrove type; in some cases the plant of xerophytic habitat, as because the area is physiogically dry for plants.

In the Sundarbans, there spread dense jungle (forests) of mangrove trees and shrubs along the banks of such tidal rivers, *khals* or canals. These *khals* known as *suti-khals* or creeks, which are closely shut in by jungle and the tide flows ins and outs and leaves a bank of mud which is as much as 3 to 5m above from the sea-level. Some of the lands are lower and are liable to be submerged and immense embankments are indispensable to its reclamation.

Climate of the District

The most remarkable features are that during this period of study (1976 to 1980), this tract along with the other southern districts of West Bengal faced the record flood-due to rain, in 1978, record drought, with minimum rain, in 1979 as well as the record temperature rises up to 42°C in Summer, 1979.

Normally, the winter season continuing from the middle of November to the middle of February, and at that time the sky is almost free from cloud and rainfall is only 17.1 mm (Table.IV). The mean maximum temperature falls from 31.2°C in the month of November to 26.1°C in the month of December and January (Table-II), and the humidity also falls down in the first stage of winter and again rises and occasional low-laying fogs appear on the first morning, which dissipates with the rising of the sun, (Table-III). In the month of February, the temperature begins to rise and the mean maximum for this month is 29.8°C. From the middle of February, the southern winds (*mousumi-wind*) becomes more frequent and this period from 15th February to 15th March is actually the transition period between the winter and summer. During this period, occasional thunder storms accompained by rain (hail type) is the characteristic features. With the advancing of the year, the temperature gradually goes up and the mean maximum temperature is 34.1°C in the month of April and after that it declines by about 0.5°C in May; at that time night

Table I

Place of Observation: Alipore: Index No.: 42-807

Rainfall (0.1mm) (1974 to 1980)

Latitude-22° 32′ N & Longitude 88° 20′ E

T.R. = Total Rain; R.D. = No. of Rainy Days;

Months	1974	1975	1976		1977		1978		1979		1980		Remarks
	T.R.	T.R.	R.D.	T.R.	R.D.	T.R.	R.D.	T.R.	R.D.	T.R.	T.R.	Average	
January	0.0	9.4	--	0.0	4	6.1	1	Tr	2	34.9	7.3	11.2	Rainfall very rare
February	0.0	38.2	1	3.3	3	48.7	7	9.0	4	2.4	69.2	24.4	Rainfall rare
March	152.6	1.4	4	9.4	2	Tr	9	57.9	4	13.9	105.6	48.7	Rainfall with thunderstorm
April	12.6	12.4	7	104.7	19	47.5	10	55.4	8	62.3	2.2	42.4	Rainfall occasional
May	145.6	114.5	17	202.7	18	164.9	14	154.0	7	59.1	115.7	136.6	Rainfall occasional
June	132.5	207.9	18	139.7	22	478.7	20	204.2	16	179.9	268.4	230.2	Rainy season started
July	369.6	385.6	27	449.6	28	469.3	26	459.2	25	397.5	241.7	396.1	Rainy season
August	363.9	342.3	27	355.6	27	333.2	28	262.9	24	270.2	361.0	327.0	Rainy season
September	279.7	215.7	25	185.5	21	244.0	23	*944.7	25	155.1	253.0	325.4	*Flooded, Rainy season
October	301.9	223.6	7	12.1	8	75.4	11	142.3	6	73.4	82.3	130.1	Rainy season declined
November	0.0	44.4	4	4.3	5	27.0	2	40.5	0	0.0	0.0	16.6	Rainfall rare
December	0.0	0.0	-	0.0	3	57.5	1	3.9	3	47.3	0.0	15.5	Rainfall very rare
Total	1758.4	1595.4	137	1466.9	160	1952.3	152	2334.0	124	1296.0	1526.4	1704.2	
Average/ Month	146.5	133.0	11.4	122.2	13.3	162.7	12.7	194.5	10.3	108.0	127.2	143 (days)	

Table II

Place of Observation: Alipore: Index No.: 42-807

Temperature (°C) - (1976 to 1978)

Latitude- 22° 32′ N & Longitude - 88° 20′E

Months	1976				1977				1978				Remarks
	Maximum	Minimum	Average Max.	Average Min.	Maximum	Minimum	Average Max.	Average Min.	Maximum	Minimum	Average Max.	Average Min.	
January	29.9	10.0	25.1	15.8	29.9	9.1	24.8	16.3	30.1	10.3	26.0	14.8	Winter
Feb.	35.0	14.2	28.7	16.9	33.9	9.7	29.6	16.6	33.3	15.9	29.8	17.2	Winter
March	38.0	13.9	34.0	22.4	38.4	18.3	33.2	22.8	38.1	19.2	35.6	23.1	Summer started
April	39.7	20.2	36.3	24.6	38.3	21.1	33.9	29.9	39.3	22.2	34.1	22.6	Summer
May	39.8	21.1	35.1	21.9	37.9	20.6	33.9	24.4	39.8	23.2	33.9	22.4	Summer
June	39.6	22.5	34.3	26.1	37.6	22.5	33.5	26.7	38.6	23.1	34.9	27.1	Summer
July	35.4	24.4	32.9	26.9	33.9	22.6	31.8	26.9	36.3	23.0	32.3	23.1	Monsoon started
August	34.4	23.2	31.7	25.6	35.5	24.6	32.0	26.1	34.6	23.9	31.6	24.8	Monsoon
Sept.	33.8	24.2	32.2	24.9	35.1	22.9	31.4	24.8	33.8	22.7	30.9	24.6	Monsoon
Oct.	35.5	20.3	33.5	20.7	34.4	21.1	32.3	23.5	33.6	23.2	32.6	22.8	Monsoon declined
Nov.	33.8	17.8	31.2	20.2	33.5	17.4	29.3	20.2	32.2	14.9	31.2	20.7	Winter started
Dec.	31.5	10.9	27.7	14.0	31.0	11.9	27.0	13.5	29.1	11.1	26.1	14.0	Winter
Total	426.4	222.7	382.7	260.0	419.2	221.8	372.3	265.7	418.8	232.7	379.0	257.2	
Average/ Month	35.5	18.5	31.9	21.7	34.9	18.5	31.0	22.1	34.9	19.4	31.6	21.4	

Table III

Place of Observation: Alipore: Index No.: 42-807

Relative Humidity-(%) (1976-1978)

Latitude - 22° 32′ N & Longitude - 88° 20′ E

Months	1976				1977				1978				Remarks
	Aver. 8-30	Aver. 17-30	Max.	Min.	Aver. 8-30	Aver. 17-30	Max.	Min.	Aver. 8-30	Aver. 17-30	Max.	Min.	
January	080	055	098	047	081	051	096	031	079	053	086	046	Low humid
February	065	047	096	021	078	048	088	025	068	049	087	061	Low humid
March	071	044	095	028	076	053	098	026	078	054	099	066	Low humid
April	062	051	083	019	075	054	091	017	079	046	098	029	Low humid
May	072	067	096	021	086	062	097	026	072	041	100	056	Med. humid
June	076	071	097	023	091	064	090	027	095	076	062	051	Med. humid
July	082	085	098	069	092	071	095	060	090	073	048	057	Max. humid
August	084	082	100	070	090	076	096	057	094	071	100	056	Max. humid
September	080	079	095	061	094	073	096	066	089	086	100	067	Max. humid
October	069	061	093	047	091	069	098	053	091	085	100	074	Max. humid
November	070	067	089	041	090	051	082	046	072	041	092	042	Low humid
December	042	052	087	045	073	046	091	054	063	050	086	032	Low humid
Total	853	761	1127	492	1017	718	1117	488	970	726	1106	637	
Average/ Month	071	063	094	041	082	060	093	041	081	060	092	053	

Aver. = Average; Max. = Maximum; Min. = Minimum.

Table IV

Comparative Chart of Climates in Three Main Seasons

	Winter Nov. to Feb.	Summer Mar. to June	Monsoon July to Oct.	Annual
A. *Temperature (°C) Average 1976-1978*				
1. Maximum temperature	28.0	34.4	32.0	31.5
2. Maximum temperature	16.7	24.0	24.0	22.1
B. *Rainfall (0.01mm) (1974-1978)*				1704.2
1. Average rainfall/month	17.1	114.5	294.6	
2. Average rainy days/month	2.5	6	21	
C. *Relative Humidty (%) From 1976-1978*				
1. Average humidity at 8.30 AM	071	078	087	
2. Average humidity at 17.30 PM	051	058	076	
3. Maximum humidity	089	092	097	
4. Minimum humidity	041	033	061	

All these Meteorological data are taken from – The Regional Meteorological Station, Alipore, Calcutta, West Bengal.

temperature increases slowly, and the highest monthly average is not reached till June and July, when it is only maximum 33.6°C and minimum 29.1°C. The mean yearly temperature during the period is maximum 31.5°C and minimum 22.1°C (Table-IV).

The monsoonic weather starts with the cyclonic storms in the month of May; and brings heavy rainfall and the average consequently rises from 42.4 mm in April to 133.6 mm in May (Table-I). And with the commencement of south west monsoon, which generally occurs in the last half of June, but sometimes deferred till the beginning of July; humidity increases 90% of saturation, while heavy cloud is lasting and daily rainfall occurs till September. The average rainfall in June is 230.2 mm, in July 396.1 mm, in August 327.0 mm and in September it was recorded to 325.4mm. The mean temperature slowly diminishes from 34.2°C in June to 32.8° in October. During the latter half of September, and through October, cloudy weather changes to bright sunshine; the bright periods lengthening until they merge into the continuous fine weather of the cold season. Heavy downpours also occur in short periods, chiefly accompanied by thunder-storms. The rainy season is more than 5-6 months in a year and the average number of rainy days is 143 per year.

The severe cyclones of the transition period from May to November, occasionally enter the south-west angle of the Bay of Bengal and move towards north. The annual average rainfall is 1704.2 mm.

The winter is colder (28.0°C - 16.7°C) but not much, as well as the summer (34.4°C - 24.0°C) is hotter but not troublesome; the rainfall also is (1704.2 mm) moderate and humidity (97% - 33%) is very high.

Agriculture And Cropping Pattern in the District

General Condition

The arable lands of the District are of three types (i) the comparatively high lands, along the villages and the banks of the rivers, (ii) the low-lying depressions that stretch away below the river banks and on the northern portions and (iii) the lands adjoining the Sundarbans which have been reclaimed very recntly. All these riparian tracts are fairly well raised (6-9 mm above m.s.l.) and the drainage passes away easily to the basins below. It has been elevated by the deposition of silt from the rivers and has attained a height which ensures it against inundation (but at the same time prevents it formerly left behind them). The low lands that constitute the greater portion of the cultivated area are mainly under rice and jute. Large areas are occupied by bills and bheries which may be either larger freshwater lakes or brackish water swamps; these remain completely or partially water logged and are suitable for aquaculture. The farmers are, generally, forced to wait till the bill lands dry up and if there is a year of heavy rainfall the cultivations are completely stopped for that year. In the reclaimed lands adjoining the Sundarbans, cultivations are only rendered possible by means of embankments, construted to keep out the salt water; and also the fairly heavy rain fall is necessary to wash out the impregnated salt from the soil before rice seeding can be grown and transplanted.

List-3—Rainfall and Cropping Pattern (Modified after National Commission on Agriculture, 1976.)

Cropping Pattern	Geographical Area	Elevation metres above sea level	Annual rainfall Total cm	Rainy days	Month of Max. Rain fall
Paddy 75%	14,136 km^2.	9m–6m	1704.2	143	7–8

It is necessary that the maximum possible quantity of rain water is conserved in ponds, pools and under dams, in soil profiles and under ground storages so that the same could be readily used to save the crops in times of water stresses. The net sown area is 46% under this District of which the main crop (75%) is paddy. Relative yield Index Values of paddy crop is 102. Irrigation is negligible in the District 24 parganas and the paddy crop is unirrigated but its yield is close to all India average (National Commission on Agriculture-1976.)

Next to paddy, jute is the important crop of the District; cotton and wheat are being a success as new crops in 24 Parganas. Maize, soyabean, sunflower and many other leguminous seeds and fodder crops should also be given attention. The high lands along the villages are mainly under the vegetables, pulses, oil-seeds and even for the wheat or jute cultivation; some of the lands are also used for orchards or nursery beds.

Game Birds and Wild Life Sanctuaries

In the Sundarbans, there were the jungle fowls; a few years back, a bird sanctuary was established in Saznakhali named as 'Pakhir- Ala'; but the

Sikaries (hunters) and others often disturbed them and it was not fully developed as it was originally planned. Only a small flights of common teal used to visit in the monsoon and winter months and the other migratory ducks – common mallard, pochard, shoveller, gadwall, teal, pintail and widgeon would also visit this area. These ducks are very often disturbed by the shikaris for that they could not stay very long. Shipes are found during the winter on the paddy fields, Golden flaver was also met when the paddy has been harvested. Partridge and quail(button) are scarce. In the Alipore Zoo, all these ducks visit during the winter. The only other game bird is the curlew, which is found on the banks of all streams. They are shy to a degree and large flocks are often seen flying off on the least sign of danger. In the Salt Lake area, there was also a proposal for setting up a bird sanctuary or bird games.

Sundarbans was a forest of barking deer, spotted deer, hog-deer, wild pig, fishing cat, civet cat, palm civet, mongoose, monkey, jackal, fox and the world famous 'Royal Bengal Tiger'. Beside these, the crocodile was found in tanks and the tidal rivers. It was practically open to all only a few years back, though these were under the jurisdiction of the Forest Department. Recently (1973), a project was undertaken, under the name 'Tiger Project' and all the wild lives are now safe by this management in the natural condition for better conservation. Along with the Project Tiger, a Crocodile Rearing and Breading Programme at Bhagbatpur of the Sundarbans has also been carried out side by side since, 1976. This result showed some good sign of natural conservation. Since 1989, these vast areas of the Sundarbans, W.B. are included under the most important 'Biosphere Reserve Project' and strict conservation measures were undertaken to protect these biota and the ecosystems in its true perspectives. It is also planned to divert the human interactions in these virgin mangrove forests through arrangements of alternate job facilities in the periphery of the mangrove forests of the Sundarbans.

3

PAST BOTANICAL ACTIVITIES OF THE TWENTY FOUR PARGANAS DISTRICT, INDIA

Since the year 1776, the Botanists were interested mostly about the 'Mangrove vegetation' of the Sundarbans; during this period, the entire southern tract of the District 24 Parganas was under the largest chunk of mangrove vegetation of the Sundarbans. Dr. Roxburgh (1796), the great pioneer Botanist of the then Royal Botanic Garden, obtained a number of interesting plants from this region with the help of Dr. Buchanan Hamilton and Dr. William Carey of Serampore and accounted them in his famous *Hortus Bengalensis* (1814) – containing a synopsis of plants indigenous to or cultivated in the vicinity of Calcutta. J.O. Voigt's *Hortus Suburbanus Calcuttansis* (1845), Andersons' *Catalogue of plants, indigenous in the neighbourhood of Calcutta with directions for examinations and preservation of plants* (1862); Schlich's *Remarks on the Sundarbans* (1875) are the other important publication of the past Botanists. Mr. Ellis (1880) collected the interesting species namely, *Porteresia coarctata* (*Oryza coarctata*) and *Kleinhovia hospita* and sent these to Calcutta. C.B. Clarke gave particular attention to the Sundarbans' flora and the first comprehensive account of Sundarbans was given in the Presidental Address to the Linnean Society, London (1895) on the tropic *'Sundarbans of Bengal'* and Heining also worked on the *Blindroot suckers of Sundarbans*.

Dr. Wallich has not attended to the Sundarbans Flora particularly but an interesting plant species obtained by him from the Sundarbans area (1817) that no one has found upto the Prain's time. Dr. Griffith collected a few specimens from this tract and deposited to the (CAL) Calcutta Herbarium.

Dr. Falconer also collected a few specimens and preserved in the (CAL) Calcutta Herbarium; Mr. Kurz, during the time of Dr. Andersons had made several collections and reported the species *Aldrovanda veciculosa* from this tract. Mr. Gamble added some new collections to the older list and Mr. Brace also worked on the Sundarbans Flora.

Mr. Heining, a forest officer, prepared a working plan for the Sundarbans and collected almost all the plants for four or five seasons during 1881-1884 and all these help a lot to Sir D. Prain for his publications (1903a, 1903b & 1905). Mr. Heining studied in detail the vegetational aspect as well as the topography and many other aspects of the Sundarbans. Mr. G.A. Gammie also visited Sundarbans with Mr. Heining.

During the early part of the 20th Century, Sir David Prain published the *'Bengal Plants'* (1903a), *'Flora of Sundribun'* (1903b), *'Notes on the Sundarbans Plants'* (1903c) and *'The vegetation of the District Hugli-Houra and 24-Pergunah*as (1905), where he gave the lists of plants and the detail of vegetational study after a careful and sincere observations. For the first time, D. Prain (1903b) described a systematic and comprehensive account for the Sundarbans Delta flora.

David Prain made collections with the help of Mr. Lace, Conservator of forests, West Bengal, during August 1902. His work also was based on the collections of the pioneor workers and his own collectors, as well.

In fact, the present Sundarbans is far different from the Prain's (1903b) Sundarbans, as the larger and richer portion of it is now under the jurisdiction of Bangladesh (erstwhile East Pakistan) since 1947; and a vast areas are now under the reclaimed cultivated lands.

With the beginning of the 20th Century, a number of workers worked and published on the flora of the 24 Parganas or Lower Ganga Delta plants, which are as follows:-

Ramaswamy, M. (1916)	: Floating adaptations of the aquatic plants of Lower Bengal;
Biswas, K. (1927)	: Flora of the Salt Lakes, Calcutta;
———— (1950)	: Study of the Flora of South Calcutta with special reference to the Flora of University College of Science Compound, Ballyguange;
Curtis, S.J. (1933)	: Working plan for the Forests of the Sundarban Division;
Griffith, J. (1936)	: Notes on a tour in South Bengal, Calcutta;
Ghosh, A.K. (1941)	: Submerged forest in Calcutta;
Benthall, A.P. (1946)	: Trees of Calcutta and its neighbourhood;

Culshaw, J.C. (1950) : Some West Bengal Plants;

Majumder, R.B. (1956) : Studies on the Grasses of 24 Parganas, West Bengal;

Majumder, R.B. (1962) : Weed Flora of the district 24 Parganas, West Bengal;

Anonymous (1957) : Proceedings of the Mangrove Symposium;

Chatterjee, D. (1957) : Symposium of mangrove vegetation;

Rao, V.S. (1959) : Mangrove forests and the problem of reclaiming saline banks;

Ghosh, R.B. (1960) : Preliminary observations of the Flora of the dilapidated walls and buildings of Calcutta and Suburbs;

Mukerjee, S.K. (1960) : The genus *Bruguiera* in Sundarbans-a new record.

———-(1961) : On *Bruguiera hainessii* Rogers from Sundribans–new record;

Patil, R.P. (1961) : A notes on the vegetation of Sazinakhali in Sundarbans;

Dutta, N.M. & D. Mitra (1962) : Three newly recorded plants from West Bengal;

Chaudhuri, S.K., N. Mallik & J. Sen (1962) : The History of the vegetation of the lower Gangatic plain;

Banerjee, A.K. (1964) : Forests of Sundarbans;

Ganguly, D.K. (1965) : Sundarbans;

Majumdar, N.C. (1965) : Aquatic and semi-aquatic Flora of Calcutta and adjacent localities;

Datta, S.C. & N.C. Majumdar (1966) : Flora of Calcutta and vicinity;

Dutta, N.M. & A.K. Ganguly (1967) : Contribution to the Flora of Calcutta and neighbourhood. I. Commonly occurring herbs & shrubs;

Datta, S.C. & J. Bhattacharya (1969) : Flowering of Calcutta Trees;

Chanda, S. & B.B. Mukherjee (1969) : Radio-carbon dating of two microfossiliferous quarternary deposites in and around Calcutta;

Bose, T.K. & D. Mukherjee (1969) : Studies of *Plumeria* in Calcutta and neighbourhood;

Guha Bakshi, D. N. and Subir Sen (1971) : A botanical exploration of Sagar Island in 24-Parganas;

Mukherjee, B.B. (1972) : Pollen analysis of a few Quarternary deposits of Lower Bengal basin;

Mitra, R.L. (1973) : Taxonomic studies on the Flora of Tollygunge and surrounding region in 24-Parganas District, W.B.

Das Gupta, R. (1973) : Contribution to botany of a portion of Salt Lake, W.B.

Caratini, C., S. Chanda & G. Thanikaimoni (1974) : Main characteristics of Indian mangroves;

Mukherjee, A.K. (1975) : The Sundarbans of India and its biota;

Choudhuri, Rai H.N. and G. N. Tribedi (1976) : On the occurrence of some medicinal plants in 24-Parganas, West Bengal;

Blasco, F. (1975) : The mangroves of India;

Chanda, S. (1977) : An Eco-Floristic survey of the mangroves of Sundarbans, West Bengal;

Mahapatra, A.K. & R.K. Chakraverty (1977) : A survey of the aquatic flora of the Sagar Island and their economic aspects with reference to integrated rural development;

Mahapatra, A.K. (1978) : A brief survey of some unrecorded less known and threatened plant species of Sundarbans, West Bengal,

Mukherjee, B.B. and J. Mukherjee, (1978) : Mangroves of Sundarbans, India;

Paria, N.D. (1978) : Flora of Ballygunge Science College Campus;

Paria, N.D. & S.K. Sahoo (1979) : Herbaceous flora of 24-Parganas;

Naskar, K.R., D.N. Guha Bakshi & G.G.Maity (1979)	: Occurrence of *Moghania macrophylla* (Willd.) O. Kuntze from the Lower Ganga Plain of West Bengal;
Mitra, R.L. and L.K. Banerjee (1979)	: The Genus *Bruguiera* Lamk. (Rhizophoraceae) in India;
Naskar, K.R. (1981)	: Flora of the 24 Parganas District (Ph.D. Thesis).
Naskar, K.R. and D. N. Guha Bakshi (1982)	: Sundarbans – the world famous mangrove forests of the District 24-Parganas in West Bengal;
Naskar, K.R. and D.N. Guha Bakshi (1983 a)	: Eco-Taxonomic studies of the Sundarbans Mangrove Forests and its impact on Socio-Economic aspects;
Naskar, K.R. and D.N. Guha Bakshi (1983 b)	: Some-Less familiar plants of the tidal mangrove forests of Sundarbans;
Naskar, K.R. (1983)	: Halophytes and their unique adaptation on the tidal mangrove forests of Sundarbans.
Naskar, K.R., D.N. Guha Bakshi (1987)	: Mangrove swamps of the Sundarbans.
Naskar, K.R. (1990)	: Aquatic & Semi-Aquatic Plants of the Lower Ganga Delta.

So it is quite evident that all these works, enumerated here, are sporadic and scattered and no systematic work was undertaken on the floristic composition of the most important district of 24-Parganas, West Bengal, including the Sundarbans Mangrove Flora till the publications of Naskar & Guha Bakshi (1987).

Roxburgh's work only consisted of brief diagnosis, which accompanied all the species, with long and lucid descriptions.

The first attempt was made by J.W. Masters (1839) on behalf of the member of the Agricultural and Horticultural Society of India, arranged to their natural families, with observations on the properties and manner of cultivating some of the most interesting species.

Voigt's, *Hortus Suburbanus Calcuttensis* included the entries in Carey's garden receipt book; it dealt with all plants that had been planted at

Serampore, but had succumbed at Sibpur, and all plants were introduced at Serampur between 1814 and 1834.

Anderson's Catalogue of 'Plants indigenous in the neighbourhood of Calcutta' is a compilation and consists of a bare list of the plants given by Voigt, so it is of no practical use except as check list.

David Prain's contributions (1903a, 1903b and 1905) were monumental, very much laborious and the only authentic base for the taxonomic study of the early 20th century; but the geographical area and situations changed so rapidly that it needs a revision in the light of the present situation. Prain's (1903b) Sundarbans is quite different with that of the present day's Sundarbans both in terms of area and vegetation; there are so many changes of river systems geographical boundary, ecology, as well as the human interferences, to the vegetation. Moreover, the changes of scientific name, using of synonym, basionym and detailed references are also used to-day and the question of addition, deletion or omission also arise to the present situations as the present 'Flora' represent quite different style of dealing and arrangements. The detailed ecology, utilization and distribution record of the species were not mentioned in the Prain's writings (1903a, 1903b and 1905). 'Bengal Plants' was a vast publication and covered a huge area; but the present area under study is only negligible fraction in comparison to the vast area covered in 'Bengal Plants'. Prain (1903b) in his 'Flora of Sundriban' listed only 334 species from the Sundarbans area but now the major portion of his (1903 b) study falls under the jurisdiction of Bangladesh. In the Flora of 'Hugli-Hourah and 24-Pergunnah' (Prain, 1905) he dealt with 1316 Phanerogams, but it was only a tentative list, which may not serve the present needs of a complete flora.

A.P. Benthall (1946) in his book 'The Trees of Calcutta and its Neighbourhood' nicely dealt with 276 common trees both indigenous and exotic and 274 illustrations, of which 69 are naturalised in the neighbourhood, 91 are indigenous in other parts of India, 50 are indigenous to elsewhere of Asia (other than India), 14 are native of Africa, 42 of American, 9 of Australian and 1 of sea Islands. The author considered the tree species only and no attempt had been made to include the numerous shrubs, herbs and woody climbers.

The works of Ramaswamy (1916), Biswas (1927 and 1950), Culshaw (1950), Chatterjee (1958), Ghosh (1960), Mukerjee (1957, 1960, and 1961), Patil (1961), Datta & Mitra (1962), Chaudhuri, Mallik and Sen (1962), Ganguly (1965), Majumdar (1965), Datta & Majumdar (1967), Datta &

Bhattacharya (1969), Bose & Mukherjee (1969), Guha Bakshi & Sen (1971), Das Gupta (1973), Mukherjee (1975), Chanda (1977), Mahapatra & Chakraborty (1977), Mahapatra (1978), Paria & Sahoo (1978, 1979), Mukherjee & Mukherjee (1978), Naskar (1983, 1984, 1985 and 1986) and Naskar and Guha Bakshi (1979, 1982, 1983a, 1983b, 1984, 1985a, 1985b and 1986) have represented various different aspects of botanical studies; some of them are very much informative though it may not serve any solution of a typical flora. All these, established some interesting observations, which is completely different from the field of flora study for a particular area like 24-Parganas District in West Bengal.

The works of Majumder (1956 and 1962) threw light on some essential data regarding the floristic components; certainly, those works, particularly the—'Studies on Grasses' was done after a thorough and minute investigation and after a detailed study. It is definitely an authentic work and reference of grasses of this region. The 'Weed Flora', where he mentioned 344 weed species under 188 genera and 54 families may also claim the superior analysis of weeds growing to this tract. But all these must not claim a typical flora of the vast area 24-Parganas—as no mention other than the weeds were made there. Moreover, no attention laid on the flora of the Sundarbans mangrove forests, which almost covered 1/3rd of the district area of 24-Parganas.

The work of Datta & Majumdar (1966) on the 'Flora of Calcutta and vicinity' is a voluminous publication, where they dealt with 657 species and 24 varieties under 414 genera and 104 families of Phanerogams; but they have not dealt with any identifying keys. This publication only elaborated a tentative list with a few lines for each regarding their diagnostic characters, phenology and brief distribution record along with the available vernacular names; no mention of field data or references regarding field collections were mentioned there. Moreover, it covered only a limited area of Calcutta and its surroundings.

The work of Dutta & Ganguly (1967) has some scientific value as it has mentioned the field number, vernacular names with the original citation and the citation from the 'Flora of British India' and 'Bengal Plants'. But this small publication only dealt with 171 species of 129 genera and under 44 families of the class Dicotyledones.

Mitra (1973) submitted his M.Sc. Thesis (University of Bombay) dealt with the 'Taxonomic studies on the Flora of Tollygunge and surrounding region of the 24-Parganas District, West Bengal'. In his Flora, Mitra (1973) covered only a limited area of the municipal town of the 'Alipore Sub-division' of North East corner of the District 24-Parganas; so the scope of this Flora is very much limited in relation to the entire 24-Parganas district Flora, entirely.

Naskar (1983,1985 and 1986) and Naskar and Guha Bakshi (1982,1983a,1983b 1984 and 1985) have published a number of papers on the mangroves of the Sundarbans and Flora of 24-Parganas, which are only the little part of the district flora. Recently, Naskar and Guha Bakshi (1987) published the 'Mangrove Swamps of the Sundarbans in West Bengal–an Ecological Perspective' which cover much about the flora of the Sundarban's tidal forest but has not covered the flora of that area which are growing on the non mangrove zones in that Sundarbans. Naskar (1990) also published the Aquatic and Semi-Aquatic flora of the lower Ganga delta. This present work of Naskar (1993) had mentioned 1175 angiosperm species under 154 families, which is the complete Flora till the present time.

4

PRESENT WORK

In this floristic work, the author deals with 1175 Angiosperm species under 680 genara, which constitute under 154 families. All those taxa are either (i) collected directly from the fields by the frequent field trips covering all the seasons, all over the year and all parts of the district, for about four years (1976-1980), and (ii) from the studies of the herbarium specimens of the Central National Herbarium, Botanical Survey of India, Howrah (CAL) as well as from the available literature. Quite a large number of collections were made to have a true representative character of the flora of the district with a special emphasis to the mangrove vegetation of the Sundarbans, in the southern part of 24-Parganas.

The detailed field conditions, habit and habitat, phyto-association or phyto-communities, trend of the vegetation nature, shape, size and the phenological data to each and every individual species and the surroundings were observed minutely and enrolled all such data to the field notes. All these characteristics or the diagnostic characters were noted and described as such observed in the laboratory and line drawings to the interesting, rare and the taxonomically important species are also drawn carefully. For keeping the records, the plant materials were collected and preserved in the general accepted methods of herbarium technique and also have been deposited these herbarium specimens to the Central National Herbarium (CAL.) and Calcutta University, Botany Department herbarium as voucher specimens. During the field trips, the climatic conditions, topography, soil type, geological condition, river systems, forest types, games or wild life sanctuaries, namely 'Tiger Reserve Project', 'Bird Sanctuary', 'Crocodile Rearing and Breeding Project' and the plant base industries were also observed and described in brief in the introductory part of this Flora. The other information like history of the district, geological information, geographical position, area, population, climate data and so many other aspects were collected and elaborated from the 24 Parganas District Gazette (1913), Census report (1981) and from many other Government Agencies. All such factors rendered a greater correlation with the floristic study. To make clear and strengthen the observations, several photographs of landscape, vegetation, interesting species, phyto-association and few line drawings are also included.

The sequence of families followed in this Flora according to Benthum & Hooker's system of classification with little modification in case where the families have been recently split up (Kenneth M.Backer,1973).

The genera and species in a family are arranged alphabetically.

A brief and distinctly opposite phenological characters are normally used to classify the families under the indented key along with some diagnostic characters of each and every family.

In the main part of the Flora, the systematic treatment have done under each family as follows:-

i) The Indented keys were used after comparing the distinctive characters of the different genera of a particular family, and those heads were differentiated by making paragraph.

ii) The generic names followed by the original author's name were mentioned.

iii) The elaborate Indented keys were also used for the different species under certain genus.

iv) The arrangement of the species were also done alphabetically under the particular genus initiated with the serial numbers.

v) The valid names of the species with the original citations according to the latest 'International Code of Botanical Nomenclature' are followed and the citations of the latest taxonomic work are also noted as far as possible.

Synonyms and basionyms have also mentioned with the correct names along with the citation of the respective Regional Flora.

Referances to the J.D. Hooker's 'Flora of British India', Roxburgh's 'Flora Indica', Watt's 'Dictionary of Economic Products of India', Prain's 'Bengal Plants', 'Flora Sundribans', Benthall's Trees of Calcutta', CSIR's 'Wealth of India' and available recent revision work, and some local publications were also mentioned.

vi) Vernacular names in Bengali are mentioned to make it known to the interested non-botanical persons.

vii) A brief description of each species was made, showing the primary characters.

viii) Phenological observations have recorded to make clear the reproductive behaviour and nature's role in the plants.

ix) A short description regarding the ecological notes and frequency along with the notes on cultivated plants.

x) Distribution and frequency data have also been mentioned in relation to Indian territory and the World. Some of the past history regarding the introducing notes have added with availability.

xi) Economic notes have also supplied from the 'Wealth of India' and from the local people.

The vegetation analysis, climate chart, the details of ecosystems and the possibilities are also discussed with the concluding remarks.

In this book, the mangrove flora of the Sundarbans, W.B. are discussed vividly alongwith sketches and photographs of most of these mangrove flora, available in the Sundarbans. The detailed description of this mangrove ecosystems is also published by Naskar & Guha Bakshi (1987); as such here only the flora of the mangrove forests are only highlighted.

5

ECOLOGICAL TYPES OF VEGETATIONS IN THE LOWER GANGA DELTA OF INDIA

The following ecological classifications of the Lower Ganga deltaic Angiosperm flora were made by Naskar (1981):

1. *Hydrophytes or Helophytes*- the freshwater loving marshy plants of the Lower Ganga delta were also described by Naskar (1991). The following types of hydrophytes were recorded:

- 1.1. Free-floating hydrophytes, either surface floating or submerged;
- 1.2. Rooted-hydrophytes with floating leaves;
- 1.3. Rooted-submerged hydrophytes;
- 1.4. Rooted-emerged hydrophytes;
- 1.5. Wet-land flora or moist loving plants.

2. *Halophytes or Mangroves*- the saltwater loving plants of the Sundarbans area were also reported by Naskar & Guha Bakshi (1987). The following types of the halophyes were recorded from the intertidal mangrove forests of the Sundarbans in India:

- 2.1. True Mangroves;
- 2.2. Obligatory mangroves or halophytic non-mangroves;
- 2.3. Mangroves Associated flora or Back Mangals;
- 2.4. Weeds or introduced flora of the mangrove forests of the Sundarbans.

3. *Mesophytic Flora*- tropical high land flora. The following types of the mesophytic flora were reported by Naskar (1981).

- 3.1. Weeds of high cultivated lands;
- 3.2. Weeds of the roadsides and waste places;

3.3. Weeds of the low but dry cultivated lands, during summer season;

3.4. Weeds on the garbages of Dhapa-Tapsia and other areas;

3.5. Weeds on the railway tracks;

3.6. Weeds of the old-houses and crevices of dilapidated walls;

3.7. Weeds of the gardens and orchards;

3.8. Common garden escapes;

3.9. Common cultivated economic herbs, shrubs and trees;

3.10. Common hedge plants;

3.11. Common climbers and twiners.

4. *Xerophytic Flora-* This type of plant associations are very meagre in the lower parts of Bengal, as the climate and weather for this tract is typical tropical type. Availability of water for the growth of plant is adequate here. Only in the coastal sandy area or in the beach forests, a few plant species grow spontaneously.

Warming (1909) classified the flowering plants into 8 groups, based on the availability or substrate water. Among these eight groups, only the hydrophytes, halophytes, mesophytes and xerophytic flora are abundant in the lower Ganga delta; while the other four types, *viz.*, Oxylophytes, psycrophytes, lithophytes and palamophytes are lacking in this region. Oxylophytes grow on the acid soil and the psycrophytes, lithophytes and palamophytes grow on the cold regions, rocks/stones and formative sand respectively. These last types of soils are normally absent in this lower part of the Ganga Delta.

Angiosperm flora of the lower Ganga delta are mixed types, as there are no physical barriers for checking the plant migration. From the ancient time, several foreign elements were introduced here by the foreign traders or visitors of Bengal. Moreover, the rivers Ganga, Brahmaputra systems also accelerated the plant migration from the Great Himalaya and from the upper parts of India. From the bay, large number of plants also migrated here from several adjacent countries. The biotic and abiotic factors afford an interesting type of floral association here in the lower Ganga Delta.

6

ANALYSIS OF THE LOWER GANGA DELTA FLORA OF INDIA

The floristic components of the lower Ganga Delta with special reference to the 24-Parganas District were studied by Naskar (1981). In this study, the total number of Angiosperm families, genera and species, both exotic and indigenous, their habits and habitats were analysed critically by the field and laboratory studies. Flora of the tidal mangrove forests of the Sundarbans were also studied and analysed in relation to their habits and habitats and economic importances (Naskar & Guha Bakshi, 1987).

Comparative estimation of the dominant plants of 24-Parganas' along with the Flora of Bengal, published by D. Prain (1903a, 1903b and 1905), Flora of the Gangetic Plain (Hooker, 1907) and Sketch Flora of the British India (Hooker, 1907) (Table-VIII) are also highlighted. By this comparative analysis, the nature of the flora and the ecological or floristic characters can be predicted critically. Here, the ecological types of vegetation clearly showed typical tropical phytozonations.

Sizes of the families and genera, on the basis of number of plants and their phenological characters (Table-VII), distributions on the other parts of the Globe (Table-VI) and detail list of the families along with genera and species (Table-X) and ecological types (Table-V) are discussed here in the tabular forms.

The biological spectrum or the 'Life Forms' of the Lower Ganga delta, the 24-Parganas in comparison with that of the Raunkier's 'Normal Spectrum (1934)' is also compared (Table-V). In this lower Ganga delta, the Phanerophytes (P), *i.e.,* Megaphenerophytes, Mesophenerophytes, Microphanerophytes, Nanophytes, Liana, Epiphytes, Parasites and the succulent plants are dominant. Therophytes are the secondary dominant.

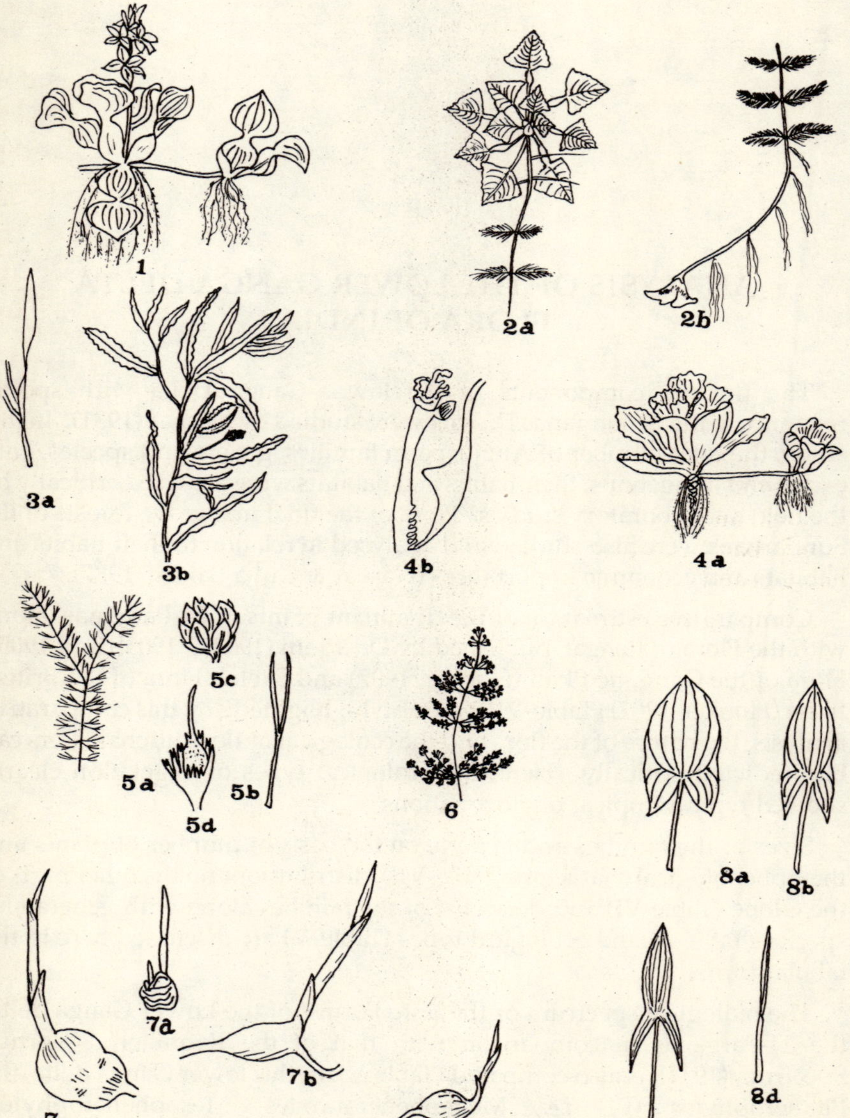

PLATE - 1

PLATE-I. Showing different aquatic flora and their variations

Fig. 1. *Eichhornia crassipes* (Mart.) Solms. — flowering shoot with swollen inflated petiole, runner and fibrous roots and root-hairs

Fig. 2. *Trapa* sp. (a) aerial twigs and (b) submerged roots with fruit

Fig. 3. *Potamogeton crispus* L. — flowering twig

Fig. 4. *Pistia stratiotes* L. — with runner and flower

Fig. 5. *Ceratophyllum demersum* L. — vegetative twig and flower

Fig. 6. *Utricularia* sp. — vegetative part

Fig. 7. Tuber of Aquatic plants —

(a) *Eleocharis* sp., (b) *Vallisneria* sp.,

(c) *Potamogeton* sp., (d) *Sagittaria* sp.,

Fig. 8. Leaf shape variations of *Sagittaria* sp. in different aquatic conditions

Table V

Biological Spectrum of the Lower Ganga Delta Flora and its Comparative Analysis with that of Normal Spectrum – Raunkiaer's Life Form, 1934.

Normal Spectrum (Raunkiaer, 1934)			Lower Ganga Delta (24-Parganas) Flora (Naskar, 1981)				Phytoclimate
Normal Biological Spectrum of the World; considering 1000 Phanerogamic Flora			Temperature (0°C) Rainfall average/annum Number of Rainy Months Relative Humidity (%) Phanerogams Flora Studied		- 18.35°C - 35.5°C - 1704 mm - 5 months/annum - 041% - 094% - 1175 species		PHANEROPHYTES & THEROPHYTES (Phanero-theophytes)
1. Phenerophytes	(P)	= 46%	1. Mega-phanerophytes 2. Meso-phanerophytes 3. Micro-phanerophytes 4. Nano-Phanerophytes 5. Liana 6. Epiphytes & Succulent 7. Parasites	(MGP) (MSP) (MIP) (NP) (L) (E & S) (P)	= 1.00% = 11.23% = 8.67% = 11.53% = 7.32% = 2.46% = 0.68%	42.89%	
2. Chameophytes	(CH)	= 09%	8. Chameophytes	(CH)	= 10.21%		
3. Hemicryptophytes	(H)	= 26%	9. Hemicryptophytes	(H)	= 12.85%		
4. Cryptophytes	(CR)	= 06%	10. Geophytes 11. Hydrophytes & Helophytes	(G) (HH)	= 11.57% = 8.02%		
5. Therophytes	(TH)	= 13%	12. Therophytes	(TH)	= 14.46%		

Table VI

Distribution of the Lower Ganga Plant Species to the Other Parts of the Globe/Country of Origin

Serial Nos.	Phyto-Geographical Regions of the Globe	Nos. of Angiosperm Flora	Percentage of Angiosperm
1	Cosmopolitan and/or Tropical Species	357	30.4%
2	South-East Asian Species (Indo-China-Malayasian-Sri Lanka-Australia-Java-Sumatra etc.)	482	41.4%
3	West Asian and Mediterranean Species	52	4.4%
4	North African Species	15	1.3%
5	Tropical African Species	90	7.7%
6	Tropical American Species	77	6.5%
7	Temperate Regions (Europe-N, America and Mexico)	57	4.9%
8	Indian and Sub-Himalayan Species	45	3.4%
Total	Angiosperm Plants Species Studied from The Lower Ganga Delta	1175	100.0%

Table VII

Phenological Characterization of the Flora of Lower Ganga Delta in West Bengal

Periods/Seasons	Average Climate			Percentage of Species Flowering & Fruiting
November - February (Winter)	Rainfall/month	-	17.1 - 20.7 mm/month	40%
	Temperature	-	16.7°C - 28°C	
	Relative humidity	-	047% - 094%	
March - June (Summer and Pre-monsoon)	Rainfall/months	-	110.0 - 114.0 mm/month	32%
	Temperature	-	24°C - 34.4°C	
	Relative humidity	-	033% - 096%	
July - October (Monsoon and Post-Monsoon)	Rainfall/month	-	281.0 - 295.0 mm/month	50%
	Temperature	-	24.5°C - 32°C	
	Relative humidity	-	061% - 097%	
Throughout the year	------------------------------			12%

Table VIII

Dominant Angiosperm Families of the Twenty Four Parganas District, Bengal, Gangetic Plain and British India

24–Parganas District Flora (Naskar, 1981)			Total Species	Bengal Plants (D.Prain, 1903a)	Flora of Gangetic Plain (J.D. Hooker, 1907)	Sketch Flora of British India (J.D. Hooker, 1907)
Fabaceae (Leguminosae)	–	11.49%	135	Fabaceae	Poaceae	Orchidaceae
Poaceae (Gramineae)	–	10.80%	127	Poaceae	Fabaceae	Fabaceae
Asteraceae (Compositae)	–	5.84%	64	Cyperaceae	Cyperaceae	Poaceae
Cyperaceae	–	5.50%	62	Asteraceae	Asteraceae	Rubiaceae
Euphorbiaceae	–	4.00%	47	Orchidaceae	Scrophulariaceae	Euphorbiaceae
Malvaceae	–	2.64%	31	Euphorbiaceae	Malvaceae	Acanthaceae
Acanthaceae	–	2.30%	27	Urticaceae	Acanthaceae	Asteraceae
Rubiaceae	–	2.30%	27	Acanthaceae	Euphorbiaceae	Cyperaceae
Convolvulaceae	–	2.20%	26	Scrophulariaceae	Convolvulaceae	Lamiaceae
Scrophulariaceae	–	2.04%	24	Convolvulaceae	Lamiaceae	Urticaceae
Total = 10 Families	–	49.11%	570	-----	-----	-----

Table IX

Angiosperm Flora of the Lower Ganga Delta (24-Parganas) in West-Bengal

Classs	*Nos. of Families*	*Nos. of Genera*	*Nos. of Species*	*Nos. of Indigenous Spp.*	*Nos. of Exotic Spp.*	*Nos. of Trees*	*Nos. of Shrubs*	*Nos. of Herbs*	*Nos. of Mangroves*	*Nos. of Terrestrials*	*Nos. of Marshy Plants*	*Nos. of Cultivated species*	*Nos. of Weeds*
Dicotyledons	124	531	874	616	258	229	261	384	63	692	119	251	623
Monocotyledons	30	149	301	245	56	6	19	276	7	140	153	38	263
Total Angiosperms	154	680	1175	861	314	235	280	660	70	832	272	289	886

Table X

Total Angiosperm Families from the Lower Ganga Delta (24- Parganas) in West-Bengal, India

Sl. No. (1)	Names of the Families (2)	Genera Nos. (3)	Species Nos. (4)	Indigenous Species (5)	Exotic Species (6)	Tree Spp. (7)	Shrub Spp. (8)	Herb Spp. (9)	Mangrove Species (10)	Terrestrial Species (11)	Aquatic Species (12)
1.	Ranunculaceae	4	4	3	1	-	2	2	-	3	1
2.	Dilleniaceae	1	1	-	1	1	-	-	-	1	-
3.	Magnoliaceae	2	2	-	2	2	-	-	-	2	-
4.	Annonaceae	5	8	3	5	7	1	-	-	8	-
5.	Menispermaceae	5	5	5	-	-	5	-	-	5	-
6.	Nymphaeaceae	2	3	3	-	-	-	3	-	-	3
7.	Papaveraceae	1	1	-	1	-	-	1	-	1	-
8.	Fumariaceae	1	1	1	-	-	-	1	-	1	-
9.	Brassicaceae (*nom. alt.* Cruciferae)	7	10	9	1	-	-	10	-	10	-
10.	Capparidaceae	2	3	2	1	1	2	-	2	2	-
11.	Cleomaceae	1	3	3	-	-	-	3	-	3	-
12.	Violaceae	1	1	1	-	-	1	-	-	1	-
13.	Bixaceae	1	1	-	1	1	-	-	-	1	-
14.	Cochlospermaceae	1	1	-	1	1	-	-	-	1	-
15.	Flacourtiaceae	1	2	1	1	1	1	-	-	2	-
16.	Polygalaceae	1	1	1	-	-	-	1	-	1	-

Contd...

(1)	(2)	(3)	(4)	(5)	(6)	(7)	(8)	(9)	(10)	(11)	(12)
17.	Caryophylaceae	4	4	4	-	-	-	4	-	3	1
18.	Portulacaceae	1	2	2	-	-	-	2	-	2	-
19.	Tamaricaceae	1	4	-	4	3	1	-	-	4	-
20.	Elatinaceae	1	2	2	-	-	-	2	-	1	1
21.	Cluciaceae (*nom. alt.* Guttiferae)	2	2	1	1	2	-	-	1	1	-
22.	Dipterocarpaceae	1	1	1	-	1	-	-	-	1	-
23.	Malvaceae	14	31	19	12	4	15	12	3	28	-
24.	Bombacaceae	3	3	1	2	3	-	-	-	3	-
25.	Sterculiaceae	12	17	13	4	13	1	3	2	14	1
26.	Tiliaceae	4	6	5	1	2	-	4	1	5	-
27.	Elaeocarpaceae	2	2	1	1	2	-	-	-	2	-
28.	Linaceae	1	1	1	-	-	-	1	-	1	-
29.	Malpighiaceae	2	2	2	-	-	2	-	-	2	-
30.	Zygophyllaceae	2	2	-	2	-	-	2	-	2	-
31.	Oxalidaceae (*nom.alt.* Geraniaceae)	3	4	2	2	1	-	3	-	3	1
32.	Balsaminaceae	2	2	2	-	-	-	2	-	1	1
33.	Rutaceae	5	10	9	1	6	4	-	-	10	-
34.	Simaroubaceae	1	1	1	-	1	-	-	-	1	-
35.	Ochnaceae	1	1	1	-	1	-	-	-	1	-
36.	Burseraceae	1	1	1	-	1	-	-	-	1	-
37.	Meliaceae	8	10	7	3	9	1	-	3	7	-
38.	Olacaceae	1	1	1	-	-	1	-	-	1	-

(1)	(2)	(3)	(4)	(5)	(6)	(7)	(8)	(9)	(10)	(11)	(12)
39.	Celastraceae	1	1	1	-	1	-	-	-	1	-
40.	Hippocrataceae	1	1	1	-	1	1	-	1	-	-
41.	Rhamnaceae	2	3	2	1	1	2	-	-	1	-
42.	Vitaceae (*nom. alt.* Ampelidaceae)	4	8	8	-	-	7	1	-	8	-
43.	Leeaceae	1	4	4	-	-	3	1	-	4	-
44.	Sapindaceae	7	9	7	2	6	2	1	-	9	-
45.	Anacardiaceae	5	6	5	1	6	-	-	-	6	-
46.	Moringaceae	1	1	1	-	1	-	-	-	1	-
47.	Papilionaceae	42	88	62	26	14	22	52	7	77	4
48.	Caesalpiniaceae	12	30	15	15	16	8	6	1	29	-
49.	Mimosaceae	9	17	7	10	11	4	2	-	4	-
50.	Rosaceae	3	4	2	2	1	2	1	-	4	-
51.	Crassulaceae	1	1	-	1	-	-	1	-	1	-
52.	Droseraceae	1	1	1	-	-	-	1	-	-	1
53.	Haloragaceae	1	2	2	-	-	-	2	-	-	2
54.	Rhizophoraceae	4	9	8	1	7	2	-	9	-	-
55.	Combretaceae	3	6	4	2	5	1	-	1	5	-
56.	Myrtaceae	5	10	5	5	8	2	-	1	9	-
57.	Lecythidaceae	1	2	2	-	2	-	-	2	-	-
58.	Melastomataceae	1	1	1	-	1	-	-	-	1	-
59.	Lythraceae	4	9	7	2	1	2	6	-	2	7
60.	Sonneratiaceae	1	3	3	-	3	-	-	3	-	-
61.	Punicaceae	1	1	-	1	1	-	-	-	1	1

Contd...

(1)	(2)	(3)	(4)	(5)	(6)	(7)	(8)	(9)	(10)	(11)	(12)
62.	Onagraceae	1	4	4	-	-	-	4	-	-	4
63.	Trapaceae	1	1	1	-	-	-	1	-	-	1
64.	Samydaceae	1	1	1	-	1	-	-	-	1	-
65.	Passifloraceae	2	2	1	1	-	1	1	-	2	-
66.	Caricaceae	1	1	-	1	-	-	1	-	1	-
67.	Cucurbitaceae	12	18	14	4	-	5	13	1	18	-
68.	Cactaceae	1	1	1	-	-	1	-	-	1	-
69.	Aizoaceae (*nom. alt.* Ficoideae)	2	3	3	-	-	-	3	3	-	-
70.	Molluginaceae	2	3	3	-	-	-	3	-	2	1
71.	Apiaceae (*nom. alt.* Umbelliferae)	9	9	7	2	-	-	9	-	4	5
72.	Alangiaceae (*nom. alt.* Cornaceae)	1	1	-	1	1	-	-	1	-	-
73.	Rubiaceae	17	27	21	6	8	11	8	-	22	5
74.	Asteraceae (*nom. alt.* Compositae)	47	64	40	24	-	58	6	2	49	13
75.	Campanulaceae (*nom. alt.* Lobeliaceae)	1	1	1	-	-	-	1	-	-	1
76.	Sphenocleaceae	1	1	1	-	-	-	1	-	-	1
77.	Plumbaginaceae	2	2	2	-	-	2	-	1	1	-
78.	Primulaceae	2	3	2	1	-	-	3	-	-	3
79.	Myrsinaceae	2	2	2	-	2	-	-	1	-	1
80.	Sapotaceae	4	4	3	1	4	-	-	-	4	-
81.	Ebenaceae	1	4	3	1	4	-	-	-	4	-

(1)	(2)	(3)	(4)	(5)	(6)	(7)	(8)	(9)	(10)	(11)	(12)
82.	Oleaceae	1	3	1	2	-	3	-	-	3	-
83.	Nyctanthaceae	1	1	-	1	-	1	-	-	1	-
84.	Salvadoraceae	1	1	1	-	-	1	-	1	-	-
85.	Apocynaceae	16	22	14	8	9	10	3	1	21	-
86.	Asclepiadaceae	13	16	15	1	-	6	10	2	13	1
87.	Periplocaceae	3	3	2	1	-	3	-	-	3	-
88.	Loganiaceae	2	2	2	-	-	1	1	-	1	1
89.	Menyanthaceae (*nom. alt.* Gentianaceae)	4	7	7	-	-	-	7	-	2	5
90.	Polemoniaceae	1	1	-	1	-	-	1	-	1	-
91.	Hydrophyllaceae	1	1	1	-	-	-	1	-	-	1
92.	Ehretinaceae	2	4	3	1	4	-	-	-	4	-
93.	Boraginaceae	3	6	5	1	-	-	6	-	5	1
94.	Convolvulaceae	11	26	19	7	-	10	16	-	23	3
95.	Cuscutaceae	1	1	1	-	-	-	1	-	1	-
96.	Solanaceae	8	21	9	12	-	5	16	1	19	1
97.	Scrophulariaceae	12	24	18	6	-	1	23	1	7	16
98.	Orobanchaceae	1	1	1	-	-	-	1	-	1	-
99.	Lentibulariaceae	1	3	3	-	-	-	3	-	-	3
100.	Bignoniaceae	7	7	1	6	6	1	-	-	7	-
101.	Pedaliaceae	1	1	-	1	-	-	1	-	1	-
102.	Acanthaceae	19	27	26	1	-	4	23	1	13	13
103.	Verbenaceae	10	15	12	3	5	8	2	1	14	-
104.	Avicenniaceae	1	3	3	-	1	2	-	3	-	-

Contd...

(1)	(2)	(3)	(4)	(5)	(6)	(7)	(8)	(9)	(10)	(11)	(12)
105.	Lamiaceae (*nom. alt.* Labiatae)	12	19	16	3	-	2	17	-	18	1
106.	Plantaginaceae	1	1	-	1	-	-	1	-	1	-
107.	Nyctaginaceae	3	4	1	3	-	1	3	-	4	-
108.	Amaranthaceae	8	18	12	6	-	1	17	-	14	4
109.	Chenopodiaceae	6	9	6	3	-	-	9	3	6	-
110.	Basellaceae	1	1	1	-	-	1	-	-	1	-
111.	Phytolacaceae	2	2	1	1	-	-	2	-	2	-
112.	Polygonaceae	3	12	11	1	-	5	7	-	6	6
113.	Aristolochiaceae	1	1	1	-	-	-	1	-	1	-
114.	Piperaceae	2	3	2	1	-	2	1	-	3	-
115.	Lauraceae	3	7	5	2	6	-	1	-	7	-
116.	Proteaceae	1	1	-	1	1	-	-	-	1	-
117.	Loranthaceae	3	4	4	-	-	2	2	2	2	-
118.	Euphorbiaceae	23	47	31	16	13	16	18	2	43	2
119.	Ulmaceae	1	1	1	-	1	-	-	-	1	-
120.	Urticaceae	4	4	2	2	-	-	4	-	4	-
121.	Moraceae	5	17	14	3	15	2	-	-	17	-
122.	Cannabinaceae	1	1	-	1	-	-	1	-	1	-
123.	Casuarinaceae	1	1	-	1	1	-	-	-	1	-
124.	Ceratophyllaceae	1	1	1	-	-	-	1	-	-	1
Total	124 Dicotyledonous Families	531	874	616	258	229	261	384	63	692	129

(1)	(2)	(3)	(4)	(5)	(6)	(7)	(8)	(9)	(10)	(11)	(12)
125.	Hydrocharitaceae	5	5	4	1	-	1	4	-	-	5
126.	Orchidaceae	4	4	3	1	-	-	4	-	3	1
127.	Zingiberaceae	7	11	10	1	-	-	11	-	10	1
128.	Cannaceae	1	1	1	-	-	-	1	-	1	-
129.	Musaceae	1	1	1	-	-	1	-	-	1	-
130.	Bromeliaceae	1	1	-	1	-	-	1	-	1	-
131.	Iridaceae	2	2	-	2	-	-	2	-	2	-
132.	Agavaceae	1	1	-	1	-	1	-	-	1	-
133.	Hypoxidaceae	1	1	1	-	-	1	-	-	1	-
134.	Amaryllidaceae	3	4	1	-	-	1	-	-	1	-
135.	Dioscoreaceae	1	8	5	3	-	-	8	-	8	-
136.	Liliaceae	6	7	4	3	-	-	7	-	7	12
137.	Smilacaceae	1	2	2	-	-	2	-	-	2	-
138.	Pontedariaceae	2	4	-	4	-	-	4	-	-	4
139.	Xyridaceae	1	2	1	1	-	-	2	-	-	2
140.	Commelinaceae	4	11	10	1	-	-	11	-	-	11
141.	Flagellariaceae	1	1	1	-	-	1	-	-	1	-
142.	Arecaceae (*nom. alt.* Palmae)	7	8	4	4	6	2	-	2	6	-
143.	Pandanaceae	1	2	2	-	-	2	-	-	2	-
144.	Typhaceae	1	2	2	-	-	-	2	-	-	2
145.	Araceae	9	14	14	-	-	-	14	-	10	4
146.	Lemnaceae	3	4	4	-	-	-	4	-	-	4
147.	Alismataceae	2	3	3	-	-	-	3	-	-	3

Contd...

(1)	(2)	(3)	(4)	(5)	(6)	(7)	(8)	(9)	(10)	(11)	(12)
148.	Butamaceae	1	1	-	1	-	-	1	-	-	1
149.	Aponogetonaceae	1	2	-	2	-	-	2	-	-	2
150.	Potamogetonaceae	1	2	1	1	-	-	2	-	-	2
151.	Najadaceae	1	1	1	-	-	-	1	-	-	1
152.	Eriocaulaceae	1	6	6	-	-	-	6	-	-	6
153.	Cyperaceae	7	62	60	2	-	-	62	-	10	52
154.	Poaceae (*nom. alt.* Gramineae)	72	128	102	26	-	8	120	6	71	51
Total	30 Monocotyledonous Families	149	301	245	56	6	19	276	8	140	153
Total	Dicotyledonous & Monocotyledonous Families = 154.	680	1175	861	314	235	280	660	71	832	272

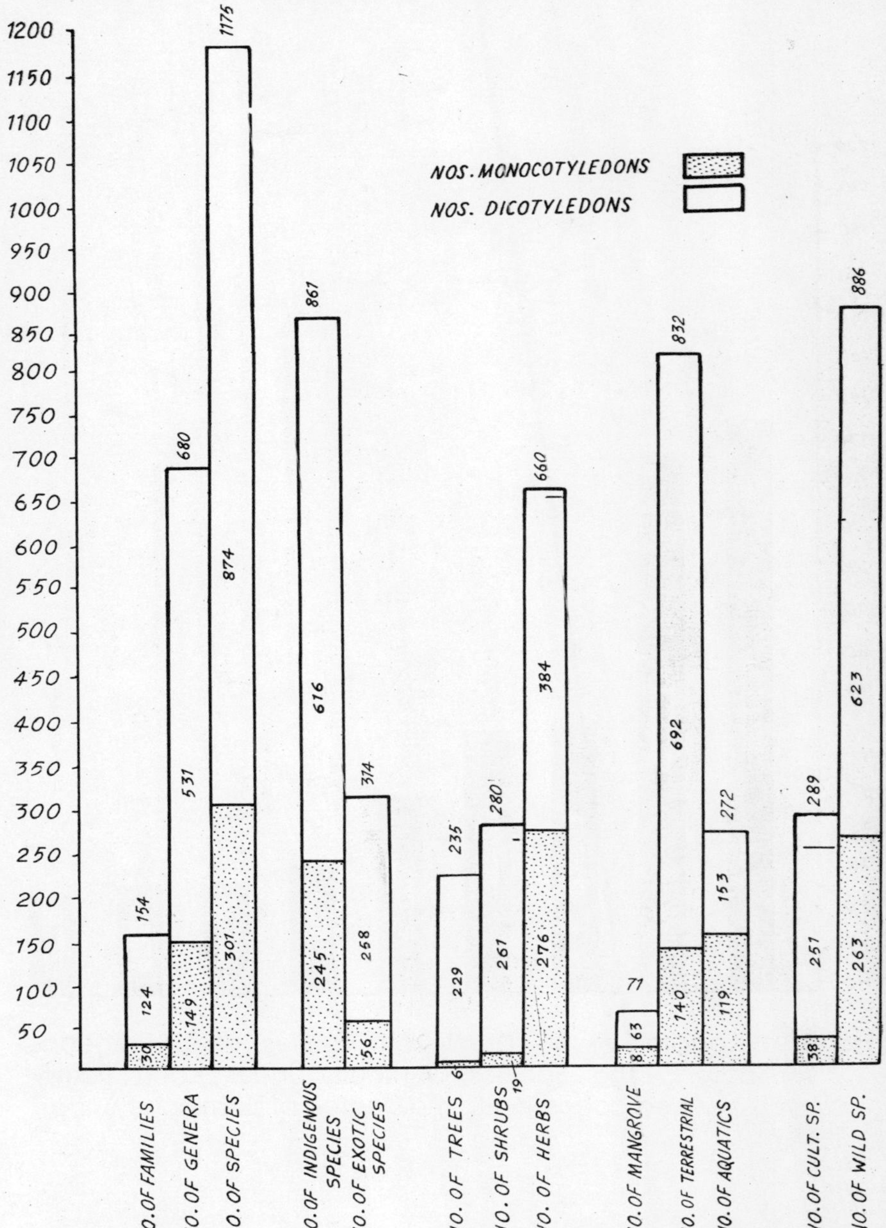

Hist.2. Showing the Total Floristic Components of 24-Parganas District

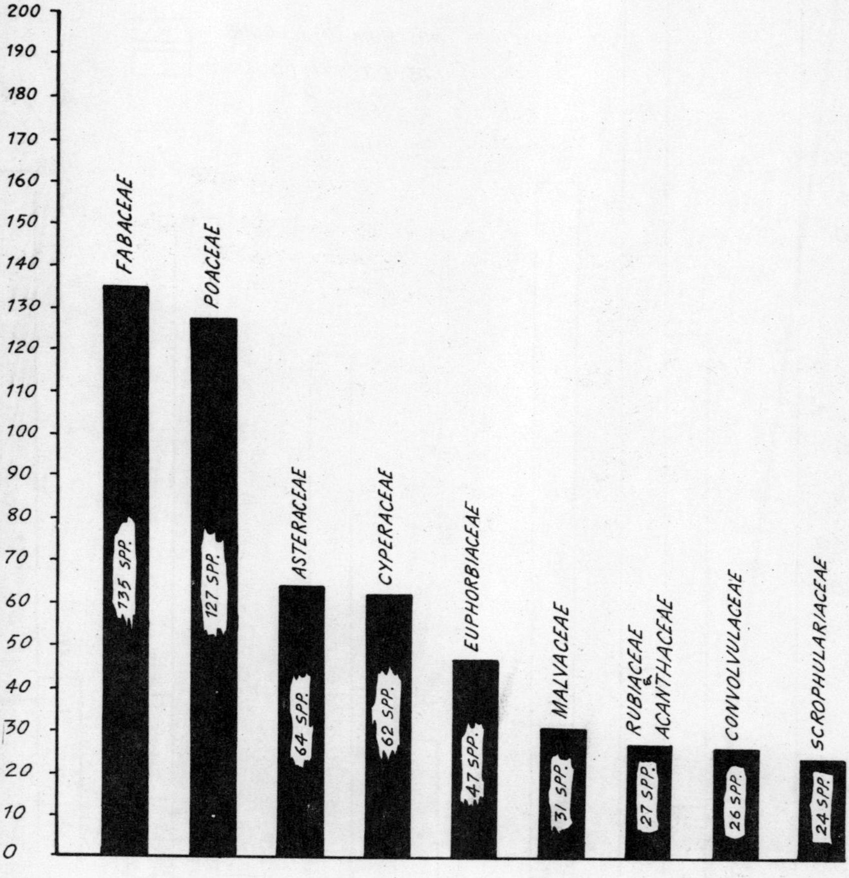

Hist. 3. Histogram of 10 Dominant Families of 24-Parganas, Showing the Species Numbers Under Each Family. (Arranged according to the Degree of Dominancy)

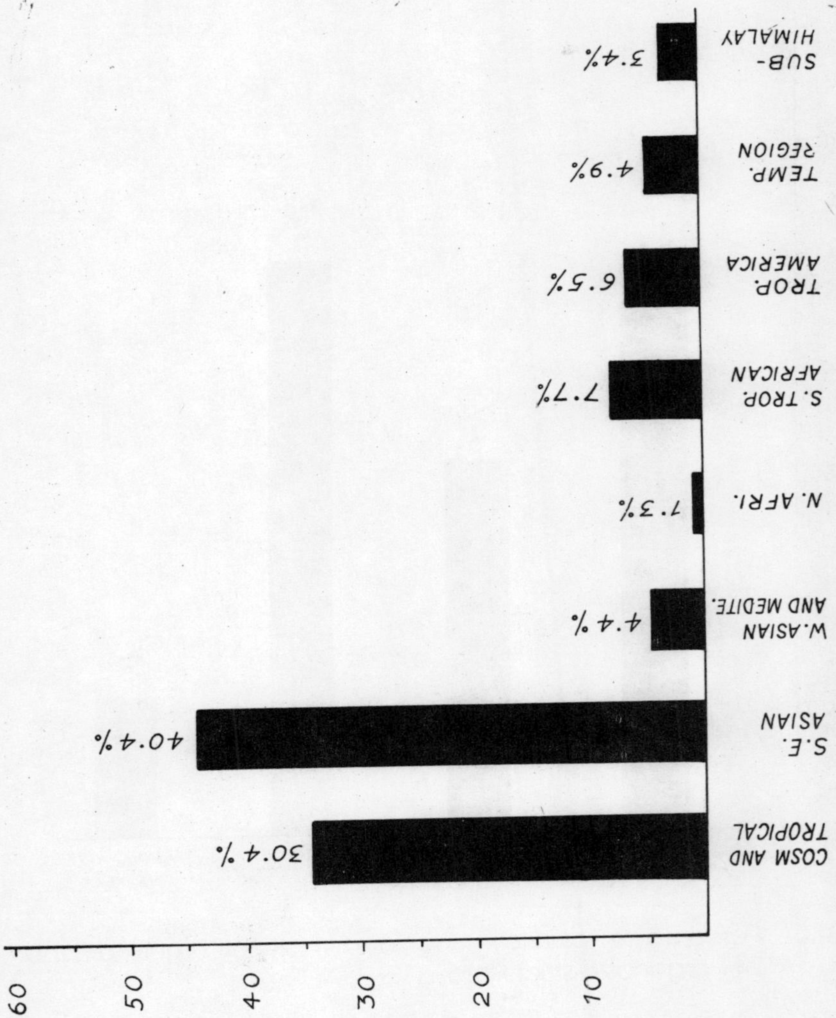

Hist.4. Percentage of Origin of 24-Parganas' Plants

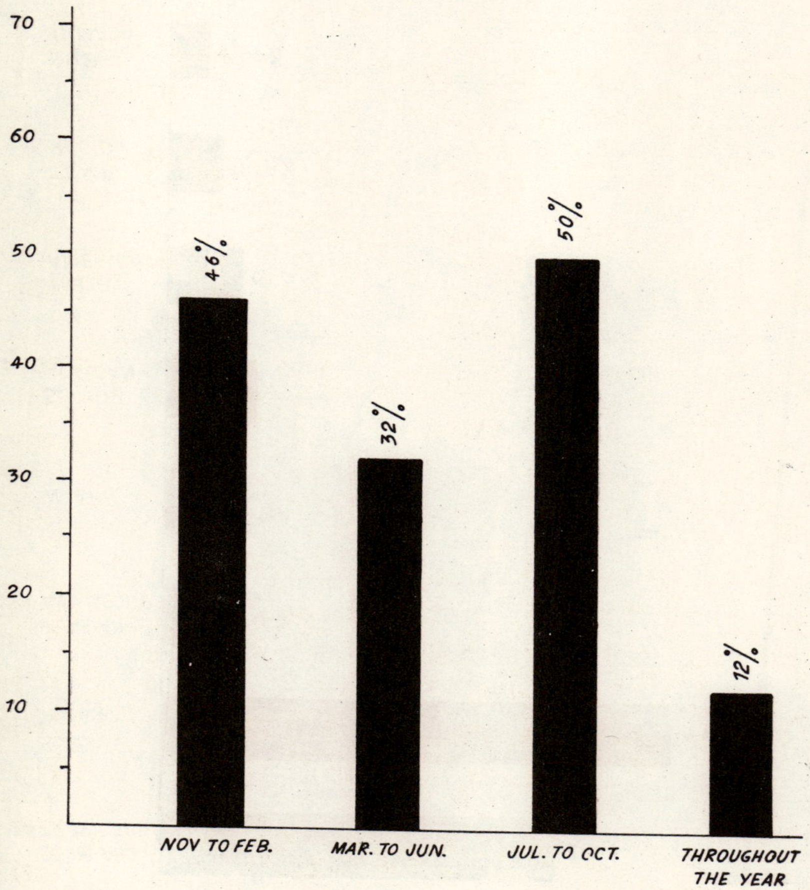

Hist.5. Percentage of Flowering and Fruiting Times of the 24-Parganas District Flora.

ANGIOSPERMEAE

Key to the Families with Brief Diagnostic Characters

1a.	Tap-roots; leaves reticulate veined; floral parts generally 4 or 5 or multiple of 4 or 5; cotyledons two	DICOTYLEDONEAE (2)
1b.	Fibrous-roots; leaves parallel veined; floral parts generally 3 or multiple of 3; cotyledon one	MONOCOTYLEDO-NEAE (124)
2a.(1)	Flowers with petals	(3)
2b.(1)	Flowers usually without petals and calyx rarely absent	(105)
3a.(2)	Perianth segmented, generally in two whrols	(4)
3b.(2)	Petals usually combined into a lobed corolla	(71)
4a.(3)	Ovary generally superior	(5)
4b.(3)	Ovary generally inferior	(58)
5a.(4)	Floral parts spirally arranged	(6)
5b.(4)	Floral parts cyclic	(12)
6a.(5)	Plants aquatic, floating or submerged herbs	(7)
6b.(5)	Plants terrestrial, rarely semi-aquatic	(8)
7a.(6)	Leaves large not dissected; but sometime serrate or dentate	

VI. *NYMPHAEACEAE* (Pages—116-119)

Aquatic herbs with rhizomatous roots; petioles and peduncles very long; flowers large acyclic, perianth segments many, free, spiral; fruits leathery, many seeded.

7b.(6) Leaves smaller, finely dissected

CXXIV. *CERATOPHYLLACEAE* (Page—610)

Branch single, arising from the nodes, flowers minute, monoecious, solitary; perianth of 6-12 narrow sub-valvate segmente.

8a.(6) Mostly trees or shrubs; oil cell present (9)

8b.(6) Mostly climbing shrubs or erect herbs; oil cell absent (11)

9a.(8) Flowers tri-merous; bisexual

IV. *ANNONACEAE* (Pages—106-111)

Buds naked; stipules absent; perianth of trimerous whroles; staments many with adnate anthers; ripe carpels free or rarely conjoined.

9b.(8) Flowers not trimerous (10)

10a.(9) Fruits partly enclosed in a hallow axis

CXV. *LAURACEAE* (Pages—561-565)

Mostly tree, barks and leaves with aromatic or camphoraceous smell; flowers small; ovary syncarpous, tricarpellary, 1-celled, ovule-1, pendulous.

10b.(9) Fruits not enclosed in the axis

III. *MAGNOLIACEAE* (Pages—104-105)

Flowers large, solitary, highly and sweet scented; torus much elongated, fruits with spicate carpels, sometime connate.

11a.(8) Erect herbs or shrubs; fruits achene or follicle

I. *RANUNCULACEAE* (Pages—101-103)

Radical or alternate leaves with twisted, sheathing petioled; flowers regular; calyx petaloid; stamens many, free, fruits achenes or follicles.

11b.(8) Climbing shrubs; fruits a drupe

V. *MENISPERMACEAE* (Pages—111-116)

Slender rarely, woody climbers; leaves simple but palmately nerved, flowers unisexual; perianth several in trimerous whorls; seeds curved.

12a.(5) Plants generally insectivorous

LII. *DROSERACEAE* (Page—317)

Copiously covered with long glandular hairs, with secreting fluids by means of which they catch and digest small animals.

12b.(5) Plants not insectivorous (13)

13a.(12) Placentation parietal; hypanthium absent (14)

13b.(12) Placentation axile; hypanthium present (19)

14a.(13) Flowers actinomorphic (15)

14b.(13) Flowers zygomorphic (18)

15a.(14) Gynoecium not elevated (16)

15b.(14) Gynoecium elevated on a gynophore (17)

16a.(15) Sepals 4, stamens many, carpels 2 to many

VII. *PAPAVERACEAE* (Pages—119-120)

Herbs with coloured or milky sap; flowers regular, bisexual, calyx caducous, stamens many.

16b.(15) Sepals 4; stamens many or 6; carpels 2

IX. *BRASSICACEAE* (Pages—121-127)

Herbs with pungent juice; leaves often lyrate; corolla cruciform with distinct limb and claw; fruits siliqua.

17a.(15) Usually herbs; replum present

XI. *CLEOMACEAE* (Pages—129-132)

Herbs, replum present; fruits a capsule siliqua.

17b.(15) Usually trees or shrubs with hairs and scales, replum absent

X. *CAPPARIDACEAE* (Pages—127-129)

Flowers bracteate, tetramerous; ovary superior, 1 or more locular; placentation parietal with a gynophore.

18a.(14) Flowers hypogynous; stamens and carpels not elevated

VIII. *FUMARIACEAE* (Pages—120-121)

Petals 4 in pairs, somewhat connivent; stamens 4, free and opposite to the petals or 6.

18b.(14) Flowers perigynous

XLVI. *MORINGACEAE* (Page—236)

Leaves pinnately compound; flowers zygomorphic.

19a.(13) Carpels 1 to many, free or united; overy surrounded by an androperianth tube (20)

19b.(13) Carpels united; androperianth tube absent (24)

20a.(19) Flowers hypogynous; plants succulent

LI. *CRASSULACEAE* (Page—317)

Stems and leaves succulent; stamens and carpels as many as petals and pistil with a scaly gland at the base.

20b.(19) Flowers perigynous (21)

21a.(20) Fruits not a legume

L. *ROSACEAE* (Pages—315-316)

Leaves alternate, simple or variously compound, flowers small, medium or large, regular; calyx 5-10, the odd sepals superior; petals free; stamens many on the disc lining the hypanthium.

21b.(20) Fruits a legume (22)

22a.(21) Inflorescence a raceme; flowers irregular; petals imbricate or vexillary; stamens usually 10 (23)

22b.(21) Inflorescence a head or spike; flowers regular; petals valvate; stamens 4 to numerous, free or united

XLIX. *MIMOSACEAE* (Pages—304-315)

Leaves 2-pinnate; flowers small regular, 4-5 merous, into dense heads and spikes.

23a.(22) Calyx distinctly united below beyond the top of the disc; posterior petals outermost largely (standard); stamens 10 often monadelphous or diadelphous (9+1).

XLVII. *PAPILIONACEAE* (Pages—237-287)

Leaves almost compound, stipulate; corolla with standard wings and keel; ovary superior.

23b.(22) Calyx divided into lobes (expt. *Bauhinia*) down to the top of the disc; posterior petals innermost and smallest; stamens 10 or less, some are abortive, free

XLVIII. *CAESALPINIACEAE* (Pages—288-304)

Leaves deeply emerginate; inflorescence a raceme; flowers slightly irregular, petals imbricate, posterior one innermost and smallest.

24a.(19) Stamens indefinite (25)

24b.(19) Stamens rarely more than twice as many as petals and alternate to them (41)

24c.(19) Stamens as many as petals and opposite to them (56)

25a.(24) Sepals valvate; placentation axile (26)

25b.(24) Sepals imbricate; placentation parietal (30)

26a.(25) Sepals free (27)

26b.(25) Sepals united

XXV. *STERCULIACEAE* (Pages—173-186)

Flowers regular or zygomorphic; calyx persistent gamosepalous; stamens usually obdiplostemous but sometimes many, monadelphous or united.

27a.(26) Stamens united into a column (28)

27b.(26) Stamens free or in separate bundles (29)

28a.(27) Carpels 5 to many

XXIII. *MALVACEAE* (Pages—148-171)

Leaves simple with palmately veined, alternate, stipulate, flowers regular, subtended by a whorl or bract and bracteoles.

28b.(27) Carpels 2-5

XXIV. *BOMBACACEAE* (Pages—171-173)

Tall tree with thick and spreading branches; stamens more or less free and pendulous; anthers 1-celled.

29a.(27) Petals thin, coloured, unguiculate, entire or sub-entire, imbricate or twisted in buds, anthers globose or oblong

XXVI. *TILIACEAE* (Pages—186-191)

Flowers regular, without an epicalyx; calyx with free sepals, deciduous; stamens many not united into a tube.

29b.(27) Petals rigid white or sepaloid, almost always lacinate, induplicate, valvate in buds, anthers linear

XXVII. *ELAEOCARPACEAE* (Pages—191-192)

Tall trees; anthers opening by terminal pores, linear; petals incised; fruit drupaceous.

30a.(25) Endosperm starchy (31)

30b.(25) Endosperm oily (33)

31a.(30) Leaves alternate (32)

31b.(30) Leaves opposite or whorled

XX. *ELATINACEAE* (Pages—143-144)

Usually small marshy herbs; leaves with paired stipules; inflorescence axillary cymose, tri-or tetra-merous flowers.

32a.(31) Leaves not scale-like; sap coloured

XIII. *BIXACEAE* (Page—133)

Woody plants with palminerved or palmately lobed, larger leaves and usually with minute caducous stipules; flowers large.

32b.(31) Leaves scale-like; sap watery

XIX. *TAMARICACEAE* (Pages—140-143)

Plants with sinuous branchlets, scale-like leaves; sepals and petals 5, rarely 4; stamens iso- or diplo-stemonous, on the margin of a crenulate disc, sub-perigynous; ovary 1-celled.

33a.(30) Leaves alternate (34)

33b.(30) Leaves opposite or whorled

XXI. *CLUSIACEAE* (Pages—144-147)

Plants with resin or oil; sepals 4 to many; stamens many, free or somewhat united into bundle.

34a.(33) Petals somewhat united (35)

34b.(33) Petals completely free (36)

35a.(34) Fruits not winged

LXVI. *CARICACEAE* (Pages—358-359)

Erect trunk with terminal crown of leves; with milky latex; petals united on the base, with double row of stamens.

35b.(34) Fruits winged

XXII. *DIPTEROCARPACEAE* (Page—147)

Plants with abundant resins; flowers panicled; sepals 5, connate below, calyx persistent; petals 5, contorted, stamens usually a multiple of 5.

36a.(34) Pistil apocarpous

II. *DILLENIACEAE* (Pages—103-104)

Leaves with strong pinnately nerved and sheathing petioles; flowers very large; sepals and petals mostly 5; stamens many in bundles, carpels 5-20.

36b.(34) Pistil syncarpous (37)

37a.(36) Flowers generally with gynophore (38)

37b.(36) Flowers without a gynophore (39)

38a.(37) Flowers with a distinct corona between petals and stamens

LXV. *PASSIFLORACEAE* (Pages—357-358)

Plants climbing with tendrils; leaves with gland on petioles; flowers with modified corona.

38b.(37) Flowers without a distinct corona between the petals and stamens

LXIV. *SAMYDACEAE* (Page—357)

Flowers axillary, fascicled, pedicels short jointed, bracts scale-like, numerous; fruits succulent, globose or ovoid or ellipsoid.

39a.(37) Sap coloured

XIV. *COCHLOSPERMACEAE* (Page—134)

Leaves deeply, digitately lobed or divided; flowers opening before leaves; capsule smooth, seeds covered with cotton.

39b.(37) Sap watery (40)

40a.(39) Flowers often perfect

XV. *FLACOURTIACEAE* (Pages—134-135)

Leaves pellucid dotted; flowers polygamous; sepals 4-several; petals small, 1-3 times as many as sepals or indefinite.

40b.(39) Flowers perfect, placentation parietal

XII. *VIOLACEAE* (Pages—132-133)

Flowers irregular, sepals 5 persistent, petals 5, hypogynous; ovary 1 celled with many anatropous ovules on 3 parietal placentae.

41a.(24) Ovule pendulous with a dorsal raphe (42)

41b.(24) Ovule pendulous with a ventral raphe (46)

42a.(41) Flowers zygomorphic (43)

42b.(41) Flowers actinomorphic (45)

43a.(42) Plants herbaceous

XXXII. *BALSAMINACEAE* (Pages—197-198)

Plants with watery stem; irregular flowers, anthers connate around ovary and capsule with explosive dehiscence.

43b.(42) Plants woody (44)

44a.(43) Leaves opposite

XL. *HIPPOCRATEACEAE* (Page—217)

Leaves pinnately compound; inflorescence thyrsoid; flowers irregular; stigma simple, coriaceous capsule and cotyledons adherent face to face.

44b.(43) Leaves alternate

XLIV. *SAPINDACEAE* (Pages—226-232)

Leaves palmately or pinnately compound; inflorescence panicles or dichotomous cymes; flowers polygamous with annular or oblique disc; stamens 5-10, free.

45a.(42) Stems with resin-ducts

XLV. *ANACARDIACEAE* (Pages—232-235)

Plants with resinous juice; inflorescence panicle; female polygamous with cup-shaped or glandular disc.

45b.(42) Stems without resin-ducts

XXXIX. *CELASTRACEAE* (Pages—216-217)

Fasciculate or cymose flowers, ovary with a disc, berry or capsule; seeds enveloped by bright coloured aril.

46a.(41) Flowers unisexual (47)

46b.(41) Flowers bisexual (49)

47a.(46) Flowers reduced, plants terrestrial, carpels 3

CXVIII. *EUPHORBIACEAE* (Pages—569-598)

Plants with milky sap; ovary superior, 3-locular, ovules collaterals, pendulous with ventral raphe.

47b.(46) Flowers not reduced (48)

48a.(47) Ovules two in each loculus

XXXVI. *BURSERACEAE* (Page—206)

Resin-duct present, leaves compound; flowers unisexual; stamens distinct.

48b.(47) Ovule one in each loculus

XXXIV. *SIMAROUBACEAE* (Page—205)

Bark bitter; leaves pinnate; flowers unisexual; filament based with scale and one ovule per loculus.

49a.(46) Flowers with secretary cells (50)

49b.(46) Flowers without secretary cells (51)

50a. (49) Leaves not gland dotted

XXXVII. *MELIACEAE* (Pages—207-214)

Leaves pinnate, exstipulate, oblique at the base; flowers hypogynous with annular or tubular disc.

50b.(49) Leaves with gland dotted.

XXXIII. *RUTACEAE* (Pages—198-211)

Flowers hypogynous with cushion-like or annular disc; ovary superior, elevated on a disc.

51a.(49) Flowers actinomorphic (52)

51b. (49) Flowers zygomorphic (55)

52a. (51) Carpels not splitting apart at maturity (53)

52b. (51) Carpels splitting apart at maturity (54)

53a.(52) Leaves alternate, simple, entire; alternate stamens usually reduced to staminodes; filaments united at the base, ovary 3-celled.

XXVIII. *LINACEAE* (Pages—192-193)

Flowers pentamerous, actinomorphic, corolla contorted of distinct and fugaceous petals; filaments basally connate, small staminodes.

53b.(52) Leaves alternate, simple, stipulate; stamens many on the large disc; ovary deeply 3-10 lobed, the lobes becoming in fruit.

XXXV. *OCHNACEAE* (Page—206)

Plants with watery sap; flowers regular, hermaphrodite, bracteate; sepals 4-5, imbricate, persisting; petals 5, rarely 4-10, free, hypogynous, imbricate; anther basifixed; deciduous; carpels united into a superior ovary.

54a.(52) Carpels splitting as mericarps

XXXI. *OXALIDACEAE* (Pages—195-197)

Leaves exstipulate; flowers actinomorphic, calyx imbricate; contorted corolla, stamens connate at the base; styles free.

54b.(52) Carpels splitting as nutlets

XXX. *ZYGOPHYLLACEAE* (Pages—194-195)

Leaves pinnate or two-foliolates; persistent paired stipules; flowers with prominent disc; free stamens bearing basal scales and carpels splitting aprat at maturity as spiny or tuberculate nutlets.

55a.(51) Anthers with length wise dehiscence

XXIX. *MALPIGHIACEAE* (Pages—193-194)

Lianous habit; leaves opposite, appressed, modified unicellular hairs, glandular calyx, clawed petals, peculiar anthers and lobed winged fruits.

55b.(51) Anthers with poricidal dehiscence

XVI. *POLYGALACEAE* (Page—136)

Flowers subtended by bracts and bracteoles. Sepals 5, the inner 2 larger, petaloid and wing-like. Petals usually 3, the lower median a dorsal frimbriate appendages.

56a. (24) Trees, shrubs or lianas, fruit drupe

XLI. *RHAMNACEAE* (Pages—217-219)

Plants spiny; inflorescence axillary cymes; flowers perigynous with disc filling the calyx tube; calyx valvate.

56b. (24) Climbing shrubs; fruits a berry (57)

57a. (56) Tendril-climbing shrubs, flowers in racemes or cymes clusters; stamens free; ovary 2-celled, ovule 2 in each cell

XLII. *VITACEAE* (Pages—219-224)

Plants climbing by leaf opposed tendrils; inflorescence arising opposite to a leaf at the nodes; flowers hypogynous with prominent disc, androecium antipetalous.

57b.(56) Erect shrubs; leaves large with sheathing petioles, stamens connate, ovary 3-6 celled

XLIII. *LEEACEAE* (Pages—224-226)

Branches striate or furrowed; flowers on leaf-opposed peduncles, in corymbose cymes; calyx 5-toothed; petals 5, connate below and adherent to the staminal tube, revolute.

58a.(4) Plants spiny, usually succulent

LXVIII. *CACTACEAE* (Page—371)

Leaves reduced to tufts of bristles or spines or glochids arranged in areoles; flowers solitary; perianth gradual transition from sepals to petals.

58b (4) Plants neither spiny nor succulent (59)

59a.(58) Flowers in umbels, ovules 1-2 in each loculus (60)

59b.(58) Flowers not in umbels; ovules several to many in each loculus (61)

60a (59) Flowers pentamerous

LXXI. *APIACEAE* (Pages—373-377)

Aromatic herbs with fistular stems; leaves sheathing base; inflorescence umbel; flowers epigynous; corolla unequal stamens free, 5.

60b.(59) Flowers tetramerous

LXXII. *ALANGIACEAE* (Page—377)

Umbellate inflorescence, inferior ovary and fleshy indehiscent drupe or berry.

61a.(59) Flowers perigynous (62)

61b.(59) Flowers epigynous (64)

62a.(61) Leaves with 3-9 longitudinal veins

LVIII. *MELASTOMETACEAE* (Page—346)

Petals 4 or 5, free, imbricate; stamens as many as or double of petals, filaments unequal and curved inwards.

62b.(61) Leaves not so veined (63)

63a.(62) Ovary 2-6 celled; embryo straight

LIX. *LYTHRACEAE* (Pages—346-349)

Plants often with 4 angular branches; calyx valvate; corolla crumpled; androecium inflexed in buds; ovary superior, 2-6 locular; ovules many.

63b.(62) Ovary more than 6 celled; embryo not straight

LX. *SONNERATIACEAE* (Pages—350-355)

Plants with pneumatophores and halophytic; flowers large, solitary, axillary; bracts -o; sepals 4-8, thickly leathery, connate below in a widely campanulate calyx.

64a.(61) Plants aquatic (65)

64b.(61) Plants terrestrial (66)

65a.(64) Petals present

LXIII. *TRAPACEAE* (Pages—356-357)

Branches assimilatory; uniseriate, multicellular hairs on petioles; pedicels and lower surface of leaves; top-shaped drupe.

65b. (64) Petals reduced or not

LIII. *HALORAGACEAE* (Page—318)

Plants monoecious; unisexual; stamens 4-8 and inferiur ovary with a solitary ovule in each loculus.

66a. (64) Loculi superposed

LXI. *PUNICACEAE* (Page—355)

Flowers large, orange-red, axillary, or several clustered; sepals 5-7, connate in a funnel shaped coriaceous calyx.

66b. (54) Loculi not superposed (67)

67a.(66) Leaves alternate (68)

67b. (66) Leaves opposite (69)

68a.(67)′ Fruits often woody

LVII. *LECYTHIDACEAE* (Pages—345-346)

Leaves not gland-dotted, stamens many; ovary inferior; berry or woody capsules.

68b.(67) Fruits not woody

LVI. *MYRTACEAE* (Pages—339-345)

Leaves gland dotted; flowers epigynous with disc living the calyx tube; stamens many, bent inwards in bud, free.

69a.(67) Generally lianas

LV. *COMBRETACEAE* (Pages—334-339)

Calyx valvate, stamens 4-5 in 1 whorl or 6-10 in whorls.

69b .(67) Not lianas (70)

70a.(69) Viviparous germination, common mangroves

LIV. *RHIZOPHORACEAE* (Pages—318-334)

Leaves coriaceous; corolla convolute or inflexed in bud; stamens 8- many; anther 4-celled; ovary inferior.

70b. (69) Non-viviparous germination and not like the mangroves

LXII. *ONAGRACEAE* (Page—356)

Flowers epigynous, tetramerous, sepals valvate; corolla convolute; stamens in 2 whorl of 4; outer whorl larger than inner; ovary 4 locular; ovules many.

71a. (3) Ovary superior (72)

71b. (3) Ovary inferior (101)

72a. (71) Stamens opposite to the corolla lobes (73)

72b. (71) Stamens alternate with the corolla lobes (77)

73a. (72) Stamens usually in 1 whorl (74)

73b. (72) Stamens in 2-whorl (76)

74a. (73) Style 1, fruit a berry, capsule a drupe (75a)

74b. (73) Style 5, fruit an achene or an utricle (75b)

75a. (74) Shrubs or tree; fruits a berry or a drupe

LXXIX. *MYRSINACEAE* (Pages—425-428)

Shrubs or small tree; ovules generally burried in the proliferating placentae; fruit drupe.

75b. (74) Herbs; fruit a capsule

LXXVIII. *PRIMULACEAE* (Page—425)

Petals 5, united; stamens as many as corolla lobes and opposite to them; placentation free central; fruit capsule.

76a. (73) Stamens as many as corolla lobes

LXXX. *SAPOTACEAE* (Pages—428-430)

Plants bearing laticiferous ducts; leaves usually exstipulate; stamens epipetalous in 2 or 3 whorls, outer antipetalous whorls reduced to staminodes or wanting.

76b. (73) Stamens twice as many as corolla lobes

LXXXI. *EBENACEAE* (Pages—430-432)

Flowers unisexual; stamens as many as corolla lobes, multilocular ovary, ovules bitegmic and typically in pairs, styles two to eight, as well as seeds.

77a. (72) Corolla herbaceous; fruit not a circumscissile capsule (78)

77b. (72) Corolla scarious; fruit a circumscissile capsule

CVI. *PLANTAGINACEAE* (Page—541)

Seemingly parallel-veined, basal sheathing at the base. Inflorescence capitate or spicate on wiry or stout scapes.

78a.(77) Stamens inserted near the corolla base; ovary 2-celled (79)

78b.(77) Stamens adnate to the corolla tube, ovary with 2-5 celled (86)

79a. (78) Plants with milky sap (80)

79b. (78) Plants without milky sap (82)

80a. (79) Filaments distinct, follicles linear

LXXXV. *APOCYNACEAE* (Pages—435-447)

Plants with latex; exstipulate, calyx imbricate; stamens 5, epipetalous, free from the stigma.

80b. (79) Filaments united follicles lanceolate (81)

81a. (80) Pollen aggregated into 2-4 waxy bodies, called pollinia

LXXXVI. *ASCLEPIADACEAE* (Pages—447-455)

Plants with latex; anthers adnate to the stigma; pollen grains cohere to form pollinia. Stigma 5-angled; fruit pair of follicles.

81b. (80) Pollen granular and in tetrads; translators spoon shaped and provided with adhesive disc

LXXXVII. *PERIPLOCACEAE* (Pages—455-456)

Twining shrubs; leaves hoary pubescent beneath; petals connate in a rotate corolla; lobes thick.

82a. (79) Shrubs or trees (83)

82b. (79) Plants generally herbaceous; style not cleft

LXXXIX. *MENYANTHACEAE* (Pages—457-460)

Leaves exstipulate, cymose, stamens as many as corolla lobes; ovary unilocular with glandular disc.

83a.(82) Stamens fewer than corolla lobes (84)

83b.(82) Stamens as many as corolla lobes

LXXXVIII. *LOGANIACEAE* (Pages—456-457)

Leaves opposite, stipulate; stamens as many as corolla-lobes; ovary superior, 2-locular.

84a. (83) Fruits generally a didynamous berry (85)

84b. (83) Fruits compound capsule

LXXXIII. *NYCTANTHACEAE* (Page—434)

Branches quadrangular, flowers in small sessile, bracteate heads disposed in terminal trichotomous cymes.

85a.(84) Stamens two; ovary free, 2-celled, seeds albuminous or not

LXXXII. *OLEACEAE* (Pages—433-434)

Leaves opposite; flowers dimerous; corolla free or connate.

85b.(84) Stamens four; ovary free, 1-2 or imperfectly 4-celled, seed exalbuminous

LXXXIV. *SALVADORACEAE* (Page—435)

Flowers small, dioecious or polygamous-dimorphic, clustered or panicled, sepals connate in free campanulate.

86a.(78) Corolla mostly regular (87)

86b. (78) Corolla irregular (95)

87a.(86) Climbing herbs or shrubs (88)

87b.(86) Erect herbs or shrubs (89)

88a.(87) Plants autotrophic with green leaves, interstaminal scale lacking; cotyledons folded

XCIV. *CONVOLVULACEAE* (Pages—465-478)

Leaves alternate, exstipulate, corolla twisted; intrastaminal disc annular or cupular, ovary superior, 2 locular with 2 ovules.

88b. (87) Plants parasitic without green leaves, interstaminal scales present; cotyledons reduced or lacking

XCV. *CUSCUTACEAE* (Page—479)

Leaflets twining, yellow or reduced parasitic herbs; flowers small, white or rose-coloured, fascicled sessile or shortly pedicelled.

89a.(87) Ovary 3-celled

XC. *POLEMONIACEAE* (Page—460)

Calyx gamosepalous; stamens inserted at different levels on the corolla tube; ovary tricarpellate; ovules indefinite.

89b. (87) Ovary not 3-celled (90)

90a.(89) Ovary not separating into nutlets at maturity (91)

90b. (89) Ovary separating into nutlets at maturity (92)

91a. (90) Style usually deeply 2-cleft

XCI. *HYDROPHYLLACEAE* (Page—460)

Helicoid cymes; imbricate aestivation, stamens without scales, between the filaments.

91b. (90) Style not deeply 2-cleft

XCVI. *SOLANACEAE* (Pages—479-489)

Leaves alternate, exstipulate, often in unequal pairs. Flowers regular, corolla paired. Anthers apparently connate.

92a. (90) Stamens 2 or 3 (93)

92b. (90) Stamens 5, carpels 2 (94)

93a. (92) Inflorescence racemose, centripetal, spicate

CIII. *VERBENACEAE* (Pages—518-526)

Leaves mostly opposite, ovary 4-lobed or entire, superior, 2-4 locular, ovules 1-2 in each loculus.

93b. (92) Inflorescence capitate cymes

CIV. *AVICENNIACEAE* (Pages—526-531)

Plants with pneumatophores and crypto-viviparous germination.

94a. (92) Style usually terminal, fruit a drupe

XCII. *EHRETIACEAE* (Pages—460-462)

Styles simple or bilobed; drupe 2-seeded or of 1-seeded, pyrenes.

94b. (92) Style gynobasic, fruit a nutlets

XCIII. *BORAGINACEAE* (Pages—462-465)

Leaves exstipulate; inflorescence scorpioid cyme; ovary 4-lobed, 2-locular having 2 ovules in each loculus.

95a. (86) Ovary separating into nutlets at maturity; stamens 2 or 4; styles gynobasic

CV. *LAMIACEAE* (Pages—531-541)

Plants often aromatic with square stems; leaves opposite; flowers often bilabiate, ovary 4-lobed superior.

95b. (86) Ovary not separating into nutlets at maturity (96)

96a. (95) Parasitic or insectivorous plants (97)

96b. (95) Green plants, hardly parasitic, not insectivorous (98)

97a. (96) Plants root-parasitic

XCVIII. *OROBANCHACEAE* (Pages—497-498)

Flowers zygomorphic, calyx valvate, stamens didynamous; ovary one-celled with 2-6 parietal placentae.

97b. (96) Plants insectivorous and parasitic

XCIX. *LENTIBULARIACEAE* (Page—498)

Leaves bearing utricles, flowers irregular, stamens 2, epipetalous, anther 1-celled.

98a. (96) Fruits commonly spiny; placentation axile

CI. *PEDALIACEAE* (Page—503)

Plants with mucilage containing glandular hairs; flowers irregular; ovary superior; 2-4 locular; fruit completely spinous.

98b. (96) Fruits not spiny; berry or capsule (99)

99a. (98) Anthers usually free (100)

99b. (98) Anthers united

CII. *ACANTHACEAE* (Pages—503-518)

Plants often with cystoliths; flowers irregular with prominant bracts and bracteoles.

100a. (99) Plants herbaceous

XCVII. *SCROPHULARIACEAE* (Pages—490-498)

Leaves exstipulate; flowers irregular petals often spread, stamens 4, didynamous.

100b. (99) Plants woody

C. *BIGNONIACEAE* (Pages—499-503)

Flowers zygomorphic, siliqua-like woody capsule.

101a. (71) Stamens distinct, rarely coherent

LXXIII. *RUBIACEAE* (Pages—337-389)

Leaves opposite, stipulate; flowers epigynous, petals 4 or 5; stamens 4 or 5, epipetalous; anthers introse. Ovules 1 - many in each loculus.

101b. (71) Stamens coherent or variously connate (102)

102a.(101) Flowers in involucrate heads

LXXIV. *ASTERACEAE* (Pages—391-422)

Capitulum inflorescence; flowers epigynous, corolla united.

102b.(101) Flowers not in involucrate heads (103)

103a.(102) Flowers generally imperfect

LXVII. *CUCURBITACEAE* (Pages—359-371)

Plants climbing by solitary extra axillary tendrils. Leaves palmi-nerved, cordate lobed; flowers epigynous.

103b.(102) Flowers generally perfect (104)

104a.(103) Flowers generally lax

LXXV. *CAMPANULACEAE* (Page—422)

Plants with watery or milky sap; leaves exstipulate, flowers epigynous, corolla valvate.

104b.(103) Flowers in dense spikes

LXXVI. *SPHENOCLEACEAE* (Page—422)

Aquatic fleshy herbs; bears alternate leaves; bracteate spike and white flowers.

105a.(2) Ovules with single embryo sac

CXXIII. *CASUARINACEAE* (Pages—609-610)

Stem jointed, verticillate, leaves whorled scale like; flowers perianthless, stamens one. Fruits enclosed by woody bracteoles.

105b.(2) Ovules with many embryo sac (106)

106a.(105) Perianth none or bract like; flowers mostly perfect in spikes

CXIV. *PIPERACEAE* (Pages—560-561)

Succulent herbs; inflorescence dense spikes; flowers minute, bracteate naked; ovules basal, orthotropous 1 in each .

106b.(105) Perianth either sepaloid or petaloid (107)

107a.(106) Ovary inferier (108)

107b.(106) Ovary superior (110)

108a.(107) Stamens opposite to the perianth-segments and united with them (109)

108b.(107) Stamens alternate with the perianth -segments and free from them

CXIII. *ARISTOLOCHIACEAE* (Pages—559-560)

Stems softly woody with broad, medullary rays; leaves exstipulate; perianth petaloid, enlarged to form a bell-like or trumpet like structure, ovary inferior.

.09a.(108) Ovary differentiated from placenta

XXXVIII. *OLACACEAE* (Page—216)

Exstipulate leaves, flowers with distinct calyx and corolla, ovary superior and placenta free ovules.

109b.(108) Ovary undifferentiated from placenta

CXVII. *LORANTHACEAE* (Page—565)

Plants aerial parasitic; receptacle cup-shaped; carpels 1, ovary inferior, unilocular; loculus ill defined.

110a.(107) Flowers perfect or imperfect; carpels 1

CXVI. *PROTEACEAE* (Page—565)

Leaves exstipulate; showy bracteate heads. Perianth tetramerous, petaloid, coloured valvate.

110b.(107) Flowers perfect; carpels 2 or more (111)

111a.(110) Leaves usually ochreate

CXII. *POLYGONACEAE* (Pages—556-559)

Herbs with swollen nodes; leaves stipulate; ovary triangular, superior unilocular, ovule solitary, erect orthotropous.

111b.(110) Leaves not ochreate; carpels 2 or more (112)

112a.(111) Placentation mostly at the apex of loculus (113)

112b.(111) Placentation free-central or basal (116)

113a.(112) Fruit not an achene (114)

113b.(112) Fruit an achene (115)

114a.(113) Fruit a drupe, nut or samara; sap not milky

CXIX. *ULMACEAE* (Page—598)

Sap not milky; filaments erect. Ovules amphitropous, anatropous, pendulous from the apex of ovarian chambers; drupe or samara.

114b.(113) Fruit multiple, sap milky

CXXI. *MORACEAE* (Pages—601-609)

Plant with milky sap; filaments inflexed or straight.

115a.(113) Plants armed with stinging hairs, not aromatic.

CXX. *URTICACEAE* (Pages—598-600)

Leaves armed with stinging hairs.

115b.(113) Plants erect or climbing, aromatic

CXXII. *CANNABINACEAE* (Page—609)

Climbing or erect aromatic herbs, fruits achene.

116a.(112) Perianth bract like (117)

116b.(112) Perianth not bract like (118)

117a.(116) Stamens free; staminodes rare

CIX. *CHENOPODIACEAE* (Pages—551-554)

Succulent herbs, inflorescence cyme or spike, without scarious bracts; ovary superior.

117b.(116) Stamens united at the base, often with alternating staminodes

CVIII. *AMARANTHACEAE* (Pages—544-550)

Inflorescence congested with scarious bracts; perianth dry, membranous; filaments connate in a cup or tube.

118a.(116) Flowers haplochamydeous (119)

118b.(116) Flowers diplochlamydeous (122)

119a.(118) Perianth segments 5, united, carpel 1

CVII. *NYCTAGINACEAE* (Pages—541-543)

Leaves often in unequal pairs; flowers involucrate; perianth petaloid; ovary superior, unilocular, fruit achenes.

119b.(118) Perianth segments 4 or 5, free or united; carpels generally more than 1 (120)

120a.(119) Leaves neither fleshy nor thread like

CXI. *PHYTOLACCACEAE* (Pages—555-556)

Leaves exstipulate, perianth small, non-petaloid, 4-9, segmented.

120b.(119) Leaves either fleshy or thread-like (121)

121a.(120) Perianth segments united from a tube; staminodes absent; ovary inferior, capsule loculicidal

LXIX. *AIZOACEAE* (Pages—371-372)

Leaves fleshy or thread like, stamens 3-5 or many, the outermost form petaloid structure, ovary inferior.

121b.(120) Perianth segments almost free; petaloid staminodes generally absent; ovary superior

LXX. *MOLLUGINACEAE* (Pages—372-373)

Flowers small, greenish, axillary, sessile or pedicelled, clustered or in cymes or racemes.

122a.(115) Sepals 2; petals 4 or 5 (123)

122b.(115) Sepals 5; petals 0 or 5

XVII. *CARYOPHYLLACEAE* (Pages—137-139)

Stem with swollen nodes; leaves opposite; inflorescence dichotomous cymes; stamens 8-10 in 2 whorls.

123a.(122) Twining herbs

CX. *BASELLACEAE* (Pages—554-555)

Climbing succulent herbs; flowers often brightly coloured.

123b.(122) Herbs or undershrubs

XVIII. *PORTULACACEAE* (Pages—139-140)

Leaves succulent with scarious stipules, flowers hypogynous with a disc.

124a(1) Carpels free (125)

124b.(1) Carpels united (135)

125a.(124) Inflorescence not a true spadix (126)

125b.(124) Inflorescence a dense glomerule or fleshy spadix (133)

126a.(125) Perianth sepaloid or petaloid (127)

126b.(125) Perianth scale-like or bristle like (131)

127a.(126) Ovary inferior; perianth made up of calyx and corolla

CXXV. *HYDROCHARITACEAE* (Pages—613-614)

Aquatic herbs; epigynous; ovary inferior, unicellular, placentation periatal.

127b.(126) Ovary superior (128)

128a.(127) Perianth sepaloid (129)

128b.(127) Perianth petaloid in 1 whorl (130)

129a.(128) Perianth of 4, concave, green valvate segments

CL. *POTAMOGETONACEAE* (Page—662)

Perennial, submerged herb; flowers floating; perianth variable; stamens 1-4; carpels 1-6, ovule solitary apical.

129b.(128) Perianth of 1-3, equal or unequal, pink or white segments or obsolate

CXLIX. *APONOGETONACEAE* (Pages—661-662)

Laticiferous, perennial aquatic herbs; leaves reticulately veined.

130a.(128) Ovule solitary in each carpel, rarely 2 or few; placentation marginal

CXLVII. *ALISMATACEAE* (Pages—660-661)

Perianth segments 6, outer 3 sepaloid and inner 3 petaloid, free; stamens 6, 9 or indefinite, free.

130b.(128) Ovules many in each carpel; placentation almost superficial

CXLVIII. *BUTOMACEAE* (Page—661)

Placentation superficial.

131a.(126) Plants woody

CXLIII. *PANDANACEAE* (Pages—651-662)

Palm like, branched shrubs or small tree, supported by stilt roots, dioecious.

131b.(126) Plants herbaceous (132)

132a.(131) Plants generally submerged; roots not rhizomatous

CLI. *NAJADACEAE* (Pages—662-663)

Flowers wholly submerged; ovary superior, unilocular, ovule solitary, erect.

132b.(131) Plants not submerged, roots rhizomatous

CXLIV. *TYPHACEAE* (Pages—652-653)

Tall aquatic or marshy herbs with creeping rhizomes, leaves erect.

133a.(125) Leaves generally plicate; carpels one ovuled

CXLII. *ARECACEAE* (Pages—645-651)

Leaves large, plaited in buds, inflorescence simple or compound spadix, enclosed in 1 or more spathes; perianth small, leathery.

133b.(125) Leaves generally plicate (134)

134a.(133) Rhizomatous or tuberous; rarely woody

CXLV. *ARACEAE* (Pages—653-658)

Herbs with acrid watery latex, leaves broad, long petioled.

134b.(134) Small, free-floating thalloid aquatic herbs, very minute

CXLVI. *LEMNACEAE* (Pages—658-663)

Minute, free-floating, freshwater herbs; without roots or with thread like roots.

135a.(124) Flowers borne in the axile or dry chaffy scales (136)

135b.(124) Flowers not in the axile of dry chaffy scales (137)

136a.(135) Culms mostly solid, triangular; leafsheath closed; fruit an achene

CLIII. *CYPERACEAE* (Pages—665-679)

Solid, triangular culms, leaves trichotomous, without ligule; leafsheath closed.

136b.(135) Culms mostly hollow, cylindric; leaf sheath like; fruit a caryopsis

CLIV. *POACEAE* (Pages—679-748)

Inflorescens panicles; culms round and hollow between internodes.

137a.(135) Plants neither saprophytic nor epiphytic; seeds not small; endosperm present (138)

137b.(135) Plants mostly epiphytic or saprophytic; seeds very small; endosperm absent

CXXVI. *ORCHIDACEAE* (Pages—614-617)

Flowers showy, bisexual, irregular and epigynous; embryo undifferentiated.

138a.(137) Endosperm mealy (139)

138b.(137) Endosperm fleshy or hoary (143)

139a.(138) Perianth variable; generally sepaloid (140)

139b.(138) Perianth not variable; generally petaloid

CXXXVIII. *PONTEDERIACEAE* (Pages—640-641)

Leaves with swollen petiole; inflorescence panicle or raceme, subtended by spathe like leaf sheath.

140a.(139) Plants usually with a grass like habit (141)

140b.(139) Plants not usually grass like habit (142)

141a.(140) Flowers inconspicous

CLII. *ERIOCAULACEAE* (Pages—663-664)

Small acaulent herbs with a dense tufted of basal linear and grass like leaves and one or more scapes.

141b.(140) Flowers conspicous

CXXXIX. *XYRIDACEAE* (Page—641)

Tufted perennial marsh plants, showy irregular flowers in bracteate heads.

142a.(140) Plants smaller

CXL. *COMMELINACEAE* (Pages—641-645)

Terrestrial with jointed stem, leaves alternate, entire, sheathing; perianth segments 6, in 2 seriate.

142b.(140) Plants larger with serrate leaves

CXXX. BROMELIACEAE (Page—624)

Inflorescence brightly coloured; flowers trimerous; perianth biseriate, anthers 2-celled.

143a.(138) Functional stamens 3-6 (144)

143b.(138) Functional stamen 1 (151)

144a.(143) Plants normally climbing or twining, not armed (145)

144b.(143) Plants normally erect or rarely twining; occasionally armed (145)

145a.(144) Perianth sepaloid; ovary inferier

CXXXV. *DIOSCOREACEAE* (Pages—629-633)

Large fleshy twining herbs with bulbils on the internodes; flowers unisexual.

145b.(144) Perianth sub-petaloid; ovary superior

CXLI. *FLAGELLARIACEAE* (Page—645)

Stem leafy erect or scandent climbing by the cirrhose leaf-tips; ovary 3-celled.

146a.(144) Ovary superior (147)

146b.(144) Ovary inferior (148)

147a.(146) Plants generally erect herbs or shrubs

CXXXVI. *LILIACEAE* (Pages—635-638)

Inflorescence racemose, sometime cymose; flowers regular, hypogynous; ovary superior, trilocular.

147b.(146) Plants normally twining shrubs with arms

CXXXVII. *SMILACACEAE* (Pages—638-639)

Climbing or erect shrubs; aerial stem leafy; fruit berry.

148a.(146) Inflorescence variable (149)

148b.(146) Inflorescence generally leafless, umbellate, enclosed by a spathe or scape bearing single flowers (150)

149a.(148) Leaves thin, flat, unarmed

CXXXIII. *HYPOXIDACEAE* (Pages—626-627)

Stamens with introse anthers.

149b.(148) Leaves large, thick and fleshy, sometime armed

CXXXII. *AGAVACEAE* (Page—626)

Inflorescence racemose or paniculate; stamens with introse anthers.

150a.(148) Stamens-6

CXXXIV. *AMARYLLIDACEAE* (Pages—627-629)

Ovary inferior, trilocular with many ovules arranged in 2-series on the axile placentae.

150b.(148) Stamens-3, alternate with the petals

CXXXI. *IRIDACEAE* (Pages—625-626)

Perennial herbs, corms or rhizomes.

151a(143) Staminode-1, not petaloid; fertile stamens-5

CXXIX. *MUSACEAE* (Pages—623-624)

Perennial herbs but tree like in appearence; leaves large, alternate; distichous or spirally arranged; inflorescence spike or panicle.

151b.(143) Staminode-1 or more; sometime or all of them petaloid (152)

152a.(151) Fertile stamens 1; anther 2-celled

CXXVII. *ZINGIBERACEAE* (Pages—617-622)

Perennial herbs with fleshy rhizomes; leaves distichous, with a ligule at the junction of blade and sheath; flowers bisexual, irregular, epigynous.

152b.(151) Fertile stamens 1; anthers 1-celled; ovules several in each loculus

CXXVIII. *CANNACEAE* (Pages—622-623)

Perennial herbs with tuberous rhizomes; leaves parallel veined, without a ligule; flowers showy, bisexual, irregular.

DICOTYLEDONS

I. RANUNCULACEAE

Key to the Genera

1a	Climbing shrubs; leaves opposite; sepals valvate	(2)
1b	Prostrate herbs; leaves alternate; sepals imbricate	(3)
2a (1)	Terminal leaflets not transformed into tendrils; flowers apetalous	1. *CLEMATIS* L.
2b (1)	Terminal leaflets usually transformed into tendrils; flowers with petals	2. *NARAVELIA* DC.
3a (1)	Leaves pinnately compound; carpels united at base, many ovules	3. *NIGELLA* L.
3b (1)	Leaves simple, tri-partite; carpels free, 1-ovuled	4. *RANUNCULUS* L.

1. *CLEMATIS* L.

1.1. ***Clematis gouriana*** Roxb. ex DC. Syst. 1: 138.1818; Hook. f. & Thoms. Fl. Brit. India 1: 4. 1872; Watt Dict. Econ. Prod. 2: 369. 1889; Prain Beng. Plants 1: 124. 1903 (Rep.ed.1963); Wlh. India 2: 231. 1950. *C. vitalba* L. var. *gouriana* (Roxb. ex DC.) Kuntze Monog. *Clematis* Verh. Bot. Ver. Prov. Brandenb. 26: 100. 1885. (*Vern. Name*: *Chhagalbati*).

Climbing shrub; leaves bi-pinnate, ovate or oblong, acuminate, strongly nerved, upper surface shining. Flowers small in much branched decompound panicles, yellowish or greenish white. Achenes ovoid, hairy, with long feathery tails. *Fls. & Frts.:* July-January.

Rare on the roadside thickets and waste lands.

Field Collections : Bhangankhali, *K.N.-163*; Birati, *K-30.*

Distributions: Common in South India and also in the hilly areas, throughout India. Cosmopolitan, particularly in Java, Peru and Philippines.

Economic Importance: A strong acrid poisonous plant, used as insecticides.

Notes: D. Prain (1903) has mentioned this species from North Bengal, the ruins of Gour, but did not mention from Lower parts of Bengal.

Other Indian species of the *Clematis L.* is *C. paniculata* Thunb.; which is an ornamental garden plant, introduced from Japan.

2. *NARAVELIA* DC.

2.1. *Naravelia zeylanica* (L.) DC. Syst. 1: 167. 1818; Hook. f. & Thoms. Fl. Brit. India 1: 7. 1872: Watt Dict. Econ. Prod. 5: 317. 1891; Prain Beng. Plants 1: 124. 1903 (Rep. ed. 1963), Rec. bot. Surv. India 2: 285. 1903 & Ibid. 3(2): 168.1905. *Atragene zeylanica* L. Sp. Pl. 542.1753; Roxb. Fl. Ind. 2: 670.1832. *(Vern. Name: Chhagalbati).*

Climbing shrub; leaves bi-foliate, the terminal one transformed into a long slender tendril; leaflets broadly ovate, acuminate. Flower small, yellowish green or whitish in axillary and terminal panicles; petals spreading. Achenes shortly stalked, hairy, with long feathery tails. *Fls. & Frts.*: September-February.

Common in the hedges and thickets or in scrub jungles. The dense mass of the greenish yellow flowers make the plant distinct during bloom.

Field Collections: Basirhat, *K.N-2160.*

Distributions: Tropical forest of the Eastern Himalayas, extending from E. Nepal eastwards to Bengal, Assam and in Eastern and Western Peninsulas. In the temperate zones, mostly in Srilanka, Java and Malaya.

Economic Importance: Tuberous roots are rarely edible, which emit smell and relieve headache. The stem used as tooth sticks to cure toothache and twisted stems are also used to make ropes and various other purposes.

3. *NIGELLA* L.

3.1. *Nigella sativa* L. Sp. Pl. (ed.1) 584. 1753; Watt Dict. Econ. Prod.5: 424. 1891: Prain Beng. Plants 1: 125. 1903 (Rep. ed. 1963) & Rec. bot. Surv. India 3 (2) :168.1905;Wlh.India 7: 63.1968.*N. indica* Roxb. Fl. Ind. 2: 646. 1824. *(Vern. Name : Kalojira).*

Erect, annual herb; leaves pinnately compound, alternate. Flowers white blue or yellowish, carpels united at base; ovules many in bi-seriate. Fruit sub-capsular. *Fls. & Frts.:* February-April.

Purely cultivated as a winter crop for spice; very occasionally found as escape of the cultivated fields.

Field collections: Subuddhipur, *K.N.-1612.*

Distributions: Cultivated in all the states of India. Native of the Mediterranean region and common on the Old World.

Plate-I

A

B

C

Plate-II

A

B

C

Plate-III

A

B

Plate-IV

A

B

C

Plate-V

Plate-VI

Plate-VII

A

B

Plate-VIII

Plate-IX

Plate-X

Plate-XI

A

Central Inland Capture Fisheries Research Institute, Barrackpore
HERBARIUM
Mangrove Flora of Sundarbans
No. 221
Name Heritiera fomes Buch.-Ham.
Family STERCULIACEAE
Remarks
Vern. Name Sundari
Locality
Date of Collection
Collected by K. P.
Identified by
CIFRI
221

camel
Classic

B

C

Plate-XII

A

B

C

Plate-XIII

A

B

C

D

Plate-XIV

A

B

Plate-XV

A

B

Plate-XVI

A

B

Economic Importance: Seeds are pungent, aromatic stimulant and used as spice and condiment. Seed powder and sesamum oil used externally for skin eruptions.

Other cultivated species in India is *N. damascena* L.

4. *RANUNCULUS* L.

4.1. *Ranunculus sceleratus* L. Sp. Pl. 551. 1753; Hook.f. & Thoms. Fl. Brit. India 1 : 19. 1872; Watt Dict. Econ. Prod. 6 (1) : 392. 1893; Prain Beng. Plants 1 : 125. 1905 (Rep. ed. 1963) & Rec. bot. Surv. India 3 (2) : 168.1905; Hand.-Mazz. Acta Horti Gothoburg.13.162.1940; Wlh. India 8 : 364. 1969. *(Vern. Name : Polica).*

Erect glabrous, annual herb upto 60 cm high; stem succulent, hollow, deeply furrowed; radical leaves reniform, tri-partite, with linear segments. Flowers small, yellow, numerous, terminating the branchlets. Achenes numerous, in oblong, turgid pilose receptacle. *Fls. & Frts.:* December-February.

Common on the marshy places, along the canal sides and water ways; grow profusely during winter season.

Field Collections: Kumarzol, K.N.-309; Brace Bridge, *K.N.-494.*

Distributions: Almost throughout India, specially in the temperate regions. Asia in general, excepting tropical regions.

Economic Importance: The plant has very acrid properties. Stem and leaf decoction is recommended for rheumatism, pneumonia, asthma and disuria. Seed powder is used as tonic for kidney disorder. In homeopathic medicine this plant is used for skin diseases.

Notes: The other important Indian members of *Ranunculus* L. are *R. arvensis* L., *R. cantoniensis* DC., *R. lingua* L., *R. muricatus* L., *R. trichophyllus* Chaix. *Delphinium ajacis* L. is commonly planted in the parks and gardens for its showy inflorescence. It is an introduced species from South Europe.

II. DILLENIACEAE

1. *DILLENIA* L.

1.1. *Dillenia indica* L. . Sp. Pl. 535. 1753; Hook.f. & Thoms. Fl. Brit. India 1: 36. 1872; Watt Dict. Econ. Prod. 3 : 113. 1890; Prain Beng. Plants 1 : 126. 1903 (Rep. ed. 1963) & Rec. bot. Surv. Ind. 3(2): 165. 1905; Benthall Trees Calc. 1. 1946; Hoogland Blumea 7 : 108. 1952; Wlh. India 3: 64. 1952. *D. speciosa*

Thunb. Trans. Linn. Soc. 1: 200. 1791; Roxb. Fl. Ind. 2: 650.1824. *(Vern. Name: Chalta).*

Medium tree with erect trunk; branches spreading. Leaves 20-30 cm long, oblong lanceolate, acuminate, sharply serrate and with very prominant leaf veins. Flowers large, solitary, 15-20 cm diam., white, fragrant; sepals thick, fleshy, concave; petals oblong; stamens many. Fruits globose, enclosed in fleshy calyx. Seeds reniform. *Fls. & Frts.* : July-November.

Commonly planted in the old orchards throughout the district 24-Parganas and self-sown in the village surroundings, road sides scrub jungles.

Field Collections : Duttapukur, *K.N.-1980;* Baruipur, *K.N.-2302.*

Distributions: Throughout India and common in Nepal tarai. Native of South East Asia, Indomalaysia and Belgium.

Economic Importance: Stem bark, leaves and fruits are much used in our indigenous system of medicine for the treatment of various ailments. The stem bark is stringent and is used as cooling beverage in fever. Fruits (fleshy sepals) are sour, edible as raw or cooked. The acrid juice sweetened with sugar forms a cooling drink. The wood is of poor quality and used as fuel, packing cases, railway sleepers and plywood.

The other two Indian species of *Dillenia* L. are *D. aurea* Sm. and *D. pentagyna* Roxb.

III . MAGNOLIACEAE

Key to the Genera

1a	Flowers terminal, sepals-3, petals 6-12, 2-4 seriate, carpels sessile	1. *MAGNOLIA* L.
1b	Flowers axillary; perianth 9-20, at least 3-seriate, looking alike, carpel stipitate	2. *MICHELIA* L.

I. *MAGNOLIA* L.

1.1. *Magnolia grandiflora* L. Syst. (ed.10) 2: 1032.1759; Benthall Trees Calc. 3. 1946; Wlh. India 6 : 224.1962. *(Vern. Name : Champa).*

Small, evergreen tree with a pyramidal, dense crown; leaves oblong, ovate or elliptic, thick, glossy green above, rusty beneath. Flowers white, scented, sepals-3, petals 6-12, ovate; filaments purple. *Fls. & Frts.:* July-November.

Not much common; it is favourite to many gardeners for its dense evergreen foliages and the large scented flowers.

Field Collections : Barrackpore, *K.N.- 1839.*

Distributions: Indigenous to South Eastern United States, especially from Florida to Texas, along the coasts. Introduced into India probably during the year 1840.

Economic Importance : It is much liked for its sweet scented and beautiful flowers, from which certain perfumes are prepared. Bark is stimulant, aromatic and tonic; it is commonly used in malaria and rheumatism.

The other Indian species of the *Mangnolia* L. are *M.camphellii* Hook.f. & Thoms., *M. denudata* Desr. and *M. pterocarpa* Roxb.

2. *MICHELIA* L.

2.1. *Michelia champaca* L. Sp. Pl. 536. 1753; Roxb. Fl. Ind. 2: 656.1824; Hook. f. & Thoms. Fl. Brit. India 1 : 42.1872; Watt Dict. Econ. Prod. 5 : 241.1891; King Ann. Roy. Bot. Gard. 3: 216. 1891; Prain Beng. Plants 1: 127. 1903 (Rep. ed. 1963) & Rec. bot. Surv. India 3 (2): 171.1905; Benthall Trees Calc. 5. 1946; Wlh. India 6 : 370. 1962. *Michelia aurantiaca* Wall. Cat. 6492, 1828.*(Vern. Name: Champa).*

Evergreen tree; branchlets pubescent; leaves ovate lanceolate, 16-22 x 7-9 cm, tapering to a long point. Flowers 3 cm diam., pale yellow or orange, vary fragrant; peduncle short, buds silky; perianth 15-20, ovary pubescent, carpel stipitate. Fruit 6-9 cm diam. *Fls.& Frts.*: March-July.

Frequently planted in lawns and ornamental gardens for its fragrant and yellowish flowers.

Field Collections : Diamond Harbour, *K.N.-1453.*

Distributions : Commonly cultivated, but wild in the forest of the temperate Himalaya. Native of South East Asia.

Economic Importance : Ornamental plants; its flowers are showy and sweet scented. The roots are used to make toy and decorative furnitures. Seed oil used in perfumery.

The other species distributed in India are *M. bailonii* Finet & Gagnep., *M. catheartii* Hook.f. & Thoms., *M. doltsopa* Buch. Ham. ex DC., *M. lanuginosa* wall., *M. montana* Blume, *M. nilagirica* Zenker, *M. oblonga* Wall. ex Hook.f. & Thoms., *M. munduana* Hook.f. & Thoms. Most of these species are reported from the hilly regions in India.

IV. ANNONACEAE

Key to the Genera

1a	Carpels united; anthers completely enclosed by over-lapping connective	1. *ANNONA* L.
1b	Carpels not united	(2)
2a (1)	Leaves tomentose; anther-cells are not completely enclosed by overlapping connective	4. *MILIUSA* Leschm.
2b (1)	Leaves shining; glabrous; anther-cells completely enclosed by overlapping connective	(3)
3a (2)	Peduncles hooked; petals connivent at concave base	2 . *ARTABOTRYS* R.Brown
3b (2)	Peduncles not hooked; petals of both series flat, lanceolate, spreading from the base	(4)
4a (3)	Ovary many, bi-serriate; fruits a head of berry	3. *CANANGA* Rumph.
4b (3)	Ovules 1-2, basal or sub-basal; fruits a cluster of druplets	5. *POLYALTHIA* Blume

1. *ANNONA* L.

Key to the Species

1a	Leaves long, acuminate; fruits reticulate areolate	1. *A. reticulata* L.
1b	Leaves oblong lanceolate, obtuse or acute, fruits tuberculate	2. *A. squamosa* L.

1.1. ***Annona reticulata*** L. Sp. Pl. 537. 1753; Roxb. Fl. Ind. 2 : 657. 1824; Hook.f. & Thoms. Fl. Brit. India 1 : 78. 1872; Watt Dict. Econ. Prod. 1 : 258. 1889; Prain Beng. Plants 1 : 134.1903 (Rep.ed.1963) & Rec. bot. Surv. India 3 (2): 169.1905; Benthall Trees Calc. 14. 1946; Wlh. India 1 : 80. 1948. *(Vern Name: Nona).*

Medium sized tree with glabrous branches; leaves membranous, oblong lanceolate, many nerved. Flowers whitish, axillary, 2-4 together. Pedicels grabrous, small pubescent; bracts inserted at the base; carpels ovoid, 1-ovuled, style oblong. Fruit subglobose, surface rough. Seeds black. *Fls. & Frts.* : October- February.

Sometime cultivated in the old orchard, but often seen as an escape in the dry waste lands.

Field Collections: Tetulia, *K.N.-516*; Bamanea, *K.N.-2123*.

Distributions : Introduced from the tropical America during the last quarter of the 17th Century; now completely naturalized in Bengal and elsewhere. Native of tropical America and widely cultivated in the tropics.

Economic Importance: Ripe fruits are low category edible and common in the village market; reported as to cure biliousness and several other blood diseases; used against dysentery and intestinal worms; the bark is powerfully astringent. Seed oil has the insecticidal properties and is also used for soap making.

1.2. ***Annona squamosa*** L. Sp. Pl. 537. 1753; Roxb. Fl. Ind. 2: 657.1824; Hook. f. Fl. Brit. India 1: 78. 1872; Watt Dict. Econ. Prod. 1: 259. 1889; Prain Beng. Plants 1: 134. 1903 (Rep. ed. 1963) & Rec. bot. Surv. India 3(2): 169.1905; Benthall Trees Calc. 12.1946; Wlh. India 1: 80.1948. *(Vern.Name: Ata).*

Small tree upto 3 m high with pubescent young branches and glabrous when fully mature. Leaves oblong, oblong lanceolate obtuse, acute, petiolate. Flowers white or yellowish, axillary and solitary; pendulous, pubescent. Fruit fleshy, ovoid, surface rough by the ends of carpels; seeds blackish, smooth, hard. *Fls. & Frts.* : August-March.

Occasionally planted in the old orchards or village surroundings; on the village thickets grow as naturalised flora.

Field Collections : Barasat, *K.N.-916*; Bongan, *K.N.-1321*.

Distributions: Frequently cultivated and naturalised almost throughout India. It was one of the first American plant introduced by the Portuguese to India during the 16th Century. Native of tropical America and widely cultivated in the tropics.

Economic Importance: Ripe fruits are edible and are very sweet and tasty; it is said to be a good tonic. Seeds are reported as valuable insecticide; the roots are also drastic purgative; the bark is astringent and is used in dysentery. Green leaves yield essential oil; seed powder possess a number of medicinal values.

The other species of *Annona* L. distributed in India are *A.cherimolia* Mill., *A. diversifolia* Saff., *A. muricata* L. These species are the native of South America.

2. *ARTABOTRYS* R. Brown

2.1. ***Artabotrys hexapetalus*** (Linn.f.) Bhandari, Baileya 12 : 147. 1964; Srivastava Fl. Gorakh. 32.1976. *Annona hexapetala* Linn.f. Supp. Pl. 270.1781. *Artabotrys odoratissimus* R. Br. Bot. Reg. t. 423.1820 (not of Blume); Hook.f. Fl. Brit. India 1: 54.1872; Watt Dict. Econ. Prod. 1: 322. 1889; Prain Beng. Plants 1: 131. 1903 (Rep. ed. 1963) & Rec. bot. Surv. India 3 (2): 168. 1905; Wlh. India 1: 119.1948. *A. uncinatus* (L.) Merril Philipp. Jour. Sci. (Bot.) 7: 234. 1912. (*Vern. Name : Kantali Champa* or *Kat Champa*).

Glabrous shrub with long branches; leaves oblong or lanceolate, shortly acuminate glabrous, shining. Flowers yellow, solitary or in pairs; petals glabrous at maturity; carpels glabrous, yellow. Seeds oblong, a little flattened deeply grooved on one side. *Fls. & Frts:* April-August.

Commonly planted as ornamental plant for its sweet scented flowers.

Field Collections: Barrackpore, *K.N.-1840.*

Distributions: Cultivated almost throughout India. Native of South East Asia, Srilanka, Java and South China.

Economic Importance: Ornamental garden plants for its highly scented flowers; the flower is useful in vomiting, biliousness, diseases of the blood and the heart, itch, sweat, foul breath, thirst, leucoderma, headache. A decoction of the leaves is given for cholera in some of the islands of Malaysia.

3. *CANANGA* Rumph.

3.1. ***Cananga odorata*** Hook.f. & Thoms. Fl. Brit. India 1 : 56.1872; Watt Dict. Econ. Prod. 2 : 93.1889; Prain Beng. Plants 1: 132. 1903 (Rep. ed. 1908) & Rec. bot. Surv. India 3 (2): 169.1905; Benthall Trees Calc. 11. 1946; Wlh. India 2: 51.1950. *Uvaria odorata* Lam. III t. 495.f.1; Roxb. Fl. Ind. 2 : 661. 1824. *Canangium odoratum* King. *(Vern.Name: Alangilang).*

Tall tree, with straight trunk; bark smooth; shoots glabrous; leaves ovate oblong, finely acuminate puberulous beneath. Flowers yellow, trinate, drooping; peduncle solitary or several from old scar; young petals silky, carpels black. *Fls. & Frts.:* May-October. Usually planted as avenue tree or in the garden.

Field Collections : Ichhapur, *K.N.-1819.*

Distributions: Cultivated or planted throughout India. Native of Malaya and in the Tropics.

Economic Importance: This tree yields a popular scent, 'ilang- ilang' or 'ylang- ylang' and an oil known as -*'cananga'* oil or *'Macassat oil'*, which are important articles of commerce in Philippines and Japan. The oil is applied externally as a cure for headache, opthalmia, gout and skin disorders.

4. *MILIUSA* Leschm.

4.1. ***Miliusa tomentosa*** (Roxb.) Sinclair, Gardn. Bull. 14: 378. 1955; Wlh. India 6: 377. 1962. *Uvaria tomentosa* Roxb. Pl. Cor. 1: 31, t. 35, 1795. *Saccepetatum tomentosum* Hook.f. & Thoms. Fl. Brit Ind. 1: 88. 1872; Watt Dict. Econ. Prod. 6(2): 381.1893. *(Vern. Name : Gandhapalasa).*

Moderate sized deciduous tree; leaves shortly petioled. Flowers brownish green; trinate; sepals 3; petals 6, outer whorl smaller than inner whorl; stamens many. Fruits succulent, stalked, globose, tapering to the base. *Fls. & Frts.:* March - August.

Common on the hedges and thickets and also planted on road sides.

Field Collections: Joynagar, *K.N.-1620.*

Distributions : Almost throughout India, mainly in the Northern parts. Tropical forest of both the Old and New World.

Economic Importance: Leaves sometime used as fodder for goat. Wood is used for cabinet work; fruit edible. Gum is a substitute for tragacanth.

The other species of the genus, *Miliusa* Leschm. distributed in India are *M. roxburghiana* Hk.f. & Thoms., and *M. velutina* (Dunal) Hook.f. & Thoms. These are mostly timber yielding plants.

5. *POLYALTHIA* Blume

Key to the Species

1a	Branches and leaves glabrous; flowers many, fascicled petals linear; carpel ovoid	2. *P. longifolia* Thw.
1b	Branches and leaves beneath pubescent; flowers few on woody tubercles; petals ovate; carpels spherical	(2)

2a (1)	Leaves lanceolate, acuminate; peduncles 1-3 fid, tubercles axillary	1 *P. cerasoides* Benth. & Hook.f.
2b(1)	Leaves oblong, obtuse; peduncles 1-2 fid, tubercles extra axillary	3. *P. suberosa* Thw.

5.1. ***Polyalthia cerasoides*** Benth. & Hook.f. Gen. Pl. 1: 25.1862 & Fl. Brit. India 1: 63.1872; Watt Dict. Econ. Prod. 6(1): 313.1892; Prain Beng. Plants 1: 133. 1903 (Rep. ed. 1963) & Rec. bot. Surv. India 3 (2): 169.1905; Benthall Trees Calc. 8. 1946. *Uvaria (Ovaria) cerasoides* Roxb. Cor. Pl. 1: 30 t. 33. 1795 & Fl. Ind. 2: 666. 1832. *Guatteria cerasoides* W. & A. Prod. 10. 1834. *(Vern. Name: Kudumi).*

Medium tree, bark greenish; old branches glabrous; leaves oblong lanceolate, acuminate, glabrous above, softly pubescent beneath, peduncles on axillary tubercles. Flowers solitary or in a 2- flowered cyme. Sepals 3, connate at base, tomentose, ovate. Fruit dark red in colour. *Fls.& Frts.:* March-August.

Occasionally in the village shrubberies, often planted as hedge tree.

Field Collection: Brace Bridge, *K.N.1574.*

Distributions: The dry forest area of India; native of India. Tropical forest of the Old World.

Economic Importance: Used as fire wood and low quality furnitures are prepared from this wood; leavs sometime used as fodder. Ripe fruits are also edible.

5.2. ***Polyalthia longifolia*** (Sonn.) Thw.Enum.398.1864; Hook.f. & Thoms. 1: 62. 1872; Watt Dict. Econ. Prod. 6(1): 313. 1892; Prain Beng; Plants 1: 133. 1903 (Rep. ed. 1963) & Rec. bot. Surv. India 3(2): 169. 1905; Benthal Trees Calc. 9. 1946; Wlh. India 8: 157. 1969. *Uvaria longifolia* Sonn. Voy. aux Indes. 2: 233. t. 131. 1782. *Guatteria longifolia* Wall. Cat. 6442. 1828 *(Verm.Name: Debdaru).*

Tall tree with straight trunk and conical crown, young branches; leaves narrowly lanceolate, glabrous, shining above, pale beneath. Flowers numerous in short peduncles; umbelliform, recemos. Fruit ovoid, one seeded, pale brown, longitudinally grooved. *Fls. & Frts.:* January-June.

Mostly planted as avenue tree, sometime planted in the hedges or gardens for beautification.

Field Collections: Gobardanga, *K.N.-530;* Fultala, *K.N.-2146.*

Distributions: Almost throughout India. Native of Sri Lanka, E. Asiatic Countries and South Asia.

Economic Importance: Most common avenue tree for its straight trunk and wavy linear dark green foliages. The leaves are used for decorative purposes, mostly in villages. Ripe fruits are edible.

5.3. ***Polyalthia suberosa*** (Roxb.) Thw. Enum.398.1864; Hook. f. & Thoms. Fl.Brit.India 1: 55. 1872; Watt Dict. Econ.Prod. 6(1): 313. 1892; Prain Beng. Plants 1: 133. 1903 (Rep. ed. 1963) & Rec. bot,. Surv. India 3 (2): 169. 1905; Benthall Trees Calc. 7.1946; Wlh. India 8: 189. 1969. *Uvaria suberosa* Roxb. Cor. Pl. 1: 31, t. 34. 1793. *(Vern. Name: Barachali).*

Small to medium tree; branchlets sparingly pubescent; leaves narrowly ovate oblong, subacute, pubescent beneath, petioles very short. Flowers pale yellow, solitary. Ripe carpels ovoid or globose, purple, fleshy. Fruits sub-spherical, red when ripe. *Fls. & Frts.:* April-December.

Frequently found as wild in the village shrubberies and waste lands; occasionally planted as avenue tree.

Field Collections: Brace Bridge, *K.N.-2237;* Baruipur, *K.N.-2257.*

Distributions: Native of India and distributed throughout. Malaya, Srilanka and Tenasserium.

Economic Importance: The whole plant is used as firewood and leaves are used as fodder, specially for goat. Root decoction is used as abortificient.

The other members of the *Polyalthis* Blume are *P.fragrans* (Dalz.) Bedd., *P. simiarum* Hook.f. & Thoms. *P. fragrans* is mainly distributed in the Western Ghat and its wood is used for cricket stumps. *P. simiarum* Hook. f. & Thoms. is common in the Eastern parts of India and Himalayas; it has a number of medicinal properties like for cure of scorpion-sting and several skin diseases.

V. MENYSPERMACEAE

Key to the Genera

1a	Inflorescence usually umbel or cyme, rarely receme; stamens connate in a column; carpel generally one	(2)

1b	Inflorescence mostly panicle, rarely raceme; stamens free; carpels 3 or more	(3)
2a (1)	Male flowers with sepals 6-10 and petals 3-5; female flowers with 3-5 sepals and petals	3. *STEPHANIA* Lour.
2b(1)	Male flowers 4-merous; female flower with 2 sepals	1. *CISSAMPELOS* L.
3a(1)	Leaves deeply cordate; style scar sub-terminal	5. *TINOSPORA* Miers.
3b(1)	Leaves not cordate, rarely sub-cordate; style either sub-basal or cylindric	(4)
4a(3)	Carpels in male-0; stamens with sub-globose, anthers dehiscing transversely	2. *COCCULUS* DC.
4b(3)	Carpels in male-3, rudimentary; stamens sub-cylindric, anthers dehiscing vertically	4. *TILIACORA* Colebr.

1. CISSAMPELOS L.

1.1. ***Cissampelos pareira*** L. Sp. Pl.1031. 1753; Hook. f. & Thoms.Fl.Brit.India 1:103. 1872; Watt Dict. Econ.Prod. 2 : 327.1889; Prain Beng. Plants 1 : 136. 1903 (Rep. ed. 1963) & Rec. Bot. Surv. India 3(2): 170. 1905; Wlh. India 2: 185. 1950. *C. convolvulacea* Willd. Sp. Pl. 4: 863. 1806; Roxb. Fl.Ind.3 :842.1832. *(Vern.Name:Akanadi* or *Ekleja).*

Twining shrub with long herbaceous tomentose branches; leaves triangular, cordate, hemispherical or sub-orbicular, usually peltate; mature leaves glabrous above, glaucous beneath. Flowers pale-green, minute. Drupe sub-globose, red when ripe. *Fls.&Frts.:* July-November.

Common in the orchards and hedges or the waste lands, generally trailing or twining on the old trees.

Field Collections: Nimpith, *K.N.-1613;* Shyamnagar, *K.N.-2984.*

Distributions: Throughout India, especially in open rocky valley. Warmer parts of Asia, Srilanka, East Africa and America.

Economic Importance: Root has the medicinal properties; Santals use the root in diarrhoea, dysentery, colic pains, cough and urinary troubles. Leaves are said to be used for sores and itches. Pelosin is derived from it.

2. *COCCULUS* DC. *nom. cons.*

2.1. ***Cocculus hirsutus*** (L.) Diels Engler's Pflanzenr. 46 : 236. 1910; Wlh. Ind.2: 258. 1950; Santapau Rec.bot. Surv. India 16(1): 4.1967. *Menispermum hirsutum* L. Sp. P. 341.1753; Roxb. Fl. Ind.3 : 814.1832. *Cocculus villous* DC. Syst. 1.525.1818; Hook.f. & Thoms. Fl.Brit. India 1: 101.1872; Watt Dict. Econ.Prod.2: 397.1889;Prain Beng. Plants 1: 137.1903 (Rep. ed. 1963) & Rec. bot. Surv.India 3 (2) : 170. 1905. *(Vern. Name: Huyer).*

Climbing shrub; branchlets villous; leaves ovate oblong, subdeltoid villous. Male flowers in short panicles, female flowers 1-3 axillary, rarely racemose. Drupes black purple, endocarp acutely keeled, tubercled. *Fls. & Frts.* November-May.

Very common on the bushes, in hedges and waste lands.

Field Collections: Magrahat, *K.N.-270;* Surjyapur, *K.N.-279.*

Distributions: Throughout India; tropical and sub-tropical tracts in India; Tropical Africa.

Economic Importance: Leaves juice after mixing with water is taken as a cooling medicine for gonorrhoea and used externally for eczema, prurigo and impetigo. Roots are said to be useful in chronic rheumatism and venereal diseases. Root and stem decoction is sedative, hypotensive, cardiotonic and spasmolytic.

The other species of the Genus *Cocculus* DC. is *C. pendulus* (Forssk.) Diels.

3. *STEPHANIA* Lour.

3.1. ***Stephania japonica*** (Thunb.) Miers, Ann. & Mag. nat. Hist. (ser.3) 18 : 14 1866, pro parte (excl. *cuming* 1160); Diels, Pflanzenr 46 : 277. 1910. Wlh. India 10 : 42. 1976. *Menispermum japonicum* Thunb. Fl. Jap. 193.1784. *Cissampelos hernandifolia* Willd. Sp,. Pl. 4 : 861.1806. *Stephania hernandifolia* (Willd.) Walp. Rep. 1 : 96. 1842; Hook.f. & Thoms. Fl. Brit. India 1 : 103. 1872; Watt Dict. Econ. Prod. 6 (3): 359.1893; Prain Beng. Plants 1: 136. 1903 (Rep.ed. 1963)& Rec. bot. Surv. India 3(2): 170. 1905. *(Vern.Name : Nimusha* or *Nimukha).*

Slender climbing shrub; branches glabrous, striate; leaves 3.5-9.2 x 8.2, 9.2 cm, broadly ovoid deltoid, peltate membranous, acute, acuminate or sub-obtuse, mucronate, rounded and /or truncate at base. Flowers yellow, umbles often head like with sessile flowers; peduncles axillary. Drupe solitary, reddish, endocarp usually with 4-longitudinal rows of tubercles. *Fls. & Frts.:* July-December.

Very common on the dry waste lands, roadside thickets, hedges and railway tracks, climbing on bushes.

Field Collections: Bhangan Khali, *K.N.-164;* Ramnagar, *K.N.-219.*

Distributions: Throughout India; Malaya, Australia and Africa.

Economic Importance: Leaves decoction if eaten during the first 4 or 5 days of menstruation cycle is said to be useful in the birth control. The root is used in fever, diarrhoea, urinary diseases and dyspepsia.

The other members of the genus is *S. glabra* (Roxb.) Miers, distributed in the Himalayas, in Nepal and Assam and in W. Ghats.

4. *TILIACORA* Colebr.

4.1. ***Tiliacora acuminata*** (Lamk.) Miers, Arun & Mag.nat. Hist. (Ser. 3) 7 : 39, 1851; Hook.f. & Thoms. Fl. Ind. 187. 1855; Wlh. Ind. 10: 247. 1976. *Menispermum acuminatum* Lamk. Encycl. 4: 101. 1797. *Tiliacora racemesa* Colebr., Trans. Linn. Soc. 13: 67. 1822; Hook. f. & Thoms. Fl. Brit. India 1: 99. 1872: Watt Dict. Econ. Prod. 6(4) 56. 1893 : Prain Beng. Plants 1: 137. 1903 (Rep. ed. 1963) & Rec. bot. Surv. India 3(2) : 170. 1905. *Menispermum polycarpum* Roxb. Fl. Ind. 3 : 816. 1832. *(Vern. Name : Bhaga lata* or *Tiliokora).*

Climbing shrub; branches glabrous. Leaves 5-10 cm x 5 cm, ovate, acuminate, glabrous, base acute, truncate, rounded or sub-cordate; thin margin undulate. Flowers yellow in lax elongate, lax, axillary, receme. Drupes red, 1.5 cm long. *Fls. & Frt.:* March-December.

Fairly abundant during rainy season on hedges, thickets and shrubberies in general; climbing on *Ficus bengalensis* and *Mangifera indica.*

Field Collections: Beliachandi, *K.N.-268;* Haroa, *K.N.-592.*

Distributions: Throughout the tropical India, from Oude and Bengal to Kankan; Sri Lanka, Singapur and Java.

Economic Importance: The bark contains two alkaloids-tiliacorine and tiliarine. Long branches are used for basket making.

Note: Diels (Pflanzenr. 4: 278. 1910) changed the name *T.racemosa* Colebr. to *T. acuminata* Hook.f. & Thoms.

5. *TINOSPORA* Miers.

5.1. ***Tinospora cordifolia*** (Willd.) Miers, Ann. & Mag. nat. Hist. (ser.3) 7: 38. 1831; Hook.f. & Thoms. Fl. Ind. 184. 1855 & Fl. Brit. Ind. 1:97.1872; Watt Dict. Econ.Prod. 6(4): 43.1893; Prain Beng. Plants 1: 137. 1903 (Rep. ed. 1963) & Rec. bot. Surv. India 3 (2): 169.1905; Wlh. Ind. 10 : 251. 1976. *Menispermum cordifolium* Willd. Sp. Pl. 4 : 826.1806; Roxb. Fl. Ind. 3: 811. 1832. *Cocculus cordifolius* DC. Syst. 1: 518. 1818 & Prodr. 1: 97. 1824. *(Vern. Name: Gulancha).*

Large climbing shrubs, with glabrous stems and leaves; stems and barks corky; leaves cordate, 5-10 cm X 5 cm, acute or acuminate. Racemes exceeding the leaves, axillary, terminal or on the old wood; bracts subulate. Flowers yellow, male fascicled; females usually solitary, glabrous. Stamens free, ripe carpels pisiform; anthers oblong. Fruit a drupe with a subterminal scar. *Fls. & Frts :* June-February.

Very common in the village thickets, hedges, waste lands and railway tracts, commonly climbing on the way side trees.

Field Collections: Ghoramara, *K.N.-406;* Santoshpur, *K.N.-499.*

Distributions: Throughout tropical India and Andamans. Srilanka and Malayasia.

Economic Importance: It is considered as cooling and sedative by Tibbi practitioners. Antiperiodic and diuretic properties have been attributed to the plant. The juice of the plant is largely used for the stomach ailments. It is considered useful in gonorrhoea by the Hindu physician and is also given in chronic diarrhoea and dysentery. The oil extracted from the seeds is used in the treatment of tuberculosis of the lungs; the dry stem powder after drying in the shade are used to reduce high blood pressure.

Notes: The species *Tinospora tomentosa* Miers *(Vern. Name: Padma Gulancha),* a climbing shrub reported by D.Prain in Bengal Plants 1: 137.1903 (Rep. ed.1963) after collecting by him from Jatta, the ruins of the Pagoda. But in the recent exploration trips, no body so far succeded to collect it from the district of 24-Parganas. I have not even seen and collected any of the *T. tomentosa* during my field trips.

The other members of *Tinospora* Miers are *T. crispa* (L) Miers ex Hook.f. & Thoms. and *T. sinensis* (Lour.) Merr.

VI. NYMPHAEACEAE

Key to the Genera

1a	Leaves and flowers raised above the surface of water; carpels irregularly scattered, sunk in pits of the turbinate disk; ovary 1- 2; seeds exalbuminous	1. *NELUMBO* Adans.
1b	Leaves floating; carpels confluent with the disk into one ovary, ovules many; seeds albuminous	2. *NYMPHAEA* L.

1. *NELUMBO* Adans.

1.1. ***Nelumbo nucifera*** Gaertn. Fruct. 1 :73.t.f.2.1788; Subramanyam Aqua. Angios.6.1962. *Nymphaea nelumbo* L. Sp. Pl. 511.1753. *Nelumbium speciosum* Willd. Sp. Pl. 2: 1258.1799; Roxb. Fl. Ind. 2:647.1824; Hook.f. & Thoms. Fl. Brit. India 1:116.1872; Watt Dict. Econ. Prod. 5 : 342.1891; Prain Beng. Plants 1: 141. 1903 (Rep. ed. 1963) & Rec. bot Surv. India 3 (2): 171. 1905. *N. nelumbo* (L.) Druce, Bot. Exch. Club.Brit. Isles. 421.1914. *(Vern.Name:Padma* or *Kamal).*

Erect large, water herb with milky juice; root stocks stout, creeping. Leaves raised high above the water label; peltate, petiole finely spinose. Flowers white or rose 10-12 cm diam.; petals elliptic concave, veined; peduncle with scattered and weak prickles; ovules 1 or 2; seeds without albumens. *Fls. & Frts.:* April- November.

Profuse on the old ponds and jheels as attached floating herb.

Field Collections: Garia, *K.N.-310.*

Distributions: Throughout India, mainly in the plain districts. Indigenous to continental Asia, Persia, Malay Islands, China, Japan and Tropical Australia.

Economic Importance: The leaves are occasionally used as food plates; flowers are used in Hindu and Bouddhist temples for religious purposes. This is the sacred lotus of Vishnu, also liked for its large sized and delicate structure. Used largely for decorative purposes. Flowers used as cardiac tonic, in fever and liver troubles. Seeds used as cooling medicine for skin diseases. Rhizomes used in piles; leaves, pedicels and embryo contain alkaloid nelumbine.

2. *NYMPHAEA* L.

Key to the Species

1a	Plants comparatively smaller; lamina margin always toothed; flowers 5-8 cm diam., white; anther without appendages; stigma rays with clubbed appendages	1. *N.nouchali* Burm.f.
1b.	Plants comparatively large; lamina margin almost entire; flowers 5-12 cm diam., greatly variable in colour; anther with appendages; stigma rays with short hornes	2. *N. stellata* Willd.

2.1. ***Nymphaea nouchali*** Burm. f. Fl. Ind. 120. 1768; Merill Philipp. J. Sc. 19: 350.1921; Subramanyam, Aqua. Angios. 6.1962; Backer & Bakhuizen.f. Fl. Java 1: 148.1963; Wlh. India 7: 71.1966. *N. lotus* L. Sp. Pl. 511.1753; Watt Dict. Econ. Prod. 5: 436. 1891; Prain Beng. Plants 1: 140. 1903 (Rep. ed. 1963) & Rec. bot. Surv. India 170. 1905. *N. pubescens* Willd. Sp. Pl.2: 1154.1799; Conard Monog. Nymphaea 198.1908; Deb Bull. bot. Soc. Bengal 29 (2): 158.1976. *N. lotus* Hook.f. & Thoms. Fl. Brit. India 1: 114. 1872 (non Linn.1753). *N.lotus* L. var. *pubescens* (Willd.) Hook.f.& Thoms. Fl. India 241. 1855. *N. rubra* Roxb. ex Salisb. Parad. Lond. 1, sub.t.14. 1805. *(Vern. Name: Saluk).*

Floating aquatic herb; rhizomes attached with the clay; leaves 20- 25 cm across, ovate to sub-sagittate, deeply cordate with sharply sinuate teeth; bright green and glabrous above, pubescent beneath. Flowers 5-8 cm wide, white or rosy; anther without appendages; stigma rays with clubbed appendages. Seeds 1 mm diam., ellipsoid or subglobose, longitudinally striate, green *Fls. & Frts:* August-November.

Very common, floating in the less cleared rice fields and shallow water ponds, roadside canals and nalas.

Field Collections: Jugdia/Benipur, *K.N.-225;* Falta, *K.N.-1299;* Salt Lake, *M.K.G.-1342.*

Distributions: Common throughout the warmer parts of India. Mostly in the Old World tropics. The native land is S.E. Asia or Africa or both.

Economic Importance: Rhizomes and seeds are commonly used by the village people and considered as chief famine food during scarcity. Most part of the plants are used medicinally. The powder of the root is used in piles, as a paste in skin diseases; the flowers are astringent and are used in the treatment of diarrhoea; the flowers are used by the Hindu for religious purposes.

Remarks: Santapau (*Rec. bot. Surv. India* 16 (1): 5.1967) stated that flower colour cannot be the basis of classification of the genus, as it varies in a wide range and he considered the two species *N. rubra* Roxb. and *N. pubescens* Willd. I have followed Santapau (1.c.) in consigning the species. Conard (1916) has stated that the Indian species of *N. lotus* is not that of Linn. though Prain (1903a & 1905) treated it as Linnean one. Plant, hitherto referred, as *N. nouchali* may be called as *N. pubescens* Willd., as it is the next earliest name for this taxon after *N. nouchali* Burm.f., which is now considered as conspecific with *N. stellata* Willd. and is corrected the name for this group of plants. But Deb (1976) suggested no alternative and separate as *N. stellata* Willd. and *N. pubescens* Willd. Here, I have followed Van Royen (1962) in treating *N nouchali* Burm.f. as correct name and *N. pubescens* Willd. as synonyms.

2.2. ***Nymphaea stellata*** Willd. Sp. Pl. 2: 1153.1799; Hook.f. & Thoms., Fl.Brit. India 1: 114.1872; Watt Dict. Econ. Prod. 5: 438. 1891; Prain Beng. Plants 1: 140. 1903 (Rep. ed. 1963) & Rec. bot. Surv. India 3 (2): 170.1905; Conard Monog. Nymphaea 140.1908; Subramanyam Aqua. Angios. 6. 1962; Wlh. India 7: 72.1966; Deb Bull. bot. Soc. Bengal 29(2): 158. 1976. *(Vern. Name:Sapla).*

Large aquatic herb; rootstock creeping, attached to the substratum, soil; leaves quite entire, sometimes sinuate toothed. Flowers 5-15 cm white, rose, red or variously coloured; floating on by long radical scapes; sepals obtuse, ribbed; petals acute or tapering at both the ends; stigma rays terminating in short horns without appendages. *Fls. & Frts.:* April-November.

Very common on the less used ponds, jheels, beels, roadside canals and also in the paddy fields during monsoon; grow profusely as attached floating aquatic herb and acts as the serious weed on the fish ponds and paddy fields.

Field Collections: Taldi, *K.N.-167;* Harinbari, *K.N.-1109.*

Distributions: Common throughout the plain and warmer parts of India; South East Asia and N. Africa.

Economic Importance: The stems, petioles and peduncles are eaten as vegetable; it is common in the daily markets during monsoon; poor villagers depend upon it as chief food during famine; they eat either raw or cooked. The roots and seeds are edible. Root stocks are given in the dyspepsia, diarrhoea and piles. Flowers are useful in palpitation of heart. The Hindu worship the goddess Manasha with the flowers.

Remarks: Van Royen (1962) and Backer & Bakhuizen. f. (1963) are treating *N. stellata* Willd. as a synonym of *N. nouchali* Burm.f. which is the correct name for the combined taxon, as the later antedate the former. But here, I have followed most of the Indian authors.

Notes: Euryale ferox Salisb. is also reported by D. Prain (1903), particularly from the district 24-Parganas, but I could not collect and study the species within my surveyed areas.

VII. PAPAVERACEAE

1. *ARGEMONE* L.

1.1. *Argemone mexicana* L. Sp. Pl. 508. 1753; Roxb. Fl Ind. 2 : 571. 1824; Hook.f. & Thoms. Fl. Brit. India 1: 117. 1872; Watt Dict. Econ. Prod. 1: 306. 1892; Prain Beng. Plants 1: 142. 1903 (Rep.ed. 1963) & Rec.bot.Surv.India 3(2): 171.1905; Wlh. India 1: 116. 1948; Van Steenis, Fl. Males 5: 114.1954; Majumdar, Ind. Agr.6(1&2): 103.1962. *(Vern. Name:Sialkanta).*

Erect, prickly, annual herbs, 0.3-0.5 m high; with yellow latex; leaves sessile, sinuate pinnatifid, variegated green and white. Flowers brighty yellow; sepals 2-3 free, horned or winged at margins; petals 4-6 style very short, stigma 4-7, lobed on a very short style. Capsule dehiscing at the top by short valves. Seeds many. *Fls. & Frts.* Almost throughout the year, mainly February-July.

A common weed on the roadside dry waste lands, railway tracks, hedges, thickets and in the agricultural fields; grow profusely; act as the serious weed for winter crops.

Field Collections: Canning, *K.N.-139;* Godkhali, *K.N.-654.*

Distributions: Introduced in the Indian sub-continent but at present naturalized throughout, mostly in the Eastern India; Native of Mexico.

Economic Importance: The plant is bitter, diuretic, purgative, destroys worms, cures itching, leprosy and various skin diseases; inflammations and bilious fevers. The root is also beneficial in some chronic skin diseases.The yellow juice of the plant is used as a medicine for dropsy, jaundice and cutaneous affections. Seeds are nacrotic, stronger than opium and possess an emetic properties. Seed oil is used for burning and is valuable for painting purposes.

Notes: *Papaver somniferum* L.-a glaucous, glabrous, hairy herbs; leaves cauline with broad stem clasping base, coarsely dentate more or less lobed with purple solitary flower- is a restricted cultivated plant. Very rare as an escape. It is an important plant, opium is the main product of the species. Beside this *'pasto'* is also an economic edible product produced from the seeds of this plant. It has numerous medicinal uses like, in snake bite and scorpion sting, dysentery and all kinds of flexes. Roasted seeds are also used as tonic, fruit enriches the blood and strengthen the body. It helps digestion, in cough and in complaints of the chest, in fever and in anaemia. It has also harmful effect on brain.

The other members of the *Papaver* L. are *P dubium* L., *P. nudicauli* L., *P. orientale* L. and *P. rhoeas* L.

VIII. FUMARIACEAE

1. *FUMARIA* L.

1.1. ***Fumaria vaillantii*** Loisel, Desv. J. Bot. 2: 358.1809. *F. indica* (Haussk.) Pugsley J. Linn. Soc. London (Bot.) 44: 313. 1919; Wlh. Ind. 4: 68. 1956; Majumdar, Ind. Agr.6 (1&2) : 104.1962. *F. vaillantii* Loisel var. *indica* Haussk. Flora 56: 443. 1873. *F. parviflora* Wt. & Arn. (non Lamk.) Prod. 18.1834. *F. parviflora* Lamk. var. *vaillantii* (Loisel) Hook. f. & Thoms. Fl. Ind. 258.1855, sensu Fl. Brit. India 1: 128.1872; Watt Dict. Econ.Prod.3 : 454.1890. *F. parviflora* Lamk. Encyc.2 : 567.1786; Prain Beng. Plants 1: 143.1903 (Rep. ed.1963) & Rec. bot. Surv. India 3 (2): 177.1905. *(Vern.Name:Ban Salpa).*

Spreading branched, glaucous, annual herbs with watery latex; leaves finely divided; segments flat, narrowly linear acute, mucronate. Flowers rosy purplish (6mm x 2 mm) in terminal or leaf opposed racemose. Capsule globose, slightly verrucose polished. *Fls. & Frts.:* December-March.

Fairly common as a weed in the cultivated fields and on the sandy soils in the winter season.

Field Collections: Mazilpur/Joynagar, *K.N.-1616.*

Distributions: Throughout India. Temperate regions of the World and Mediterranean regions.

Economic Importance: Said to be diuretic, diaphoretic and aperient.

Notes: Fumaria vaillantii Loisel (1809) is the only species in India, wrongly referred to as *F. officinalis* L. and *P. parviflora* Lamk. (Sarkar & Mitra, *Bull. bot. Soc. Bengal 28:* 12.1974).

IX. BRASSICACEAE

(non. alt. Cruciferae)

Key to the Genera

1a	Pods dehiscent	(2)
1b	Pods indehiscent	(6)
2a (1)	Pods narrow and long	(3)
2b (1)	Pods broad and short	(5)
3a (2)	Pods with seed almost throughout the length and dehiscing throughout	(4)
3b (2)	Pods without seed in the beak and not dehiscing thoughout	1. *BRASSICA* L.
4a (3)	Pods almost cylindric, seeds globose; flowers yellow	7. *RORIPPA* Scop.
4b (3)	Pods flattened; seeds compressed; flowers white	3. *CARDAMINE* L.
5a(2)	Pods sub-globose, seeds many, small compressed; flowers whitish	4. *COCHLEARIA* L. *nom.con.*
5b (2)	Pods flattened, compressed laterally at right angles; flowers white	2. *CAPSELLA* Medik.
6a (1)	Pods short globose; 2-celled, each cell with one seed; flower white	5. *CORONOPUS* L.
6b (1)	Pods elongated, terete, hollow or transversely septate; flowers yellow, white or lilac	6. *RAPHANUS* L.

1. *BRASSICA* L.

Key to the Species

1a	Leaves of the stem narrowed to their bases and not stem clasping; most of the stem lyrately lobed	1. *B. juncea* Czern.

1b	Leaves of the stem auriculate and stem clasping; without hairs, green above and faint beneath, less lobed	2. *B. napus* var. *dichotoma* Prain

1.1. ***Brassica juncea*** (L.) Czern. Consp. Pl. Chark. 8, n.5. 1859; Coss., Bull. Soc. Bot. Fr. 4: 609. 1859; Hook.f. & Anders. Fl.Brit. India 1: 157. 1872 (in part); Wlh. India 1: 217. 1948; Backer & Bakh. f. Fl. Java 1: 189. 1963; Sarkar & Mitra Bull. bot. Soc. Bengal 23(2): 96. 1969. *Sinapis juncea* L. Sp. Pl. 668. 1753, excl. Synm. *Brassica juncea* sensu Hook.f. & Thoms. Linn. Soc. (Bot.) 5 : 170. 1861, in part; Watt Dict. Econ. Prod. 1 : 528. 1889; Prain Beng. Plants 145. 1903 (Rep. ed. 1963) & Rec. bot Surv. India 3 (2): 172. 1905; *(Vern. Name: Rai Sarisha).*

Erect, branching annual herb, 100-160 cm tall; stems often tinged with purple, especially at the nodes; leaves large, lower one petiolate. Flowers bright yellow in short corymbose. Pods narrow, beak of siliqua seedless. Seeds small round, blackish. *Fls.& Frts.:* December - March.

Normally an important winter crop in both high and low lands; but it is often found as an escape in the dry waste lands, roadsides or in the railway tracks.

Field Collections: Madarhat, *K.N.-65;* Benipur, *K.N.-242.*

Distributions: Cultivated almost throughout India; South and South East Asia; probably wild in C. Asia (Asiatic Origin).

Economic Importance: Some edible oil is extracted from the seeds; also used as an illuminant; the young shoots often used as vegetable.

Notes: Brassica juncea (Linn.) Czern & Coss. var. *laevis* (Prain) Schulz was collected by R.N. Banerjee, from a 'broken old brick wall, Calcutta' on 12th January 1966 under the field number *99* (CAL); but I have not made any collection during my exploration tours in the district. Some authors (Bailey, 1949) attributed the binomial *B. juncea* to Cosson (Bull. Soc. Bot. France 6: 609. 1859).

1.2. ***Brassica napus*** Linn. var ***dichotoma*** Prain, Agricult. Ledger 1: 47. 1898; Beng. Plants 1: 146. 1903 (Rep. ed. 1963) & Rec. bot. Surv. India 3 (2): 171. 1905; Wlh. Ind. 1: 218. 1948; Datta & Majumdar Bull. bot. Soc. Bengal 20(2): 57. 1966; Sarkar & Mitra Bull. bot Soc. Bengal 23(2): 96. *Brassica napus* Linn. Sp. Pl. 666. 1753. *B. compestris* sub sp. *napus* (L.) Hook f. & Andr. F1. Brit. India 1: 56 (Partly) 1872. *(Vern.Name: Sarisha).*

Erect herb; root fusiform; leaves auriculate and stem-clasping, without hairs, green above, with a faint bloom beneath, less lobed. Flowers yellow; petals deciduous. Seeds globose, brown. *Fls. & Frts.:* December-March.

Commonly cultivated throughout the district during winter months; sometime as an escape in the waste lands and garden edges.

Field Collections: Garia, *K.N. -317;* Subhashgram, *K.N.-1343.*

Distributions: Almost throughout India. Native of Eurasia.

Economic Importance: Oil obtained from the seeds is largely used for cooking; the young leaves are eaten as vegetable.

Other cultivated species of *Brassica* L. are *B.alba* (L.) Rabenh, *B. chinensis* Juslen., non Duthie & Fuller, *B. napus* L. var. *glauca* (Roxb.) Schulz, *Brassica napus* L. var. *napus, B. nigra* (L.) Koch., *B. oleracea* L. var. *acephala* DC., *B. oleracea* L. var. *botrylis* L., *B. oleracea* L. var. *camitata* L., *B. oleracea* L. *gemmifera* DC., *B. oleracea* L. var. *gongylodes* L., *B. pekinensis* (Lour.) Rupr., *B. rapa* L.

2. *CAPSELLA* Medik. *nom. cons.*

2.1. ***Capsella bursa-pastoris*** (Linn.) Medikus, Pflanzen. Catt. 1: 85. 1792; Hook. f. & Anderson F1. Brit. India 1: 159. 1872; Moench, Method. Pl. 271. 1794; Watt Dict. Econ. Prod. 2 : 134. 1889; Prain Beng. Plants 1: 147. 1903 (Rep. ed. 1963) & Rec. bot. Surv. India 3 (2): 173. 1905; Wlh. Ind. 2: 68. 1950; Sarkar & Mitra, Bull. bot. Soc. Bengal 23(2): 99. 1969. *Thlaspi bursa-pastoris* L. Sp. Pl. 647. 1753. *(Vern Name: Shepherd's Purse).*

Erect, glabrous or hairy, small, branched annual herbs. Flowers minute, white. Pods triangular or obcordate, glabrous, broad at the top; valves smooth, seeds oblong, punctate. *Fls. & Frts.:* December-March.

Very common in the cultivated fields; sometime in the waste lands.

Field Collection: Belghoria, *K.N. -1399.*

Distribution: Almost throughout the temperate India. Asia, Europe and Persia; native of most temperate countries.

Economic Importance: Extract used in dropsy as a diuretic; seeds used in chest trouble, and as hemostatic; considered a most reliable medicine for staying fluxes of blood. Seeds contain about 15-20% oil.

Remarks: 'Shepherd's purse' is a very variable species, both in size and shape of leaves, siliquae and hairiness.

3. *CARDAMINE* L.

3.1. ***Cardamine hirsuta*** L. Sp. Pl. 655. 1755; Sarkar & Mitra, Bull. bot. Soc. Bengal 23 (2): 102. 1969. *C. debilis* D. Don; Hook.f. & Anderson Fl. Brit. India 1: 138. 1872; Prain Beng. Plants 1: 144. 1903 (Rep. ed. 1953) & Rec. bot. Surv. India 3 (2): 171.1905.

Annual herb, 10- 30 cm tall; basal leaves 2-7 jugate, 2-10 X 1- 2.5 cm glabrous or hairy with simple hairs; leaflets ovate orbicular, except the terminal one. Racemes 10 - 30 flowered, upto 10 cm long in fruit; sepals 2-2.5 mm long; petals 3 mm long. Siliquae linear, compressed, 15-25 mm long, 1 mm broad, straight, glabrous. *Fls. & Frts.:* December-February.

Very rare, a garden weed.

Field Collection: Subuddhipur, *K.N.-2255.*

Distributions: Almost throughout India. Native of most temperate countries.

Economic Importance: Stimulant and diuretic.

Other species of the genus is *C. pratensis* L., distributed in Kashmir.

4. *COCHLEARIA* L.

4.1. ***Cochlearia cochlearioides*** (Roth) Santapau & Maheshwari, J. Bombay nat. Hist. Soc. 54. 804. 1957; Sarkar & Mitra, Bull. bot. Soc. Bengal 23 (2): 99. 1969. *Alysum cochlearioides* Roth, Nov. Pl. Sp. 322. 1821. *Cochlearia flava* Buch. Ham. ex Roxb. Hort. Beng. 48. 1814 *nom. nud;* Hook. f. & Anders. Fl. Brit. India 1 : 145. 1872; Prain Beng. Plants 1 : 146. 1903 (Rep. ed. 1963.)

Annual, glabrous, fleshy herb; leave pinnatifid, toothed. Flowers in elongated racemes, whitish; sepals spreading. Pods sub-globose. Seeds compressed; rugose, cotyledons accumbent. *Fls. & Frts.:* September-February.

Very rare in the dampy places and dry ditches.

Field Collection: Santoshpur, *K.N.- 1958.*

Distribution: Gangetic valleys from Punjab to Bengal; Tropics.

Economic Importance: Used in fever.

5. *CORONOPUS* L. *nom. cons.*

5.1. ***Coronopus didymous*** (L.) Sm. Fl. Brit. 2: 691. 1804; Stewart Annot Cat. Pl. Pakistan 306. 1972. *Lepidium didymus* L. Syst. 2: 433 & Mant. 92. 1767.

Senebiera didymus (L.) Pers. Syn. Pl. 2: 185. 1807. *S. pinnatifida* DC. Mem. Soc. Hist. Paris (1799) : 144. t.9.1799; Prain Beng. Plants 1: 148. 1903 (Rep. ed. 1963), Rec. bot. Surv. India 2 : 286.1903 & Ibid. 3 (2) : 173.1905. *Coronopus pinnatifida* (DC.) Schultz.; Sarkar & Mitra, Bull. bot. Soc Bengal 23 (2): 98.1969.

Prostrate, slightly hispid, branched, annual herb; leaves 1-2 pinnatifid, leaves small, obovate, spreading. Flowers minute, white, sepals spreading. Pods separating into 2 indehiscent transeversely reticulate lobes; seeds reniform, punctate striate, solitary in each chamber. *Fls. & Frts.:* December-April.

Scanty in distribution, a weed of garden, waste lands, roadsides and river sides.

Field Collection: Bhebe/Basirhat, *K.N.-704.*

Distribution: Throughout India but scanty. Cosmopolitan, native of Europe.

6. *RAPHANUS* L.

6.1. ***Raphanus sativus*** L. Sp. Pl. 699. 1753; Hook.f. Fl. Brit. India 1: 166. 1872; Watt Dict. Econ. Prod. 6 (1): 393.1893; Prain Beng. Plants 1 : 148.1903 (Rep. ed.1963) & Rec. bot. Surv. India 3 (2) : 173. 1905. *(Vern. Name: Mula).*

Annual or biennial, diffusely branched herbs; roots fusiform, pungent; leaves radical cauline, usually lyrate, pinnatifid. Flowers in long bractless racemes; white, petals small. Pods indehiscent, didynamous, laterally compressed. Seeds solitary in each chamber. *Fls. & Frts.:* December-March.

Commonly cultivated throughout the agricultural fields.

Field Collections: Garia, *K.N.-322;* Budge Budge, *K.N.-753.*

Distributions: Commonly cultivated throughout India; Pantropic.

Economic Importance: The fusiform roots are edible.

Remarks: It looks similar to *Eruca* sp., when without fruits; but can be distinguished by longer pedicels and shorter sepals and 1-seriate ovules.

Other species, *Raphanus caudatus* L. is a native of Java, also distributed in India.

7. *RORIPPA* Scop. (*Roripa* Adans.)

Key to the Species

1a.	Leaves deeply incised; pods terete, long linear	1. *R. indica* Hiern
1b.	Leaves pinnatifid, pods short thick	2. *R. islandica* Borbas

7.1. ***Rorippa indica*** (L.) Hiern, Cat. Afr. Pl. Wel. Pt. 1:26. Addit. & Corr. 1896; Raizada Ind. For. 92: 322. 1966. *Sisymbrium indicum* L. Mant. 1: 93. 1767. *Nasturtium indicum* (L.) DC. Syst. 2: 199. 1821; Hook.f. & And. Fl. Brit. India 1 : 134. 1872; Prain Beng. Plants 1 : 114. 1903 (Rep. ed. 1963) & Rec. bot. Surv. India 3 (2): 171. 1905. *(Vern. Name: Raisarisha).*

Erect or sub-erect, glabrous herb, 10-15 cm height. Leaves 8 X 3 cm, elliptic-oblong or oblanceolate; lower leaves long- petioled, entire-lyrate, pinnatifid or sessile. Flowers in terminal raceme, yellow. Pods usually 6-13 cm long, straight or slightly curved. Seeds numerous, 2-serriate, minute, elipsoid. *Fls. & Frts.* : January-July.

Rarely found in shady and waste lands, occasionally as field weeds.

Field Collections: Haroa, *K.N.-598;* Bhebe Choumatha, *K.N.-705.*

Distributions: Northern India especially, Tropical Asia.

Economic Importance: In Malayasia, this plant is edible as salads and soups; it is considered to be diuretic, stimulant and antiscorbatic. It is also applied in diarrhoea and dysentery.

7.1/A. ***Rorippa indica*** (L.) Hiern var. ***bengalensis*** (DC.) Deb, Taxon 14: 111. 1965; Majumder Ind. Agr. 6 (1 & 2): 106.1962 (Rep.ed.). *Nasturtium indicum* var. *bengalensis* DC. Syst. 2: 199.1821; Hook.f. Fl. Brit. India 1: 134.1872; Prain Beng. Plants 1: 144. 1903 (Rep. ed. 1963) & Rec. bot. Surv. India 3 (2): 171.1905. *N. bengalensis* DC. Syst. 2: 198.1821; Hook.f. &Anders. Fl. Brit. India 1: 134.1872. *(Vern. Name: Bil rai* or *Rai Sarisha).*

Closely allied to the typical variety, but is easily distinguishable by bracteate racemes. *Fls.& Frts.:* April-August.

A field weed, common in waste lands and old walls, everywhere.

Field Collection: Gobindapur, *K.N.-1237.*

Distributions: Throughout India, Srilanka, Malaya, Philippines China and Japan.

7.2. ***Rorippa islandica*** (Oeder) Borbas, Balaton Tav. 2: 392. 1900. *Sisymbrium islandicum* Oeder, Fl. Dan. 3 :7, t. 409.1753. *S. palustre* Leysser Fl. Hal. 166.1961. *Nasturtium palustre* (L.) DC. Syst. Nat. 2 : 191.1821; Hook.f.& Anderson Fl. Brit. India 1: 138. 1872; Prain Beng. Plants. 1 :144. 1903 (Rep.ed.1963) & Rec. bot. Surv. India 3(2): 173.1905. *Rorippa palustris* (Leyss.) Besser Enum. - Pl. Volhyr.27.1822.

Annual or biennial herb, 20-80 cm tall, erect, branched, glabrous or slightly hairy. Lower and middle leaves lyrate- pinnatifid, 3-5 jugate, stalked, 5-15 cm long, 1-3 cm broad. Racemes 20-30 flowered; flowers 2 mm across, pale yellow; sepals 2 mm long; petals as long as the sepals. Siliquae oblong ellipsoid, 5-10 mm long, 2-3 mm broad, glabrous. Seeds several, crowded, pale brown. *Fls. & Frts.:* April-August.

A rare field weed in the rabi crop fields.

Field Collections: Kadampur / Nabagram, *K.N.-635.*

Distributions: Eastern India, Kerala and temperate Himalaya; Europe and temperate Asia, widely introduced.

Economic Importance: Possesses antiscorbutic, tonic, stimulant and diuretic properties.

Some other cultivated/most economic species of the family- Brassicaceae in the Lower Ganga Delta regions are—

Brassica nigra (L.) Koch. *(Vern. Name: Kalo Sarisha)* -Oil yielding

B. compestris L. var. *sarson* Prain *(Vern.Name: Sarisha)* -Oil seeds

B. oleracea L. var. *botrytis (Vern.Name: Kapi)* -Winter vegetable

B. oleracea L.var. *capitata (Vern. Name: Bhandha Kapi)* -vegetable

B. rapa L. (Vern. name : Salgam)- vegetable.

X. C APPARIDACEAE

Key to the Genera

1a.	Leaves simple; sepals biseriate free, closed in bud	1. *CAPPARIS* L.
1b.	Leaves compound; sepals distinct, open in bud, inserted on the edge of the disk	2. *CRATAEVA* L.

1. *CAPPARIS* L.

Key to the Species

1a	Woody climber; flowers in short peduncled umbles	1. *C. sepiaria* L.
1b	Erect shrub; flowers in axillary clustered inaxis	2. *C. zeylanica* L.

1.1. ***Capparis sepiaria*** L. Syst. Nat. 2: 1071. 1759; Watt Dict. Econ. Prod. 2: 132. 1889; Hook. f. & Thoms. Fl. Brit. India1: 177.1872; Prain Beng. Plants 1: 150.1903 (Rep. ed. 1963), Rec. bot. Surv. India 2: 287.1903 & Ibid. 3(2): 174.1905; Jacobs, Fl. Males.6: 79.1960 *(Vern. Name: Kanta gurkamai* or *Kalia Kara)*

Spreading, wiry, zig zag branched, bushy; thorns recurved. Leaves ovate oblong or oblong lanceolate, sub-acute or retuse, pinninerved downy beneath. Flowers white in many flowered sessile or shortly peduncled umbels; sepals oblong; petals narrow, ovary glabrous. Fruit black. *Fls. & Frts.:* April-August.

Very common in the hedges, thickets and scrub jungles in the village waste lands; always climbing on the bushes.

Field Collections: Dighirpar/Canning, *K.N.-200*, Santoshpur, *K.N.-743;* Garia, *Mitra 21* (Cal.)

Distributions: Throughout the dry places in India; Philippines, Tropical Africa to Australia.

Occasionally planted as garden hedges and as goat fodder.

1.2. ***Capparis zeylanica*** L. Sp. Pl. 2:720 (ed.2).1762; Dunn Kew Bull. 1962 : 62.1962; Wlh. Ind. 2 :68.1950; Jacobs, F. Males.6: 87.1960. *C. horrida* L.f. Suppl. 264.1781; Hook.f. & Thoms. Fl. Brit. India 1:178.1872; Watt Dict. Econ. Prod. 2: 131.1889; Prain Beng. Plants 1: 150.1903 (Rep. ed. 1963) & Rec. bot. Surv. India 3 (2): 173.1905. *(Vern. Name : Kalo Kara).*

Rigid much branched, erect or rambling shrub. Stipular spines short, recurved. Leaves 2.4 X 1.9.-4. 1 cm oblong; acuminate. Flowers sub-globose to ellipsoid, seeds many. *Fls. & Frts.:* March-September.

Common on waste lands, roadsides hedges and the thickets of the railway tracts; climbing on the bushes of *Acacia nilotica, Ziziphus oenoplia, Lantana camera, Calotropis gigantea* and other hedges.

Field Collections: Dhanyakuria, *K.N.-565;* Kadampur, *K.N.-630.*

Distributions: Almost throughout India; China and Malaya.

Economic Importances: Leaves are used as a counter irritant and as a cataplasm in boils, swellings and piles; root bark bitter, sedative, stomachic, antihydrotic and used in cholera. Also used as hedge plant and fruits are edible after cooking.

2. *CRATAEVA* L.

2.1. ***Crataeva roxburghii*** R. Br. Denh. & Clapp. Travels, App. 224. 1824; Benthall Trees Calc. 15.1946. *C. religiosa* Forsst. Prodr. 35.1786; Hook.f.& Thoms.Fl.Brit. India 1: 172.1872; Watt Dict. Econ. Prod. 2 : 583.1889; Prain Beng.Plants 1: 151.1903 (Rep.ed.1963) & Rec. bot. Surv. India 3 (2): 174.1905. *C. religiosa* Forsst. var. *roxburghii* (R.Br.) Hook.f. & Thoms. Fl. Brit. India 1:172. 1872 *(Vern. Name: Barun).*

Spreading, unarmed, deciduous tree. Leaves small, ovate lanceolate, abruptly acuminate. Flowers yellow or purplish, polygamous. Berry globose. *Fls. & Frts.:* March- June.

On the river bank hedges in the Sundarbans area.

Field Collections: Bakkhali, *K.N.-1901.*

Distributions: Throughout the coastal India; native of South East Asia and Tropical Africa.

Economic Importance: The bark is demulcent, sedative, antipyretic, expectorant; it also promotes apetite, decreases the secretion of the bile and removes disorders of such urinary organs as kidney and bladder stones. It is also beneficial in the diseases of the chest and tubercular glands.

XI. CLEOMACEAE

1. *CLEOME* L.

Key to the Species

1a	Stamens inserted on the gynophores; petals white	1. *C. gynandra* L.
1b	Stamens sessile on the disk; petals violet or yellow	(2)
2a (1)	Low herbs with erect twigs; leaves 3-foliolate; flowers pinkish-violate	2. *C.rutidosperma* DC.

2b (1) Mostly erect herb; leaves 3-5 foliolate; flowers yellow 3. *C. viscosa* L.

1.1. ***Cleome gynandra*** L. Sp. Pl. 2 :671.1753; Jacobs. Fl. Males. 6 : 101.1960. *C. pentaphylla* L.Sp. Pl. 2:938 (ed.2).1763. *Gynandropsis pentaphylla*(L.) DC. Prodr.1: 238.1824; Hook.f. & Thoms. Fl. Brit. India 1: 171.1872; Watt Dict. Econ. Prod. 4: 190.1890; Prain Beng. Plants 1: 150.1903 (Rep.ed.1963), Rec.bot.Surv. India 2: 286. 1903 & Ibid. 3 (2) : 173.1905. *Gynandropsis gynandra* (L.) Briq. Ann. Cons. Jard. Bot.Geneve 17 : 382.1914. *(Vern. Name: Sada Hurhuria or Swet Hurhuria).*

Erect, annual 0.75-1.25 m high, branched, pubescent herb; leaves palmately compound, leaflets 5, sessile, broad-obovate, acute or obtuse, entire or serrulate. Racemes terminal, elongated, glandular, pubescent. Flowers white or purplish in glutinous racemes. Capsule cylindric. Seeds numerous, small. *Fls. & Frts.* : May-August.

Grow profusely on the roadside waste lands, railway tracks and cultivated fields as weeds.

Field Collections: Joynagar, *K.N.-876;* Garia, *Dutta-8* (Cal).

Distributions: Abundant throughout the warmer parts of India; and in S.E. Asia.

Economic Importance: Decoction of the root used in fever; leaves used in rheumatism by the rural people. In Africa, varieties of the species with large capsules are cultivated as vegetable.

Remarks: Cleome gynandra has been treated as belonging to a separate genus *Gynandropsis*. Iltis *(Brittonia 12:*283-290) 1960) has given convincing reasons for restoring this species as *Cleome* L.

1.2. ***Cleome rutidosperma*** DC. Prodr. 1: 241.1824; Iltis Brittonia 12 : 290.1960; Jacobs Fl. Males 6(1): 104.1960; Mukherjee Ind. For. 95 : 237.1969; Sarkar & Mitra Bull bot. Soc. Bengal 28 : 16.1974. *C.ciliata* Schum. & Thoms. Dansk Vld. Selsk. Afh. 4 : 68.1828; Balk. & Bakn. Fl. Java (em.ed.) 4A (1942) fam. 45.3.1942 (Fig. 1).

Twining herb with erect twigs, 30-100 cm height; stem angular. Leaves 3-foliolate, leaflets 1-4.5 x 0.5 - 1.8 cm, subsessile, oblong or ovate, acute, shortly ciliate. Flowers pinkish violet in leafy terminal racemes. Capsule linear, beaked, 2-valved, 5-7 cm long. *Fls. & Frts.:* May-August.

Rarely found on the waste lands and railway tracks.

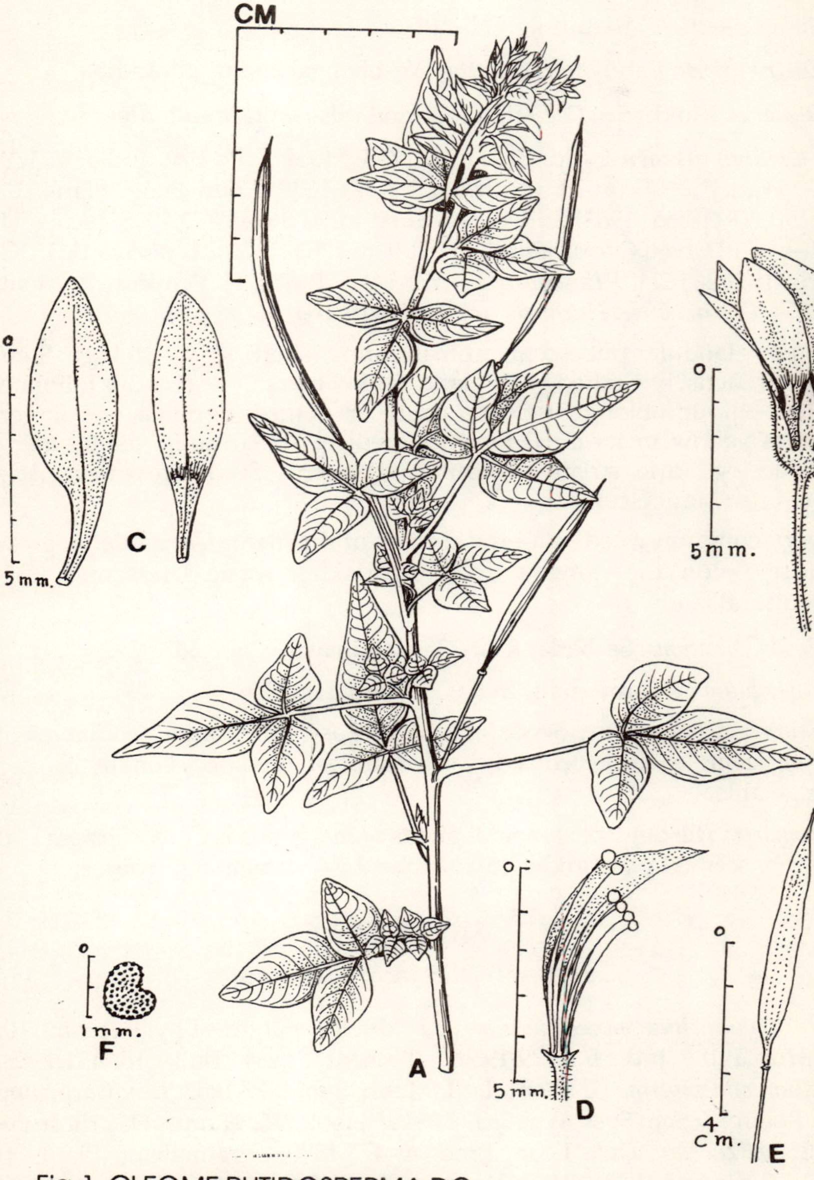

Fig. 1. CLEOME RUTIDOSPERMA DC

A. Flowering & Fruiting twig, B. Single Flower, C. Petals
D. Androecium, E. Single Fruit, F. Seed.

Field Collections: Baruipur, *K.N.-2147.*

Distributions: Rarely found in the West Bengal and upper India.

Remarks: Mukherjee (1.c.) first reported this taxon from India.

1.3. ***Cleome viscosa*** L. Sp. Pl.2: 672. 1753; Hook f. Fl. Brit. India. 1: 170. 1872; Watt Dict. Econ. Prodr. India 2 : 371.1889; Prain Beng. Plants 1 : 149.1903 (Rep. ed. 1963) & Rec. bot. Surv. India 3 (2) : 173.1905. Jacobs Fl. Males. 6: 103.1960.*C.icosandra* L. Sp. P. 938. 1753. *Polanisia viscosa* (L.) DC. Prodr. 1: 242. 1824. *P.icosandra* (L.) W. & A. Prodr. Fl. Pen Ind. Orient. 22.1834. *(Vern. Name: Hurhuria* or *Halde Hurhuria).*

Erect, glandular pubescent, aromatic annual herb, 0.5-1.0 m high. Stem grooved, densely clothed with gland tipped hairs. Leaves 3-5 foliolate; leaflets elliptic oblong or ovate, acute or obtuse, terminal one larger. Flowers yellow in terminal racemes; pedicel 1-2 cm long, terete, hairy. Capsule cylindric, striate, glandular pubescent. Seeds brown or black. *Fls. & Frts.:* May- October.

Very common weed in all the corners of the lower Ganga delta; grow profusely on the railway tracks, roadsides, waste lands and high agricultural lands.

Field Collections: Sarberia, *K.N.-235;* Saktinagar, *K.N.- 507.*

Distributions: Throughout India; Pantropic species.

Economic Importance: Seeds considered as anthelmintic, rubifaveient and vesicant; leaves juice used in ear troubles. Seed occasionally cooked as vegetable.

Remarks: The other ornamental plants of this genus is *Cleome spinosa* L. It is highly scented plant and cultivated for the ornamental purposes.

XII. VIOLACEAE

1. *HYBANTHUS* Jacq. *nom. cons.*

1.1. ***Hybanthus ennessperma*** (L.) Muell., Fragm. Phyt. Austr. 10: 81.1876; Wlh. Ind. 5 : 139.1959; Tennent Kew Bull. 16: 431.1963. *Ionidium ennessperma* (L.) Vent. Jard. Malm. Sub. t. 27.1803. *I. suffruticosum* (L.) Roem. & Sch. Syst. Veg. 5: 394.1819; Hook.f. & Thoms. Fl. Brit. India 1: 185.1872; Watt Dict. Econ. Prod. 4: 475.1890; Prain Beng. Plants 1: 152.1903 (Rep.ed.1963) & Rec. bot. Surv. India 3 (2). 174.1905. *(Vern. Name: Nunbora).*

Perennial undershrub; branches diffuse, woody. Leaves linear or lanceolate, serrate or toothed. Flowers solitary, red, pedicels shorter than the leaves. Capsule sub-globose. Seeds striate, ellipsoid. *Fls. & Frts.:* July-October.

Very common, growing profusely in open lands after rains.

Field Collections: Champahati, *K.N.827;* Belghoria, *K.N.-1397.*

Distributions: Abundant throughout India; Tropical Asia, Africa and Australia.

Economic Importance: The entire plant is bitter and acrid; used in dysentery, vomitting, burning, urethral discharges, blood troubles, asthma, cures cough and gives tone to the breasts. The Santals employ the root in bowel complaints.

Notes: Viola tricolour is a common garden plant in the lower Ganga Plains.

XIII. BIXACEAE

1. *BIXA* L.

1.1. ***Bixa orellana*** L. Sp. Pl. 512.1753; Hook.f. & Thoms. Fl. Brit. India 1: 190.1872; Watt Dict. Econ. Prod. 1: 454.1889; Prain Beng. Plants 1: 153. 1903 (Rep. ed. 1963) & Rec. bot. Surv. India 3 (2): 174.1905; Wlh. India 1: 196. 1948, *(Vern. Name: Latkan).*

Small, evergreen tree. Leaves cordate, acuminate, 14 X 10 cm, glabrous, petiole slender. Flowers in terminal panicles; pink, purple or white. Capsule ovoid or globose, base intruded, shortly prickles. Seeds many, funicle thick; testa pulpy; albumen fleshy, embryo large, cotyledons flat. *Fls. & Frts.:* August- November.

Planted as an ornamental garden plant.

Field Collection: Barrackpore, *K.N.-1841.*

Distributions: Cultivated throughout India; Native of South America.

Economic Importance: The red dye is obtained from the red pulp of the furits. It is extensively used for colouring cotton and silk, orange and also butter, cheese, confectionary, hair oil, shoe polishes, floor polishes and pharmaceutical ointments. A fairly good fibre is also suitable for cordage, which is extracted from the bark. The root are used to flavour meats. Leaves and many other parts of the plant are used as medicines.

XIV. COCHLOSPERMACEAE

1. *COCHLOSPERMUM* Kunth ex A.P. de Candolle *nom. cons.*

1.1. ***Cochlospermum religiosum*** (L.) Alston, Trimen Handb. Fl. Ceylon 6: 14. 1931. *C. gossypium* (L.) DC. Prodr. 1: 527.1824; Hook.f. & Thoms. Fl. Brit. India 1: 190.1872; Watt Dict. Econ. Prod. 2 : 412. 1889; Prain Beng. Plants 1: 153. 1903 (Rep. ed. 1963); Benthall Trees Calc. 19.1946. *Bombax gossypium* L. Syst. (ed.12) 2: 457. 1867. *(Vern. Name: Gabdi* or *Golgol).*

Medium sized or small, deciduous tree; bark thick, fibrous, light coloured, deeply furrowed, young branches tomentose, marked with scars of fallen leaves. Leaves old ones glabrous; petiole slender. Flowers in few flowered terminal panicles, golden yellow; sepals silky, petals obovate, notched. Capsule 5-lobed; pedulous; pear shaped. Seed numerous, brown, covered with soft silky wool. *Fls. & Frts.:* January-May.

Commonly planted as ornamental garden plant.

Field Collections: Nilgunj/Barrackpore, *K.N.-2107.*

Distributions: North West Himalaya, Bihar, Madhya Pradesh, Orissa, Gujarat and South India also; introduced elsewhere. A native of Malaya.

Economic Importance: White gum exudes from the bark and is used in the manufacture of cosmetics, for thickening ice-cream. The bark yields – cordage fibre; sometimes the soft wood is eaten. The gum has medicinal values.

XV. FLACOURTIACEAE

1. *FLACOURTIA* L' Heritier

Key to the Species

1a	Leaves ovate or orbicular, obtuse, tomentose beneath; often wild	1. *F. indica* Merr.
1b	Leaves ovate or lanceolate, acuminate, quite glabrous; often planted	2. *F. jangomas* Raeusch

1.1. ***Flacourtia indica*** (Burm.f.) Merrill, Interpret, Rhump. Herb. Amb. 377.1917; Benthall Trees Calc. 23.1946; Sleumer, Fl. Males. 15 (1): 76.1954. *Gmelina indica* Burm.f. Fl. Ind. 132.t.39, f.5. 1768. *Flacourtia ramontchi* L'Herit. Stirp. Nov. 3 : 59, t. 30, 30B. 1785; *F. sepiaria* Roxb. Pl. Cor. 1:48, t. 68.1795; Hook.f. & Thoms. Fl Brit. India 1 : 194.1872; Watt Dict. Econ.

Prod. 3 : 399. 1890; Prain Beng. Plants 1 : 154. 1903 (Rep. ed. 1963) & Rec. bot. Surv. India 3 (2): 174.1905. *F. latifolia* Cooke Fl. Pres. Bomb. 1: 59.1901 (Rep. ed.). *(Vern. Name : Bainchi).*

Much branched, deciduous shrub or small tree. Branches often spinous; spines 2-3 cm long. Leaves ovate, orbicular, oblong, with a cuneate base, obtuse, or crenate, glabrous. Flowers in short racemes, few flowered, greenish yellow, small, deciduous. Fruits globose, dull scarlet red; 8-16 seeded. *Fls. & Frts.:* February-July.

Very common in the scrub jungles, village thickets and roadside waste dry lands or railway tracks.

Field Collections: Amtala, *K.N.-385;* Harinbari, *K.N.-429.*

Distributions: Almost throughout India, particularly in the drier regions; Tropical and sub-tropical countries in Africa, South East Asia, Polynesia, Malaya and Java.

Economic Importance: Leaves are used as fodder; ripe fruits are edible and rather tasty but slightly acrid; immature fruits are very astringent. Village boys and rural people are used to eat the ripe fruits. Fruits are also medicinal and are useful for enlarged spleen.

1.2. ***Flacourtia jangomas*** (Lour.) Raeusch, Nomenol.bot. (ed.3): 290.1797; Benthall Trees Calc. 22.1946; Sleumer Fl. Males. 1,5(1): 1954; Wlh. Ind. 4: 42.1956. *Stigmarata jangomas* Lour. Fl. Cochinch. 22: 634.1790. *Flacourtia cataphracta* Roxb. ex Willd. Sp. Pl. 4: 830. 1806; Hook.f. & Thoms. Fl. Brit. India 1: 193. 1872; Watt Dict. Econ. Prod. 3 : 398. 1890; Prain Beng. Plants 1: 154.1903 (Rep. ed.1963). *(Vern.Name: Paniala).*

Small evergreen tree, armed with spines. Leaves 5-10 cm long ovate or oblong lanceolate, acuminate, crenate serrate, membranous and shining on both the surfaces. Flowers small, greenish, in short irregular axillary and terminal racemes. Fruits dark purple when ripe, globose. Seeds 8-12, flattened. *Fls. & Frts.:* May-December.

Rarely cultivated for its fruits,which are edible.

Field Collection: Falta, *K.N.-2000.*

Distributions: Cultivated almost throughout India; native of Chattagram.

Economic Importance: Fruits are edible but become sweeter on ripening; various parts of the plants are used medicinally; the seeds yield an oil.

Remarks: Flacouritia inermis Roxb. an ornamental tree in India.

XVI. POLYGALACEAE

1. *POLYGALA* L.

1.1. ***Polygala chinensis*** L. Sp. Pl. 704.1753; Bennett, Hook. f. Fl. Brit. India 1: 204.1872; Watt Dict. Econ. Prod. 6 (1): 315.1892; Prain Beng. Plants 1: 157.1903 (Rep. ed. 1963) & Rec. bot. Surv. India 3 (2): 174.1905; Wlh. India 8: 190.1969. *P. arvensis* Willd. Sp. Pl. 3 (2): 876.1802. *P. rothiana* W. & A. Prodr. 37.1934. *(Vern. Name: Meradu).*

Slender, erect, simple or branched annual herbs, 25-30 cm high. Leaves linear oblong to oblong lanceolate, with an acute base, apiculate, short petioled. Flowers light yellow, usually in short extra axillary, few flowered racemes. Capsule 3-4 cm wide, ciliate, oblong, obliquely obcordate at apex. Seeds hairy. *Fls. & Frts.:* June-December.

Common on the pastur lands and roadsides; common in rainy season on the grassy grounds, often mixed with grasses.

Field Collections: Basirhat, *K.N.-967;* Taki, *K.N.-1007.*

Distributions: Almost throughout India; China, Malaya and Tropical Austalia.

Economic Importance: Root is useful in fever and dizziness. Leaves are useful in asthma, chronic bronchitis and catarrhal affections.

Notes: The other Indian species under this genus *Polygala* L. are *P. abyssinica* R. Br. ex Fresen, *P. acuminata* Turez, *P. arillata* Buch. Ham. ex D.Don, *P. elongata* Klein, *P. glomerata* Lour., *P. senega* L. and *P. sibirica* L. Most of these species are distributed in the Himalayas. Most of these Himalayans species have the medicinal values:

(a) *P. abyssinica* – Root contain saponin and used as substitute of Senega.

(b) *P. arillata* – Root used as purgative, for headache and for fermenting *murwa* beer.

(c) P. elongata – Plant used biliousness, constipation and specific for snake bite.

(d) *P. glomerata* – Plant used in asthma, diarrhoea, chronic bronchitis and in inflammatory conditions.

(e) *P. senega* – Rootstock is used in chronic bronchitis and asthma.

(f) *P. sibirica* – Root used in inflammation of urinary bladders and mammary abscesses.

XVII. CARYOPHYLLACEAE

Key to the Genera

1a	Stipules scarious; style 2-3 fid; capsule small and sub-globose or longer	(2)
1b	Stipules; style 3-5 fid; capsule globose	4. *STELLARIA* L.
2a(1)	Capsule valve 3, style 3, united	(3)
2b(1)	Capsule valve 5, style 5, free, petals entire	3. *SPERGULA* L.
3a(2)	Erect herb; sepals not keeled	1. *POLYCARPAEA* Lamk.
3b(2)	Prostrate or sub-erect herb. sepals keeled	2. *POLYCARPON* L.

1. *POLYCARPAEA* Lamk. *nom. cons.*

1.1. ***Polycarpaea corymbosa*** (L.) Lamk. Tab. Encycl. 2: 129.1797; Edge. & Hook.f. Fl. Brit. India 1: 245.1874; Watt Dict. Econ. Prod 6(1) : 314. 1892; Prain Beng. Plants 1: 160. 1903 (Rep. ed. 1963)& Rec. bot. Surv. India 3 (2): 175.1905; Wlh. Ind. 8: 189. 1969. *Aschyranthes corymbosa* L. Sp. Pl. 205.1753 *(Vern. Name: Okharad).*

Erect or decumbent annual herb; densely tomentose; often glabrescent; simple or dichotomously branched. Leaves pseudo-verticillate, linear or sublate, acute or acuminate or obtuse at apex; stipules lanceolate or subulate. Cymes irregular, terminal dense, much branched. Flowers upto 4 mm in diam. Capsule oblong. Seeds brown ovoid. *Fls. & Frts.:* September-December.

Common on the sandy soil, river banks and on the dry paddy fields.

Field Collections: Shyamnagar, *K.N.-2076.*

Distributions: Common throughout India; Asia, Africa, Australia and America.

Economic Importance: Used in strangury, urinary calculi, boils, inflammatory swellings and ulcers. Leaves used as poultice, worms and cold over boils and inflammatory swellings. Also useful in jaundice.

2. *POLYCARPON* L.

2.1. ***Polycarpon prostratum*** (Forssk.) Asch. & Schwein.f. Ost. Bot. Zeitsch. 39: 128.1889; Wlh. Ind. 8: 189.1969. *Alsine prostrata* Forssk. Fl.

Aggypt Arab 207. 1775. *Pharnaceum depressum* L. Mant. Alt. 2: 562.1771 (non *P. depressum* Nutt. ex Torr. & Cay, 1838). *Loeflinsia indica* Retz. Obs. Bot. 4: 8. 1786. *Polycarpon loeflingiae* (W. & A.) Benth. & Hook.f. Gen. Pl. 1: 153. 1862 in note *(Loeflingii);* Edge. & Hook. f. Fl. Brit. India 1 : 245.1874; Prain Beng. Plants 1: 160.1903 (Rep. ed. 1903) & Rec. bot. Surv. India 3 (2): 175. 1905. *P. indicum* (Retz.) Merr. Philippine J. Sci. 10: 302.1915. *(Vern. Name: Gima).*

Much branched, prostrate weed. Leaves acute or obtuse, linear oblong or spathulate, scarious white stipules. Cymes fascicled or panicled. Flowers 3 mm in diam., white; sepals acute and fairly, strongly keeled at the midrib, which is green, the rest being usually whitish; petals entire or notched. Seeds many, sub-cylindric; helium lateral and sub-basal; embryo straight. *Fls. & Frts.:* March-September.

Very common to all parts of the lower Ganga delta plain; generally on the roadside ditches, ponds, village thickets and paddy fields.

Field Collections: Kholapota/Basirhat, *K.N.-544;* Haroa, *K.N.-582.*

Distributions: Throughout the warmer parts of India; in the fields and waste places. Tropical Asia and Africa.

Economic Importance: Numerous medicinal properties of the plants exibit it very economical; also edible as vegetable. Infusion of roasted leaves is given for cough, fever and in Indo-China it is used in meals.

3. *SPERGULA* L.

3.1. ***Spergula arvensis*** L. Sp. Pl. 440. 1753; Edgeworth & Hook. f. Fl. Brit. India 1 : 243.1874; Watt Dict. Econ. Prod. 6(3); 320. 1893; Prain Beng. Plants 1: 158.1903 (Rep. ed. 1963); Wlh. Ind. 10: 3. 1976.

Glabrous, diffused, annual herb. Stem branched from the base. Flowers white, sepals obtuse; style 3, free to the base. Capsule ovoid glabrous. Seeds keeled or narrowly winged; granular or papillose black. *Fls. & Frts.:* November-February.

Rare in cultivated land during the winter season.

Field Collections: Harinbari/Sagar Island, *K.N.- 487.*

Distributions: Temperate regions of India; cultivated as fodder in Europe and South Africa.

Economic Importance: Said to be employed against pulmonary tuberculosis and also used as diuretic.

4. *STELLARIA* L.

4.1. ***Stellaria media*** (L.) Cyrill. Charac. Comm. 36.1784; Edge. & Hook.f. Fl. Brit. India 1: 230.1874; Watt Dict. Econ. Prod. 6(3): 358.1893; Prain Beng. Plants 1: 159.1903 (Rep. ed. 1963); Wlh. Ind. 10 : 39.1976. *Alsine media* L. Sp. Pl. 272.1753.

Much branched, procumbent, flaccid, annual herb. Leaves long petioled, ovate, elliptic or lanceolate. Flowers white in axillary or terminal cymes; pedicles slender, glabrous. Capsule ovoid; longer than the sepals. Seeds brown, acutely tubercled. *Fls. & Frts.:* November-April.

Generally grow on the roadsides or waste lands; also found in the cultivated fields as weed. Propagate vegetatively by the nodal roots.

Field Collections: Gosaba, *K.N.-657;* Sarsuna, *Datta-12.*

Distributions: Throughout the temperate regions of India; Arctic and N. temperate regions; doubtful native of Asia and northern temperate regions. Probably introduced with the crop plants from Europe.

Economic Importance: Said to be eaten as vegetable.

Notes: Stellaria aquatica (L.) Scop. is another herb, occurring in the Temperate Himalayas.

The other members of the family Caryophyllaceae are *Dianthus chinensis* L., *D. caryophyllus* L., *D. barbatus* L., *Gypsophila elegans* L. and *G. paniculata* L.,the common season flowers of India.

XVIII. PORTULACACEAE

1. *PORTULACA* L.

Key to the Species

1a	Nodes without appendages, flowers in terminal clusters	1. *P. oleracea* L.
1b	Nodes with ring of hairs; flowers in solitary terminals	2. *P. quadrifida* L.

1.1. ***Portulaca oleracea*** L. Sp. Pl. 445.1753; Thiselton Dyer in Hook.f. Fl. Brit. India 1: 246.1874; Watt Dict. Econ. Prod. 6 (1): 329.1892; Prain Beng. Plants. 1: 161.1903 (Rep. ed. 1963) & Rec. bot. Surv. India 3(2): 175.1905. *(Vern. Name: Bara Laniya).*

Prostrate, annual herb with spreading branches; stem tinged with pink or reddish, swollen at nodes. Leaves rounded truncate; petiole short. Inflorescences solitary or dichotomous cymes.

Flowers yellow and sessile with ovate pointed scarious scales. Capsule ovoid. Seeds tubercled, punctate. *Fls. & Frts.*: June-October.

Common weed in the waste lands, roadsides and railway tracts.

Field Collections: Suparibhanga Sundarbans, *K.N.- 420.*

Distributions: Throughout the hotter parts of India; both Tropical and Sub-Tropical regions. A well known weed of Europe.

Economic Importance: An excellent salad plant; leaves are eaten in scurvy and liver diseases. Occasionally edible as vegetable. Plant is also used for the treatment of cardiovascular diseases, spleen troubles, kidney troubles, dysentery and acts as blood purifier.

1.2. ***Portulaca quadrifida*** L. Mant. 1: 73.1767; Thiselton Dyer, Hook. f. Fl. Brit. India 1: 247.1874; Watt Dict. Econ. Prod. 6(1): 330.1892; Prain Beng. Plants 1: 161.1903 (Rep.ed. 1963) & Rec. bot. Surv. India 3: 175.1905; Geesink, Blumea 17 : 290.1967; Wlh. India 8: 220.1969. *(Vern. Name: Chhota Laniya* or *Nuniya).*

Common on the roadside waste lands and railway tracts.

Field Collections: Manirampur/Barrackpore, *K.N.- 1838.*

Distributions: Throughout the hotter parts of India; Srilanka, Malaya, Tropical Asia and Africa.

Economic Importance: Occasionally planted for the decorative purposes.

Notes: Portulaca grandiflora Hook, is a showy succulent pot-herb in the ornamental gardens: which is a native of South America and introduced into India by the Portuguese, during the last part of the 16th Century.

XIX. TAMARICACEAE

1. *TAMARIX* L.

Key to the Species

1a	Flower bisexual	(2)
1b	Flower monoecious in short close, cylindric spikes	2. *T. dioica* Roxb
2a (1)	Medium sized tree	(3)
2b (1)	Glaucous shrub; branches slender articulated; leaves minute or scale like, acute; inflorescences panicle	3. *T. gallica* L.

3a (2) Branchlets articulate at base of the sheath, slender, cylindric, jointed; leaves reduced to small sheath with minute tooth; inflorescences in long slender spikes, sometime in loose panicles 1. *T. aphylla* Kanza

3b (2) Branches pendulous; leaves subulate, with triangular semiamplexicaul base, panicled, speciform racemes 4. *T. troupii* Hole

1.1. ***Tamarix aphylla*** (L.) Kanza Boll. Orto. Bot. palermo 8: 82.1909; Wlh. Ind.10.123.1976. *T. articulata* Vahl Symb. Bot. 2 : 48, t. 32.1791; Thiselton Dryer, Hook.f. Fl. Brit. India 1: 249.1874; Watt Dict. Econ.Prod. 6(3): 409.1893. *Thuya aphylla* L. Cent. Pl. 1: 32.1755. *Tamarix orientalis* Forssk. Fl. Aegypt. 206.1755. *(Vern. Name: Lal-Jhau).*

Medium sized tree; branches articulate at base of sheath, often greyish, fastigate, slender, cylindric, jointed 5 cm long or more. Leaves reduced to a small sheaths with a minute tooth. Flowers pink, 0.25 mm diam., bisexual or monoecious, 5-merous, sessile, scattered on long slender spikes, sometimes in loose panicles at the ends of branches; stamens 5, inserted in alternate notches of 10 lobed disk. Capsule 0.25 cm long. *Fls. & Frts.:* July-October.

Well adapted in sandy and salty areas; occasionally planted in the ornamental gardens.

Field Collections: Ichhapur, *K.N.-1820.*

Distributions: Punjab and north India, cultivated everywhere. Native of Sind, Baluchistan and countries westwards to Egypt.

Economic Importance: Bark bitter and astringent, has tonic properties, and also employed to cure skin diseases.

Remarks: Prain (1903a, 1903b and 1905) did not mention this taxon from Bengal.

1.2. ***Tamarix dioica*** Roxb. Hort. Beng. 22.1814; *nom. nud.* & Fl. Ind. 2: 101.1824; Thiselton Dryer, Hook.f. Fl. Brit. India 1: 249.1874; Watt Dict. Econ. Prod. 6(3): 410.1893; Prain Beng. Plants 1 : 162.1903 (Rep. ed. 1963) & Rec. bot. Surv. India 3 (2): 176. 1903; Wlh. Ind. 10.123.1976. *(Vern. Name : Lal Jhau).*

Shrub to small tree, 3-5 m tall with drooping branches. Leaves minute, adpressed, sheathing; sheath tubular, apex acuminate. Flowers purple or pink in compact cylindrical, loosely panicled spikes. Capsule oblong, tapering. *Fls. & Frts.:* July-October.

Frequently found on the river banks and the edge of marshy lands.

Field Collections: Bamanpukuria / Minakhan, *K.N.- 1125.*

Distributions: Coastal regions of India and also in Punjab; Burma, Baluchistan to Egypt, native of Sind.

Economic Importance: Bark is bitter and astringent; tonic properties and also employed to cure skin diseases.

1.3. ***Tamarix gallica*** L. Sp. Pl. 270. 1753; Thiselton Dryer, Hook.f.Fl. Brit. India 1:249.1874; Watt Dic. Econ. Prod. 6(3): 411.1893; Prain Beng. Plants 1: 162.1903 (Rep.ed.1963) & Rec. bot. Surv. India 3(2): 176.1905. *T. indica* Willd., Ges. Natur.f.Fr. Neue Schr. 4: 214.1803. *(Vern.Name: Jhau* or *Banjhau).*

Glaucous shrub; branches slender, articulated. Leaves minute, not sheathing smooth, subulate or scale like, acute. Flowers bisexual 0.25-0.5 cm diam., white or pink, crowned in slender spikes, collected in dense panicles at the ends of branches; bracts shorter than flowers. Capsule 0.25-0.30 cm long. *Fls. & Frts.:* August - September.

Well adopted to grow in sandy places and flourishes on lands of the Sundarbans area; impregnated with salt area is suitable for the growth of this species. Occasionally planted as garden plants.

Field Collections: Sagar Mela Premises/Sundarbans, *K.N.-441.*

Distributions: Most of the coastal area in India; indigenous from the Mediterranean to Japan, Europe and Africa.

Economic Importance: Stem galls and bruished twigs used for dyeing and tannins. The galls also employed as an astringent in the cure of dysentery and similar diseases. The fruits and leaves are useful in skin diseases.

Notes: Common on the sandy islands and in the large river banks.

1.4. ***Tamarix troupii*** Hole Ind. For. 45: 248.1919; Troup Silvic. Ind. Trees 1: 16.1921; Wlh. Ind. 10: 124.1974. *T.gallica* auct. *(Non* L.). *(Vern. Name: Jhau).*

Small tree with pendulous branches, 2-6 m high. Leaves upto 2.5 x 3.5 mm, subulate, with triangular semiamplexicaul base, acute. Flowers bisexual in panicled spiciform racemes; peduncle upto 1 mm long, bracts ovate, acute. Capsule upto 2.3 mm long, trigonous conical, glabrous. Seeds white. *Fls. & Frts.:* September- March.

Common on the river banks and river beds of the Sundarbans.

Field Collections: Sagar Island, *A.K.M.-8719* (CAL).

Distributions: Most of the coastal area of India; indigenous from Mediterranean to Japan.

Notes: Hole (l.c.) wrote, the *T. gallica* L. does not occur in India and the species so far described as *T. gallica* L. from the plains of West India is *T. troupii* Hole.

The other Indian species is *Tamarix erioides* Rottl. -a shrub, also distributed in the S. India. The twigs of this species are used as brushes and also for fire wood.

XX. ELATINACEAE

1. *BERGIA* L.

Key to the Species

1a	Erect herb with lower branches decumbent, without roots; leaves 10-30 cm long; flowers pedicelled, rosy; stamens 3-5	1. *B. ammannioides* Roxb.
1b	Procumbent, glabrous herb; rooting from nodes; leaves 2.5 cm or more long; flowers sessile, white, sub-whorled; stamens 10	2. *B. capensis* L.

1.2. ***Bergia ammannioides*** Roxb.(Hort. Beng.34.1814) *nom. nud.* & Fl. Ind. 2: 456. 1824; Thistelton Dryer, Hook.f. Fl. Brit. India 1: 252.1874; Prain Beng. Plants 1: 163.1903 (Rep. ed. 1963) & Rec. bot. Surv. India 3 (2): 176.1905; Backer, Fl. Males. 4 : 205. 1951. *Lechea verticillata* Willd. Sp. Pl. 1 : 495.1797. *Elatine ammannioides* W. & A. Prodr. 41.1834. *(Vern. Name : Lal-keshuria).*

Erect or sub-erect, much branched annual herb, 15-40 cm high; branches often glandular or elliptic, acute, glandular, ciliate, serrate, finely hairy or glabrous above, beneath hairy on veins; stipules membranous, linear acute. Flowers red in dense long pedicels. Capsule 2-3 cm long ovoid dehiscing from the centre. Seeds numerous. *Fls. & Frts.:* December-May.

Common on the canal sides, paddy fields, riverbanks and marshy places; on loam and sandy soils.

Field Collections: Shyamnagar, *K.N. 2078;* Sagar Islands, *K.N.2375.*

Distributions: Throughout India; cosmopolitan.

Economic Importances: Rural people occasionally cook as vegetable and it is reported that this plant has immence medicinal values.

1.2. ***Bergia capensis*** L. Mant. 2 : 241.1771; Deb, Bull. bot. Soc. Bengal 29(2): 59.1978. *B. verticillata* Willd. Sp. Pl. 2: 770.1799; Thiselton Dryer, Hook.f. Fl. Brit. India 1: 252.1874; Prain Beng. Plants 1: 163.1903 (Rep. ed. 1963) & Rep. bot. Surv. India 3(2): 176.1905 *B. aquatica* Roxb. Corom. Fl. 2: 22. t.142.1795; Boiss. Fl. Orient. 782.1867. *(Vern. Name: Kesuria).*

Procumbent, glabrous, annual, succulent herb, 20-30 cm high; branches decumbent; rooting from the nodes. Leaves simple, opposite, elliptic lanceolate, serrate; stipules triangular acute, caducous. Flowers sessile, white, sub-whorled, 5-merous. Capsule globose, apiculate. Seeds minute, cylindric, dark brown. *Fls. & Frts.:* November - May.

Rarely on the wet places in the paddy fields, roadside nalas.

Field Collections: Nilgung/Barrackpore, *K.N.-2106.*

Distributions: Throughout India; in the Tropics.

Economic Importance: Occasionally edible as vegetable.

Notes: The other species is *B. suffruticosa* (Del.) Fenzl., distributed in Rajasthan and Gujarat.

XXI. CLUSIACEAE (*nom. alt.* Guttiferae)

Key to the Genera

1a	Stigma on a slender style; embryo of two fleshy, free cotyledons with a small radicle	1. *CALOPHYLLUM* L.
1b	Stigma sessile; embryo of a solid tigellus with minute cotyledons	2. *GARCINIA* L.

1. *CALOPHYLLUM* L.

1.1. ***Calophyllum inophyllum*** L. Sp. Pl. 513.1753; Anderson, Hook.f. Fl. Brit. India 1: 273.1874; Watt Dict. Econ. Prod. 2: 29.1889; Prain Beng. Plants

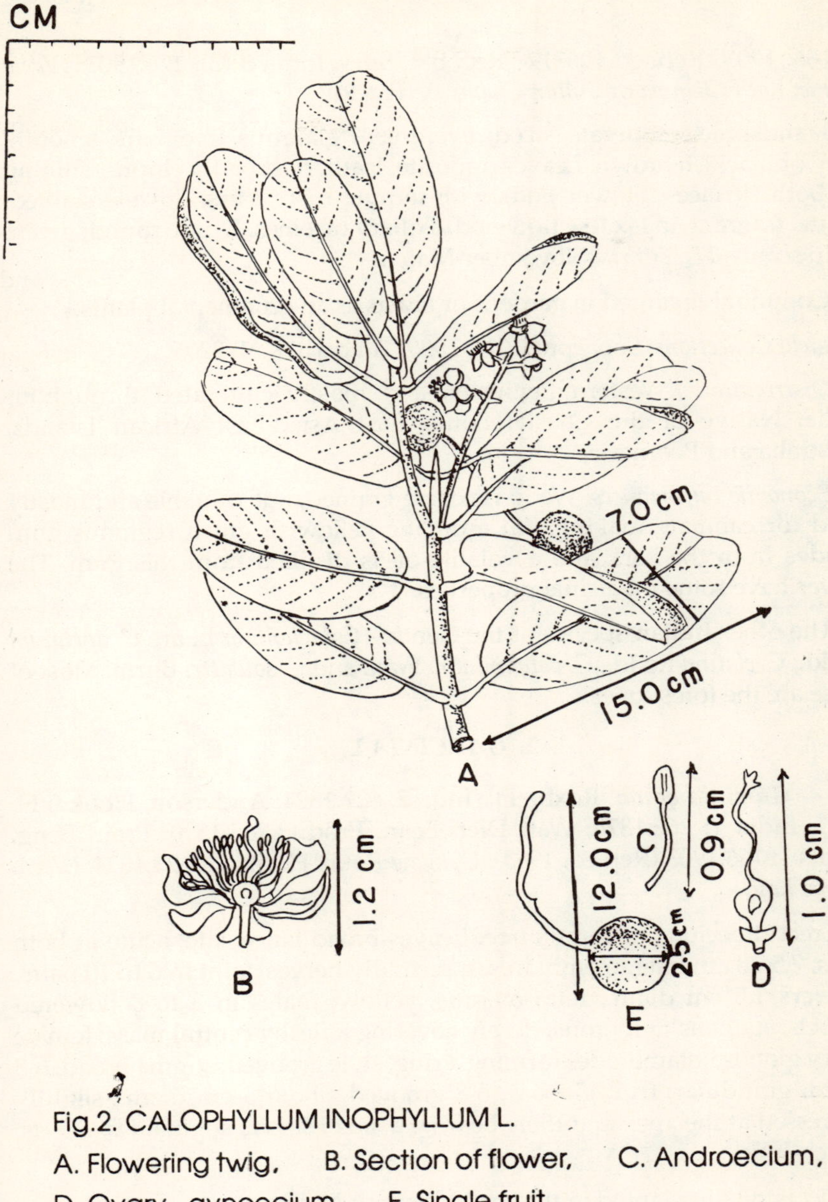

Fig.2. CALOPHYLLUM INOPHYLLUM L.

A. Flowering twig, B. Section of flower, C. Androecium, D. Ovary - gynoecium, E. Single fruit.

1: 166. 1903 (Rep. ed.1963) & Rec. bot. Surv. India 3 (2): 176.1903. *(Vern. Name: Kath Champa* or *Sultan Champa).* (Fig. 2.).

Handsome, moderate sized, evergreen, glabrous tree; bark smooth, grey or blackish brown. Leaves opposite, leathery, broadly elliptic, shining on both surfaces. Flower buds with minute rusty hairs; flowers stalked, white, fragrant, in lax few flowered. axillary racemes. Drupe round, green, pulp scanty. *Fls. & Frts:* November-March.

Commonly planted in gardens or lawns as an ornamental plants.

Field Collections: Sonarpur, *K.N.-1900;* Titagarh, *K.N.-2205.*

Distributions: Western Peninsula and mostly cultivated throughout India. Native of the Coastal South East Asia, East African Islands, Australia and Polynesia.

Economic Importances: Wood is coarse grained, but valuable and mostly used for cabinate making. An aromatic yellowish green reginous gum exudes from the bark and a volatile oil is distilled from this gum. The leaves have some medicinal properties.

The other Indian species of the Genus - *Calophyllum* L. are *C. apetalum* Willd., *C. elatum* Redd., *C. polyanthum* Wall. and *C. soulattri* Burm. Most of these are the forest trees.

2. *GARCINIA* L.

2.1. ***Garcinia cowa*** Roxb. Fl. Ind. 2: 622.1824; Anderson, Hook.f. Fl. Brit. India 1: 262.1874; Watt Dict. Econ. Prod. 3:265.1890: Prain Beng. Plants 1: 166.1903 (Rep.ed. 1963). *G. kydia* Roxb. Fl. Ind. 2: 623.1824. *(Vern. Name: Kau).*

Erect dioecious, glabrous tree. Leaves broad lanceolate, acute at both ends, 7.5-10 cm long, membranous, secondly nerves, faint in 6 to 10 pairs. Flowers 1.5 cm diam., tetra-merous, yellow; males in 3 to 8 flowered umbels, stamens numerous, desely covering a fleshy central mass; female ovary globose, staminodes forming a ring, style grooved, stigma broad, 6-8 lobed, grandular; fruit globose, 6-8 grooved, about 5 cm diam., slightly depressed at the apex and there bearing a small sessile stigma. *Fls. & Frts.* March-June.

Occasionally planted in the gardnes or roadsides.

Field Collection: Baruipur, *K.N.-2252.*

Distributions: Native of India, distributed in the eastern and southern India and Andaman Islands; Burma and Srilanka.

Economic Importances: Resinous gum used as yellow varnish in Burma and also used in medicine for dysentery. Fruits are edible, timber hard but heavy coarse grained.

Notes: The other Indian species of the genus-*Garcinia* L. are *G. atroviridis* Griff. ex T.Anderson, *G. cambogia* Dest. var. *cambogia, G. dulcis* (Roxb.) Kurz., *G. echinocarpa* Thw., *G. indica* Choisy, *G. lanceaefoliam* Roxb., *G.livingstonei* Anderson, *G. mangostana* L., *G. morella* Desr., *G. paniculata* Roxb., *G. pedunculata* Roxb., *G. pictoria* Roxb. and *G. xanthochymus* Hook. f. ex Anderson. Most of these species distributed in the foot hills and on the forests.

XXII. DIPTEROCARPACEAE

1. *SHOREA* Roxb.

1.1. ***Shorea robusta*** Gaertn. f. Fruct. 3: 48.t. 186.1805: Thiselton Dryer, Hook.f. Fl. Brit. India 1: 306.1874; Watt Dict. Econ. Prod. 6(2): 678. 1893; Prain Beng. Plants 1: 172.1903 (Rep. ed. 1963); Wlh. India 9: 313.1972 . *(Vern. Name: Sal).*

Large deciduous tree. Leaves oblong or ovate-accuminate, base rounded or cordate, entire glabrous shining; stipule falcate, pubescent, caducous. Flowers yellowish or green coloured, short, pedicelled and usually unilateral on the branches, lax terminal and axillary panicles. Fruits ovoid, densely pubescent with a thin outer shell; wings five sub-equal. *Fls. & Frts.:* March-July.

Very rare in this district, sometimes cultivated in the gardens or as avenue tree.

Field Collection: Nilganj, *K.N.-2121.*

Distributions: Sub-Himalayan tracts, Cromandol Coast, Pachmari Hills and Godavari River area. Both Tropical and Temperate regions of the world.

Economic Importances: Yield valuable timber and bark is used for tannin; stem extraction is 'dhoona', used for many purposes.

XXIII. MALVACEAE

Key to the Genera

1a	Carpels separating from the axis at maturity as dehiscent or indehiscent cocci	(2)
1b	Ripe carpels form a capsule	(9)
2a (1)	Number of style and carpel same	(3)
2b (1)	Style double than carpel	(7)
3a (2)	Bracteoles absent, mature carpels after separating dehiscent	(4)
3b (2)	Bracteoles 3, mature carpels after separating indehiscent	(6)
4a (3)	Carpels without a false dissepiment	(5)
4b (3)	Carpels with a transverse false dissepiment	14. *WISSADULA* Medik.
5a (4)	Carpels two or more seeded	2. *ABUTILON* Mill.
5b (4)	Carpels one seeded	11. *SIDA* L.
6a (3)	Leaves orbicular, palmately veined and lobed, stigma linear	8. *MALVA* L.
6b (3)	Leaves ovate lanceolate, pinnately veined, stigmas capitate	9. *MALVASTRUM* Gray
7a (2)	Bracteoles 6-9; flowers white	3. *ALTHAEA* L.
7b (2)	Bracteoles 5 or absent	(8)
8a (7)	Bracteoles connate 5; flowers pink, carpels spinescent or unarmed	13. *URENA* L.
8b (7)	Bracteoles intermixed with flowers or absent; flowers yellow or white in dense heads, carpels unarmed	7. *MALACHRA* L.
9a (1)	Staminal tube divided about the middle into 5 bundles of stamens	6. *KYDIA* Roxb.
9b (1)	Staminal tube slightly divided at the apex	(10)
10a (9)	Herbs or shrubs	(11)

10b (10)	Small to medium tree	12. *THESPESIA* Solan.
11a (10)	Bracteoles 5 or more	(12)
11b (10)	Bracteoles 3, large and persistent	4. *GOSSYPIUM* L.
12a (11)	Calyx spathaceous, deciduous; fruit more that 7 cm long	1. *ABELMOSCHUS* Medik.
12b (11)	Calyx 5-cleft, persistent, fruit less than 7 cm long	5. HIBISCUS Medik.
12c (11)	Bracteoles 7; staminal column excerted	10. *MALVAVISCUS* Dill.

1. *ABELMOSCHUS* Medik.

Key to the Species

1a	Mostly cultivated; involucral bracts equalling the calyx	1. *A. esculentus* Moench.
1b	Mostly wild; involucral bracts shorter than the calyx	2. *A. moschatus* Medik.

1.1. ***Abelmoschus esculentus*** (L.) Moench. Meth. Pl. 617.1794. *Hibiscus esculentus* L. Sp. Pl. 696.1753; Masters, Hook.f.Fl. Brit. India 1:343.1874; Watt Dict. Econ. Prod. 4:237.1890; Prain Beng. Plants 1: 180.1903 (Rep. ed. 1963) & Rec. bot. Surv. India 3(2): 178.1905; Wlh. Ind. 5:75.1959. *Abutilon hirtum* (Lamk.) Sweet Hort. Brit. 53. 1826. *A. graveolens* Wt. & Arn. Prodr. 56.1834. *(Vern. Name: Bhindi* or *Dherosh).*

Shrub or undershrub; stem covered with stellate hairs. Leaves 12 x 10 cm, orbicular cordate, shortly acuminate at apex, sub-entire or irregularly dentate; nerves 7 or 9, from the base. Flowers axillary, solitary; 10-12 cm diam., yellow; petals with purple centre. Fruits 7 angled, elongated 12-20 cm long, conical capsule. Seeds 2-3 in each, dark brown or white pubescent sub-reniform. *Fls. & Frts.:* Almost throughout the year, normally September-July.

Largely cultivated as vegetable throughout.

Field Collections: Barrackpore, *K.N.-1837.*

Distributions: Extensively cultivated throughout India; doubtful native of Ethiopia, but presently naturalized and cultivated in most Tropical countries.

Economic Importance: Fruits and the seeds are the common vegetable; decoction of the fruits are emollient and demulcent; mucilage is said to be useful in gonorrhoea.

1.2. ***Abelmoschus moschatus*** Medik. Malv. Fam. 46.1787. *Hibiscus abelmoschus* L. Sp. Pl. 696.1753; Masters, Hook.f.Fl. Brit. India 1:342.1874; Watt Dict. Econ. Prod. 4: 229.1890; Prain Beng. Plants 1: 181.1903 (Rep. ed. 1963) & Rec. bot. Surv. India 3(2): 178.1905. *(Vern. Name: Kalkastari* or *Mushak Dana).*

Tall 1-1.5 m high annual undershrub, clothed with minute bristles. Leaves variable in shape, 5-7 lobed, lobes spreading, oblong lanceolate, coarsely toothed. Flowers axillary, 10-12 cm in diam., yellow with crimson centre. Capsule oblong lanceolate, slightly hairy, 3-8 cm long. Seeds musky scented. *Fls. & Frts.:* October- March.

Rarely grow on the cultivated fields as a weed plant.

Field Collection: Nilgunj/Barrackpore, K.N.-2108.

Distributions: Common throughout India; Tropics of the Old World.

Economic Importance: Stem fibre used as a substitue for jute; seeds have a musky odour, used as perfumery under the trade name of *'grains d. ambrette'* , also used in cosmetics and to scent the hairs. The seed oil has some medicinal properties.

2. *ABUTILON* Miller

2.1. ***Abutilon indicum*** (L.) Sweet, Hort. Brit. 1: 54. 1826; Masters, Hook.f. Fl. Brit. India 1: 327.1874; Watt Dict. Econ. Prod. 1: 16.1889; Prain Beng. Plants 1: 176.1903 (Rep. ed. 1963) & Rec. bot. Surv. India 3 (2): 277.1905; Wlh. Ind. 1:4.1948; Hu Fl. China 32.1955 *Sida indica* L., Torner Cent. Pl. 2: 26.1756. *(Vern. Name: Petari* or *Jhampi).*

Herb or undershrub, 0.75-1.0 m high; stem stellate tomentose with intermixed simple short hairs. Leaves ovate or orbicular, pale beneath, acuminate; irregularly crenate dentate, cordate at base, 7-11 nerved. Flower solitary in the axile, peduncle 5-10 cm long, geniculate. Fruits subglobose. Seeds 2-4 in each carpel; reniform scrobiculate. *Fls. & Frts.:* March-October.

Very common throughout the lower Ganga delta; generally found in the loam soil, in the hedges, thickets and roadside waste lands.

Field Collection: Kakdeep/Sundarbans, *K.N.-493.*

Distributions: Throughout the hotter parts of India; in the Tropics.

Economic Importance: Stem yield fibre, bark astringent; leaves demulcent; infusion of the roots used in fever and as a cooling medicine: seeds aphrodisiac and laxative.

Notes: Abutilon graveolens G. Don var. *hirsutum* Masters was also reported by the earlier workers from the Sundarbans areas.

3. *ALTHAEA* L.

3.1. ***Althaea rosea*** (L.) Cav. Diss. 2 : 91.f.3.1785; Masters, Hook.f.Fl.Brit. India 1 :319.1874; Bailey Man. Cult. Plants 658.1949. *Alcea rosea* L. Sp. Pl. 687.1753. *(Vern.Name: Hollyhock).*

Annual garden herb, 1.5-3 m tall. Leaves large, rough, rugose, long stalked, rounded heart shapped. Flowers large 8-10 cm diam., in a long raceme or spike, colour varies from purple, rose, yellow or white; epicalyx 6-9 segmented. Ovary many celled, one ovule in each cell. *Fls. & Frts.:* December-March.

Extensively cultivated in the ornamental gardens throughout.

Field Collection: Shyamnagar, *K.N.-2077.*

Distributions: Cultivated ornamental garden plants throughout India; introduced into India about 1835. Native of China.

Economic Importance: Showy ornamental garden plants.

4. *GOSSYPIUM* L.

Key to the Species

1a	Capsule ovoid; cotton easily separable from the seeds	1. *G.barbadense* L.
1b	Capsule sherical; cotton adherent to the seeds	2. *G. herbaceum* L.

4.1. ***Gossypium barbadense*** L. Sp. Pl. 693.1753; Masters, Hook.f. Fl. Brit. India 1: 347.1874; Watt Dict. Econ. Prod. 4: 15.1890; Wlh. Ind. 4: 170.1956. *(Vern.Name: Kapastula).*

Much branched shrub, 2-3 m high, glabrous leaves cordate, black dotted, deeply 5 lobed, lobes acuminate. Pedicels short, stout often glandular at the top, axillary. Corollar yellow with purple base. Capsule 3-5 celled, ovoid, acuminate. Seeds free, black, covered with readily separably cotton. *Fls.& Frts:* October-January.

Cultivated in the low lands during summer months.

Field Collections. Ghatur More/Amtala, *K.N.-381.*

Distributions: Cultivated throughout India; Tropics of the World. Native of Tropical America.

Economic Importance: Produce a fine cotton fibre of commerce. Seed oil is edible and oil cake used as fertilizer.

4.2. ***Gossypium herbaceum*** L. Sp. Pl. 975.1753; Masters, Hook.f. Fl.Brit. India 1: 340.1874; Watt Dict. Econ. Prod. 4: 25.1890; Prain Beng. Plants 1: 184.1903 (Rep. ed. 1963) & Rec. bot. Surv. India 3 : 179.1905. *(Vern. Name : Kapastula).*

Coarse, much-branched shrub. Leaves 3-5 lobed, lobes short triangular; base cordate; Flowers yellow to pinkish purple, bracteoles quite free. Capsules spherical. Seeds with a greyish, rusty or green fuzzy cotton. *Fls. & Frts.:* August-April.

Commonly cultivated, but occasionally found as a garden escape.

Field Collections: Gosaba, *K.N.-645;* Joygram/Minakhan, *K.N.1352.*

Distributions: Cultivated almost throughout India. Native of South East Asia.

Economic Importance: Cultivated for its seed and fibre, which are the chief source of cotton of commerce. Seeds demulcent, laxative, expectorant and nervic tonic. Oil cake is useful fodder and manure.

Remarks: Other species of the genus *Gossypium* L. are *G. arboreum* L., *G. arboreum* L. var. *nadam* (Watt) Prokh, *G. hirsutum* L. etc.

5. *HIBISCUS* Medik.

Key to the Species

1a	Calyx 5-cleft or sepals 5, nearly free	(3)
1b	Calyx spathaceous	(2)
2a (1)	Flowers yellow with purple eye	6. *H.tetraphyllus* Roxb.
2b (1)	Flowers red, pendulous	5. *H.schizopetalus* Hk.
3a (1)	Not grown for ornamental purposes	(5)
3b (1)	Large garden shrubs, grown for their showy flowers	(4)
4a (3)	Bracteoles 10; flowers white to pink	2. *H. mutabilis* L.
4b (3)	Bracteoles 6-8; flowers generally red, but variable to a great extent except lilac and purple	3. *H. rosa-sinensis* L.
5a (3)	Herbs or undershrubs, not woody	(6)

5b (3)	Trees or large shrubs	7. *H.tiliaceus* L.
6a (5)	Calyx lobes with a prominant midrib and with usually thickened indurated borders	(7)
6b (5)	Calyx lobes equally 3-5 nerved, margins not thickened	8. *H. vitifolius* L.
7a (6)	Bracteoles not adnate	1. *H. cannabinus* L.
7b (6)	Bracteoles adnate to the base of calyx	4. *H. sabdariffa* L.

5.1. ***Hibiscus cannabinus*** L. Syst. Nat. (ed.10) 1149.1759; Masters,Hook.f. Fl.Brit. India 1: 339.1874; Watt Dict. Econ. Prod. 4: 231. 1890; Prain Beng. Plants 1: 267. 1903 (Rep. ed. 1963) & Rec. bot. Surv. India 3(2): 179.1905: Wlh. India 5 :77.1959. *H. cannabinus* L. var. *genuinus* Hochr. Ann. Cons. Jard. Bot. Geneve 4: 115.1900. *(Vern. Name: Mestapat).*

Prickly herb, 0.5m-1.0 m high. Leaves cordate, entire. Flowers large, yellow, axillary; pedicel long. Epicalyx 7-10 segmented. Calyx dry, horny in furit. Corolla with crimson spot inside. Fruit globose, bristly capsule. Seeds glabrous. *Fls. & Frts.:* August-November.

A common cultivated crop during the rainly season, to the high lands and often found as an garden escape in the waste places.

Field Collection: Kazipara/Barasat, *K.N.-920.*

Distributions: Throughout India, upto an elevation of 1500m; apparently indigenous to India; but some botanists consider it to be of African Origin; Old World Tropics.

Economic Importance: Yield fibre crop, as a substitute for jute. Seed oil is edible and also used for burning lamp. Oil cake is used as manure or as cattle feed.

5.2. ***Hibiscus mutabilis*** L. Sp. Pl. 694.1753; Masters, Hook.f. Fl. Brit.India 1: 344.1874; Watt Dict. Econ. Prod. 4:242.1890; Prain Beng. Plants 1:183.1903 (Rep. ed.1963) & Rec.bot. Surv. India 3 (2): 179.1905; Hu, F. China Malvaceae 49.1955; Wlh. Ind.5: 91.1959. *Ketmia mutabilis* Moench. Method 617. 1774. *(Vern. Name: Sthalpadma).*

Shrub or small deciduous tree upto 4 m high. Leaves cordate 3-7 lobed. Flowers axillary and sub-corymbose, white; epicalyx of 10 segmented. Capsules sub-globose, 2.0 cm diam., hirsute, endocarp with dense white hairs. Seeds reniform, brown, densely bearded. *Fls. & Frts.:* October-December.

Planted in the ornamental gardens, throughout.

Field Collection: Dhanyakuria, *K.N.-559.*

Distributions: Throughout India; native of China and distributed in most hot countries.

Economic Importance: The bark yield strong fibre; the flowers are said to be the remedy for chest diseases and the leaves for swelling; also used as an emollients. An infusion of the flowers is given in pectoral and pulmonary complaints. Bark fibre used for ropes and cordages.

5.3. ***Hibiscus rosa-sinensis*** L. Sp. P. 694. 1753; Masters, Hook.f. Fl. Brit. India 1: 344.1874; Watt Dict. Econ. Prod. 4: 242.1890; Prain Beng. Plants 1: 182.1903 (Rep. ed. 1963); Hu, Fl. China, Malvaceae 45.1955; Wlh. Ind. 5:91.1959. *H. fragilis* DC. Prodr. 1: 446. 1824. *H. rosa-sinensis* L. var. *genvinus* Hochr.Ann.Cons. Jard. Bot. Goneve 4: 134.1900. *(Vern. Name: Jaba).*

Evergreen large shrub. Leaves 7-12 cm long glabrous, broad ovate to narrow ovate, shruply pointed or acuminate, not lobed, margins toothed, notched or almost entire except near apex. Flowers red or scarlet; bracts 7-9, linear, staminal tube red exceeding the corolla. *Fls. & Frts.:* Throughout the year.

Commonly cultivated as ornamental plants throughout.

Field Collection: Subuddhipur/Baruipur, *K.N.-354.*

Distributions: Almost throughout India; native of China.

Economic Importances: Ornamental garden plants. The Hindu worship the Goddess Kali and Durga with the red flowers; also used as hair dye in China. Flowers are emollient, demulcent, aphrodisiac and emmenagogue. Decoction of flowers is used in bronchial catarrh. Root decoction is useful in veneral diseases.

Notes: A large number of varieties were found in respect of flower colour, shapes and sizes of the petals.

5.4. ***Hibiscus sabdariffa*** L. Sp. Pl. 695.1753; Masters, Hook.f. Fl. Brit. India 1: 349.1874; Watt Dict. Econ. Prod. 4: 243. 1870; Prain Beng. Plants 1: 267.1903 (Rep.ed.1903); Hu, Fl. China Malvaceae 58.1955; Wlh. Ind. 5: 92.1959. *H. digitatus* Cav. Diss. 3:151.1787. *H. sanguineus* Griff. Notul. 4.520.1854. *(Vern. Name: Chukar* or *Mesta).*

Erect, glabrous, annual shrub. Stem often reddish. Lower leaves undivided; upper palmately 3-5 lobed; lobes dentate or serrate, acuminate, petioles about as long as leaf. Flowers yellow with a dark crimson eye, on solitary, axillary peduncles; sepals united into a deep red, fleshy cup. Capsules ovoid, pointed, shorter than calyx. *Fls. & Frts.:* August-February.

Sometimes cultivated and very rare as a garden escape.

Field Collections: Falta, *K.N.-1289.*

Distributions: Cultivated in the hotter parts of India; Sri Lanka and throughout the Tropics; a native of the Tropical Africa and Asia.

Economic Importance: Fleshy calyx and the capsules are edible; used in beverage and medicine. A strong silky fibre is obtained from the stem. Fibre from the bark is used as jute substitute. Fruits are used in jellies and confections. Calyx infusion helpful for digestion, intestinal antiseptic and mild laxative.

5.5. ***Hibiscus schizopetalus*** Hook.f. Fl. Brit. India 1: 345.1874; Haines Bot . Bih. & Orissa 1: 72.1921 (Rep. ed.1961). *(Vern. Name: Jhumka Jaba).*

Erect shrub; sarmentose with narrow ovate or sub-rhomboidally elliptic shining leaves, crenate serrate except at base, with minute, sublate caducous stipules. Flowers axillary, dropping and fuchsia-like on long slender, jointed peduncles. Calyx spathaceous and 2-lobed. Petals scarlet, deeply laciniate and reflexed. Staminal column long, filiform, pendulous. Style branches long. *Fls. & Frts.:* Throughout the year.

Common in the village gardens often as an gaden escape and also used as hedge plants in the village gardens.

Field Collection: Gocharan, *K.N.-241.*

Distributions: Planted as an ornamental plant almost all over India; native of China.

Notes: Prain (1903a, 1903b and 1905) did not mention this taxon from the lower Ganga delta regions.

5.6. ***Hibiscus tetraphyllus*** Roxb. Fl. Ind.3: 211.1832; Masters, Hook.f. Fl. Brit. India 1: 341.1874:Watt Dict. Econ. Prod.4: 246.1890; Prain Beng. Plants 1: 181.1903 (Rep. ed.1963) & Rec.bot. Surv. India 3 (2): 179.1905; Hu, F. China Malvaceae 44.1955; Wl. Ind. 5: 97.1955. *(Vern.Name: Ban-vendi).*

Shrub, 1.5-2.0 m high, annual, hairy shrub. Leaves scabrid with short stiff hairs, cordate, serrate, acutely angled. Stipules linear lanceolate. Pedicels 2 cm long, axillary and clustered at the ends of the bracts, stout. Involucral bracts 4, longer or shorter than the calyx, ovate lanceolate.

Calyx softly villous, corolla 5.0-7.5 cm across, yellow with purple centre. Capsule 4 cm long, ovoid, 5-angled, hispid, cuspidate. Seeds faintly pubescent. *Fls. & Frts.:* August-February.

Sometime on the village thickets and waste lands.

Field Collection: Birati/Saptagram, *K-30* (CAL).

Distributions: Almost throughout India; Pantropic.

Economic Importance: Some medicinal properties is attributed in this plants.

5.7. ***Hibiscus tiliaceus*** L. Sp. Pl. 694.1753; Masters, Hook.f. Fl. Brit. India 1: 343.1874; Watt Dict. Econ. Prod. 4: 247. 1890; Prain Beng. Plants 1: 183.1903 (Rep. ed. 1963), Rec. bot. Surv. India 2 : 289.1903 & Ibid.3 (2): 179.1905; Hu, Fl.China, Malvaceae 44.1955; Wlh. Ind. 5:97.1959. *H.tiliacefolius* Salisb. Prodr. 383.1796. *H. tortuosus* Roxb. (Hort. Beng. 51.1814) & Fl. Ind.3: 192. 1832. *Paritium tiliaceus* W. & A. Prodr. 1: 52.1834. *H. tiliaceus* L. var. *tortuosus* Masters, Hook.f. Fl. Brit. India 1: 343.1874. *H.tiliaceus* L.var. *genuinus* Hochr. Ann. Cons. Jard. Bot. Geneve 4: 63. 1900. *(Vern Name: Bhola).* (Fig. 3).

Much-branched tree, upto 8 m height. Leaves alternate, crenulate, cordate, shortly acuminate, pubescent below. Flowers in axillary or terminal, few flowered racemes, yellow with crimson at the centre; epicalyx 10-segmented; corolla campanulate, 7-10 cm diam. Fruits ovoid, tomentose capsule; seeds black glabrous with pale dots. *Fls. & Frts.:* October-June.

Common on the beach and tidal forests of the lower Ganga delta, forming dense thickets near the shore and very often grow in the tidal flooded deltaic lands.

Field Collection: Chemagar/Sagar Island, *K.N.-446.*

Distributions: Common throughout the beach forests in India. Also in the Tropical coasts in the Old World Tropics.

Economic Importance: Stem barks yield coarse fibre, which is used in cordage; stem wood is important fire wood. Root is regarded as a febrifuge and also employed in the oparation of embrocations. Flowers after boiling with milk used for the remedy of earache; fresh bark decoction useful for dysentery. Root powder is diuretic, suodorific, emollient and laxative. Leaves are edible and leaves infusion is also used in ulcers and wounds.

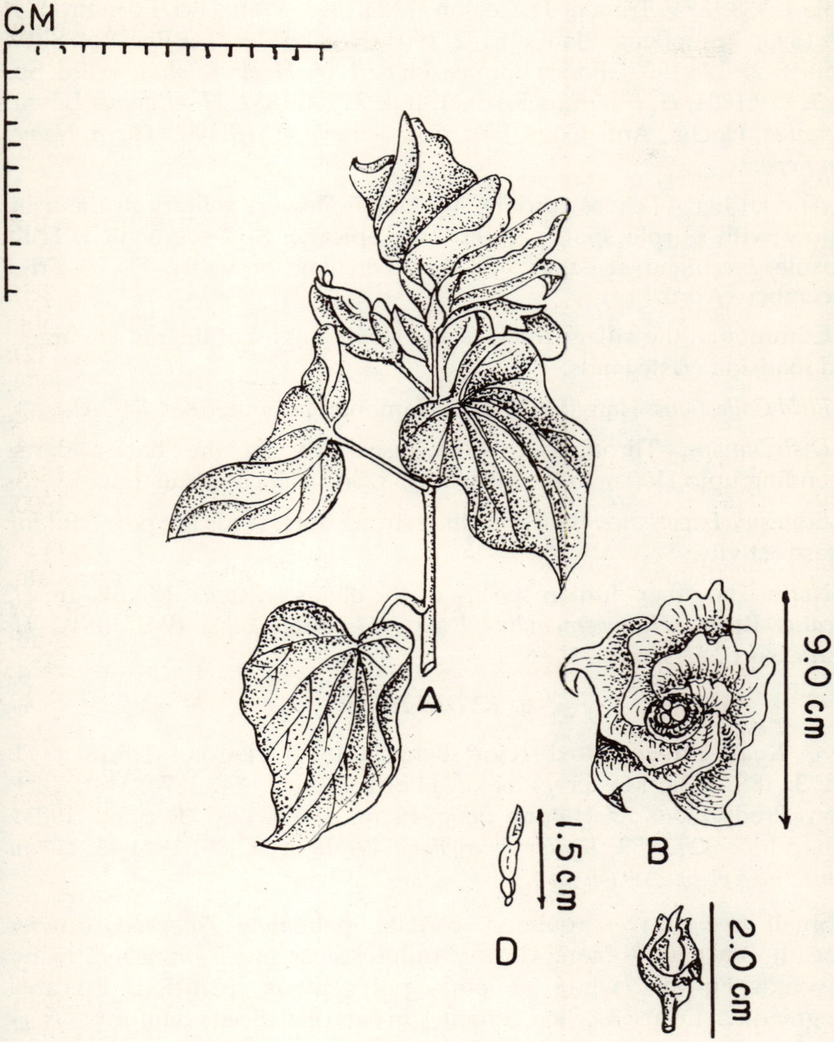

Fig.3. HIBISCUS TILIACEUS L.

A. Flowering twig, B. Flower, C. Ovary, D. Androecium.

5.8. ***Hibiscus vitifolius*** L. Sp. Pl.696.1753; Masters, Hook.f. Fl. Brit. India 1: 338.1874; Trimen, Fl. Ceylon 1: 154.1893; Watt Dict. Econ Prod. 4: 248.1890; Prain Beng. Plants 1: 182. 1903 (Rep. ed. 1963) & Rec. bot. Surv. India 3(2): 178.1905; Borss, Blumea 14: 82.1966. *H. obtusifolius* Willd. Sp. Pl. 3: 829.1801. *H. truncatus* Roxb.Fl. Ind. 3: 200.1832. *H. vitifolius* L. var. *genunius* Hochr., Ann. Cons. Jard. Bot. Geneve 4: 169.1900. *(Vern. Name: Ban Kapas).*

Annual herb. Leaves cordate, 3-5 lobed. Flowers solitary in the axis, yellow with purple spot; pedicel long; epicalyx 8-12 segmented. Fruit capsule, 5-winged at edges. Seeds tuberculate or warty. *Fls. & Frts.:* December - April.

Common in the village thickets and scrub jungles of the railway tracts and roadside waste lands.

Field Collections: Haroa, *K.N.-596;* Diamond Harbour, *SKM-8* (CAL).

Distributions: Throughout India, especially in the hotter parts; ascending upto 1100 m in the hills. Burma, Sri Lanka and Australia.

Economic Importance: Barks produce strong fibre; seeds are powerful for urease activity.

Notes: The other Indian species of the genus *Hibiscus* Medik. are *H. furcatus* Roxb., *H. macrophyllus* Roxb., *H. platanifolius* (Willd.)*SW, H. surattensis* L., *H. syriacus* L. etc.

6. *KYDIA* Roxb.

5.1. ***Kydia calycina*** Roxb. (Hort. Beng. 50.1814), Pl. Cor. t. 215,181 & Fl. Ind. 3: 188. 1832; Masters, Hook.f. Fl.Brit. India 1: 348.1874; Watt Dict. Econ. Prod. 4: 568.1890; Prain Beng. Plants 1: 178.1903 (Rep. ed. 1903); Wlh.Ind. 5: 329.1959. *Kydiaroxburghiana* Wight Ic. t. 881.1944-45. *(Vern. Name : Pola or Bankapas).*

Small tree. Leaves rounded, cordate, palmately 7-nerved, downy beneath, bracteoles oblong downy. Inflorescence much branched, many flowered. Flowers white or pink, polygamous, panicled. Epicalyx 6-segmented. Ovary 3-celled, 2 ovules in each cell. Seeds reniform. *Fls. & Frts.:* July-December.

Occasionally planted as hedge plant and also as tree for its sweet scented flowers.

Field Collection: Shyamnagar, *K.N.-2093.*

Distributions: Warmer parts of India; most of the Tropics.

Economic Importances: Inner bark fibre is used for making ropes and used for purifying (gur) molasses. The wood is soft and used as fuel; paste of the leaves is applied in body pains, also used in poultice for skin diseases. Bark fibres are used for cordage and ropes and mucilaginous substance obtained from stem is used for clarifying sugar.

7. *MALACHRA* L.

7.1. ***Malachra capitata*** L. Syst. (12 ed) 2: 458. 1767; Masters, Hook.f. Fl. Brit. India 1: 329.1874; Watt Dict. Econ. Prod. 5: 109.1891; Prain Beng. Plants 1: 178.1903 (Rep. ed.1963) , Rec. bot. Surv. India 2(4): 288.1903 & Ibid. 3 (2): 178.1905; Wlh. Ind. 6: 228.1962; *Sida capitata* L. Sp. Pl.685.1753. *(Vern. Name:Ban-Bhindi* or *Ban-Dheros.)*

Erect stiff, fleshy undershrub. Leaves cordate, acute, petioled, with a white spot at the base. Flowers aggregated in terminal or axillary clusters, yellow, shortly stalked. Carpels 5, 1 ovuled in each, rigidly hairy, styles 10. *Fls. & Frts.: May-December.*

Gregariously grow on dry waste lands, along the railway tracts and roadside hedges and village thickets.

Field Collection: Canning/Malirdhar, *K.N.-159.*

Distributions: Almost throughout India, introduced in the middle of the 19th Century as fibre yielding plant. W. Tropical Africa and a native of Tropical America.

Economic Importance: Fibre suitable in cordage, gunny bags and coarse textiles. The mucilaginous plant parts possess emollient and pectoral properties.

8. *MALVA* L.

8.1. ***Malva rotundifolia*** L. Sp. Pl. 688.1753; Masters, Hook.f. Fl.Brit. India 1: 320.1874; Watt Dict.Econ. Prod.5: 141.1891; Prain Beng. Plants 1: 174.1903 (Rep. ed. 1963); Wlh. India 6: 250.1962. *(Vern. Name: Lapha).*

Much branched herb, sparingly villous leaves sub-orbicular, lobed crenate; petioles 15 - 18 cm long. Flowers violet; branches lanceolate, half the length of the sepals. Ripe carpels downy, flat and wrinkled. *Fls. & Frts.:* December-March.

Planted in gardens for its beautiful flowers but now commonly found as a garden escape.

Field Collection: Nilgang/Barrackpoe, *K.N.-2122.*

Distributions: Throughout India; Europe, N. Africa, N. & W. Africa.

Economic Importances: Seeds are used in native medicines as a demulcent. It is used as cooling agent. Seeds are used in cough, bronchitis, ulceration of bladder and haemorrhoids and skin diseases. Leaves are used in piles.

Notes: Other species found in India are *Malva parviflora* L., *M. sylvestris* L. and *M. verticillata* L. Both these plants are useful in cough.

9. *MALVASTRUM* Gray *nom cons.*

Key to the Species

1a	Hairs simple; carpels with three small projecting points	1. *M. coromandelianum* Garcke
1b	Hairs stellate; carpels rounded not beaked	2. *M. spicatum* Gray

9.1. ***Malvastrum coromandelianum*** (L.) Garcke Bonplandia 5: 295. 1857; Wlh. Ind. 6: 251. 1962; Hu, Fl. China Malvaceae 11.1955. *Malva coromandeliana* L. Sp. Pl. 687. 1753. *Malvastrum tricuspidatum* (R. Br.) Gray Pl. Weight. 16.1852; Masters, Hook.f. Fl. Brit. India 1.: 321. 1874; Watt Dict. Econ. Prod. 5: 143.1891; Prain Beng. Plants 1: 174.1903 (Rep. ed. 1963) & Rec. bot. Surv. India 3(2): 177. 1905. *Malva tricuspidata* R. Br. W. Ait. Hort. Kew (ed.2) 4: 22. 1849.

Annual erect or sub-erect, suffruticose herb. Stems clothed with simple and stellate hairs. Leaves ovate oblong or lanceolate acute or rarely obtuse at apex, serrate truncate or obtuse at base. Flowers axillary, solitary. Fruits upto 6 mm diam. *Fls. & Frts.:* April -August.

Common weed on the waste lands and village shrubberies.

Field Collection: Gobra, *K.N.-1621.*

Distributions: Throughout India, an introduced weed. Pantropic, probably an American species.

Economic Importance: Fibre obtained from the stem bark; plant decoction is given in dysentery. Leaves used in inflamed sores and wounds as a cooling and healing salve.

9.2 ***Malvastrum spicatum*** A. Gray, in Mem. Am. Acad. N.S. 4: 22.1849; Masters, Hook.f. Fl. Brit. India 1: 321.1874; Watt Dict. Econ. Prod. 5: 144.1891; Prain Beng. Plants 1: 174.1903 (Rep. ed. 1963) & Rec. bot. Surv.

India 3 (2): 177.1905. *Malva spicata* L. Syst. (ed.10) 1146; DC. Prodr. 1: 430.1824.

Erect herb. Leaves ovate, acute, crenate serrate, sometime lobed. Flowers in a dense terminal leafy spikes; epicalyx 3- segmented; carpels rounded or angular. Ovary 1- seeded, one ovule in each cell. *Fls. & Frts.:* June-November.

Occasionally on the dry waste lands and village shrubberies.

Field Collection: Bamanghata/Salt Lake, *K.N. 1744.*

Distributions: Throughout India; Tropics of the both hemispheres, native of America.

Economic Importance: Stem bark produces low quality fibres.

10. *MALVAVISCUS* Dill.

10.1. ***Malvaviscus conzattii*** Greenm., Publ. Field Mus. Nat. Hist. Chicago (Bot.) 2: 339. 1912; Maheswari Delhi Fl. 78.1963. *(Vern. Name: Lanka Jaba).* (Fig. 4).

Erect or straggling shrub; leaves ovate, crenate-serrate, palmately nerved; stipules narrow. Bracteoles 7. Flowers red or orange red, on axillary peduncles, opening slightly at the top. Petals erect. Staminal column excerted. Fls. & Frts. Throughout the year.

Occasionally planted as hedge plants in the village gardens.

Field Collections: Joynagar, *K.N.-869,* Panihati, *K.N.-1609.*

Distributions: Cultivated almost throughout India. Native of Mexico.

Economic Importance: Common as an ornamental plant.

Notes: According to Rothmaler in Fedde. Report 53(1944): 17 the author is Guesttard (in Mem. Ac.R. Sc., Paris 1749 (1753): 403. The generic name was used also by Fabricius, En. Meth. Pl. ed. 1 (1759): 1955 and Boehmer in Ludw., Defin (1760): 77.

11. *SIDA* L.

Key to the Species

1a	Prostrate trailing herb	6. *S. veronicaefolia* Lamk.
1b	Erect or sub-erect herbs or undershrub	(2)
2a (1)	Leaves cordate at the base	(4)
2a (1)	Leaves at the base cuneate	(3)

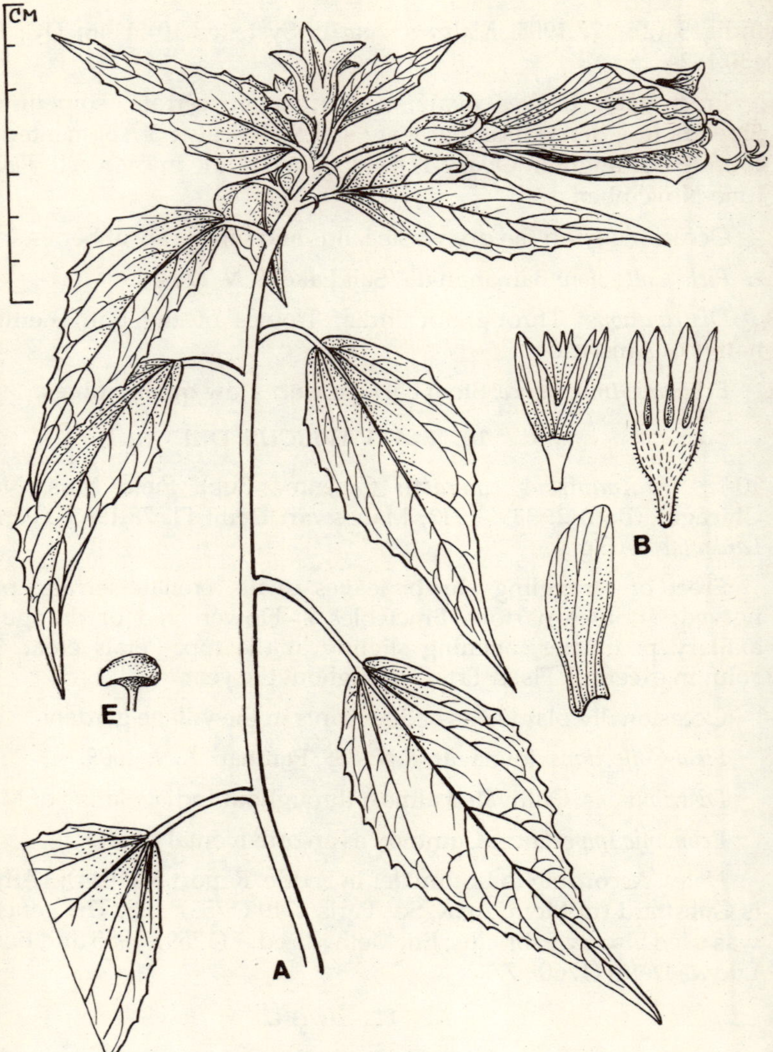

Fig.4. MALVAVISCUS CONZATTII Greenm.

A. Flowering Branch, B. Calyx tube,

C. Andoecium, ovary & stigma head, D. Stigma head.

3a (2)	Leaves nearly glabrous, narrow, lanceolate, acuminate, serrate; peduncle as long as petiole	1. *S. acuta* Burm. f.
3b (2)	Leaves pubescent to hoary beneath, obovate cuneate; peduncle longer than petiole	4. *S. rhombifolia* L.
4a (2)	Pedicels jointed in the middle; leaves and stoutish branches glutinous; viscous hairy	3. *S. glutinosa* Comm.
4b (2)	Pedicels jointed under the flower; stellately hairy	(5)
5a (4)	Leaves usually acute but sometime obtuse at the tip; petiole not spiny and as long as leafblades, carpels-10	2. *S. cordifolia* L.
5b (4)	Leaves always obtuse at the tip; petiole spiny and shorter than leaf blades; carpels 5	5. *S. spinosa* L.

11.1. ***Sida acuta*** Burm.f. F. Ind. 147.1768; Roxb. Fl. Ind. 3: 171.1832; Prain Beng. Plants 1: 176. 1903 (Rep. ed. 1963) & Rec. bot. Surv. India 3(2): 177.1905; Hu, Fl. China 153.18.1955; Whl. India 9: 322. 1972. *S. carpinifolia* L. f. Suppl. 307. 1781; Masters, Hook.f. Fl. Brit. India 1: 323.1874; Watt Dict. Econ. Prod. 6(2): 679.1893. *S. stipulata* Cav. Diss. 1:22, t.3. f.10. *(Vern. Name: Kureta).*

Undershrub, leaves nearly glabrous, narrow, acuminate serrate, stipules linear-subulate, longer than the petiole. Peduncle jointed in the middle as long as the petiole. Sepals triangular acute, calyx tube sub-globose. Carpels 5-9, rugose armed. *Fls. & Frts.:* January-July.

Farely common in the dry waste lands, village thickets, roadsides, railway tracts and in cultivated lands.

Field Collections: Dhapdhapi, *K.N.-35;* Amtala, *K.N.-380.*

Distributions: Almost throughout India; Tropics in general.

Economic Importances: An excellent fibre is obtained from the plant bark; leaves are used in rheumatic affections, the leaf juice is given for relief in chest pain and as an anthelmintic.

11.2. ***Sida cordifolia*** L. Sp. Pl. 684. 1753; DC. Prodr. 1: 464. 1824; Masters, Hook.f. Fl. Brit. India 1:324. 1874; Watt Dict. Econ. Plants 6(2): 680.1893;

Prain Beng. Plants 1: 175.1903 (Rep. ed. 1963) & Rec. bot. Surv. India 3(2): 177. 1905; Borss., Blumea 14: 199. 1966; Wlh. Ind. 9: 323.1972. *(Vern. Name: Berela).*

Hairy herb. Leaves simple, cordate at the base, 2.5-3.0 cm long, petiole as long as lamina. Inflorescence pedicilate, jointed under the flower and as long as lamina; carpels 10. Fruits with two awned tips. *Fls. & Frts.:* July-November.

Commonly found during rainy seasons and in the cold seasons; as a weed in the hedges, waste grounds and dry shady areas.

Field Collection: Subuddhipur/Baruipur, *K.N.-5.*

Distributions: Almost throughout in India. A tropical weed.

Economic Importance: The mucilaginous leaves are used as a demulcent and their infusion is given in fever as a refrigerant. Roots cooling, astringent stomachic and has tonic properties; useful in nervous and urinary diseases; ephedrine present in the seed minutely. Entire plant is useful to the Aurvedic medicine as a cardiac tonic, generally in combination with 'Musk' and 'makaradwaja'.

11.3. ***Sida glutinosa*** Commerson ex Cav, Diss. 1: 16, t.2, f. 8. 1785; Prain Beng. Plants 1: 175.1903 (Rep. ed.1963); Hu, Fl. China 153 1955. *S. mysorensis* Herb Madr. ex W. & A. Prod. 59. 1834; Masters, Hook.f. Fl. Brit. India 1: 322. 1874.

Suberect, hairy. Pedicels axillary, several crowded together. Flowers axillary, calyx widely campanulate, yellow. Carpels 5-6, smooth, grey, puberulous, not reticulated. Seed solitary, ovoid, faintly trigonous. *Fls. & Frts.:* November-April.

Occasionally grow on the village shrubberies.

Field Collections: Barasat, *K.N. 389;* Barasat, *Aziz- 83*(CAL).

Distributions: Almost throughout India; widely spread throughout the Tropics of both hemispheres.

11.4. ***Sida rhombifolia*** L. Sp. Pl. 684.1753; Roxb. Fl. Ind. 3: 176. 1832; Masters, Hook.f. Fl. Brit. India 1: 232. 1874; Prain Beng. Plants 1: 175.1903 (Rep. ed. 1963) & Rec. bot. Surv. India 3(2): 177.1905; Hu, Fl. China Malvaceae 20. 1955; Bross, Blumea 14: 193. 1966. *S. rhombifolia* L. var. *rhomboidea* Roxb. (Hort. Beng. 50.1814) & Fl. Ind. 3: 176. 1832; Masters, Hook.f. Fl. Brit. India 1: 232. 1874; Prain Beng. Plants 1:176.1903 (Rep.ed.1963). *S. rhomboidea* Roxb. ex Flom. As. Res. 9: 178. 1810. *(Vern. Name: Lal Berala).*

Undershrub; leaves rhomboid, hoary beneath, peduncles jointed at the base. Carpellary awans very short inflected. Flowers solitary in clusters at the ends of small axillary branches. *Fls. & Frts.:* May-September.

Common in the waste lands and roadside thickets; associated with the other weed flora.

Field Collections: Magrahat, *K.N.-261;* Barasat, *K.N.-913.*

Distributions: Widely distributed throughout India; an Indian origin. A cosmopolitan species.

Economic Importance: Leaves are pasted and applied on swellings. Stem is mucilaginous and used internally in skin disease and as diuretic; root extract with zinger is a remedy for fever; root itself considered valuable in rheumatism; root powder alongwith milk and sugar is used for white discharges of women.

11.4./1. ***Sida rhombifolia*** L. var. ***retusa*** (L.) Masters, Hook.f. Fl. Brit. India 1: 324. 1874; Watt Dict. Econ. Prod. 6(2): 682. 1893. *S. retusa* L. Sp. Pl. (ed 2.) 960.1763. *(Vern. Name: Jangli Methi).*

Erect shrub; branches minutely stellately hairy. Leaves ovate or truncate, sometimes retuse, cuneate at base, often coarsely toothed at top, dark green and glabrous above, more or less tomentose beneath; petioles very short. Pedicels axillary and crowded towards the ends of the branches; calyx angular, lobes triangular, acute. Corolla yellow, more than twice as long as the calyx; carpels 9-10, reticulated with 2 short awans. Seeds smooth black. *Fls. & Frts.:* September-March.

Occassionally on the village thickets.

Field Collection: Barasat, B. Sen Gupta - 153 (CAL).

Distributions: Throughout India. Tropics of the Old World.

11.5. ***Sida spinosa*** L. Sp. Pl.683.1753; Masters, Hook.f. Fl. Brit. India 1: 323.1874; Watt Dict. Econ. Prod. 6(2): 686.1893; Prain Beng. Plants 1: 172.1903 (Rep. ed. 1963); Borss., Blumea 14.191. 1966; Wlh. India 9: 325.1972. *S. alba* L. Sp. Pl. (ed 2.) 960.1763. *(Vern. Name: Swet Berela).*

Sub-erect, branched shrub. Stems with minute, hairy. Leaves ovate to oblong or laceolate to linear, rarely orbicular. Stipules liner, 2-4 mm. Flowers axillary, solitary or in cluster of 2-5 flowers. Pedicels 2-5 mm. Calyx widely campanulate, segments triangular, acute to acuminate. Seeds ovoid, obscurely triangular. *Fls. & Frts.:* November-January.

Sometimes in the waste lands and along the road-sides.

Field Collection: Kulti, *B.Gupta, s.n. (Cal).*

Distributions: Common throughout India; a Tropical Species.

Economic Importance: Leaves possess demulcent and refrigerant properties and useful in gonorrhoea, gleet and scanding urine. Root is used as a tonic and diaphoretic and is given in mild cases of debility and fever.

11.6. ***Sida veronicaefolia*** Lamk. Encyc 1:5.1785; Masters, Hook.f Fl. Brit. India 1: 322.1874; Prain Beng. Plants 1: 175.1903 (Rep. ed.1963) & Rec. bot. Surv. India 3 (2): 177.1905; Hu, Fl. China 23.1955; Wlh. India 9:325.1972. *Melochia cordata* Burm .f. Fl. Ind. 143.1768. *Sida multicaulis* Cav. Diss. 1: 10.t. 1. fig. 6.1785; Hu, Fl. China Malvaceae 24.1955. *S. humilis* Willd. Cav. Diss 5: 177. t. 134.fig. 2.1788; Masters, Hook.f. Fl. Brit. India 1:322. 1874. *S. cordata* (Burm.f.) Borss., Blumea 14: 182.1966. *(Vern. Name: Junka).*

Much branched herb. Leaves cordate, sparingly hispid. Flowers usually solitatry. Pedicel jointed at middle and longer than petiole. Carpels 5; seeds trigonous. *Fls. & Frts.:* May- October.

Fairly abundant during monsoon and post monsoon periods, at the waste lands and cultivated fields.

Field Collections: Baruipur, *K.N.-112;* Narikeldanga, *Dutta-13.*

Distributions: Almost throughout India. Native of Tropical America; and cosmopolitan in distributions.

Economic Importance: Yields an excellent fibre; leaves are edible. Decoction of the plant is used to prevent swelling of joints due to arthritis.

12. *THESPESIA* Soland ex Corr. *nom cons.*

Key to the Species

1a	Leaves sparingly pilose above, tomentose beneath; unripe capsules densely hirsute	1. *T. lampas* Dalz. & Gibs.
1b	Leaves glabrous; unripe capsules covered with rusty coloured scales	2. *T. populneoides* Kost.

12.1. ***Thespesia lampus*** (Cav.) Dalz. & Gibs. Bomb. Fl. 19.1861; Masters, Hook.f. Fl. Brit. India 1: 345.1874; Watt Dict. Econ. Prod. 6(4): 121. 1893;

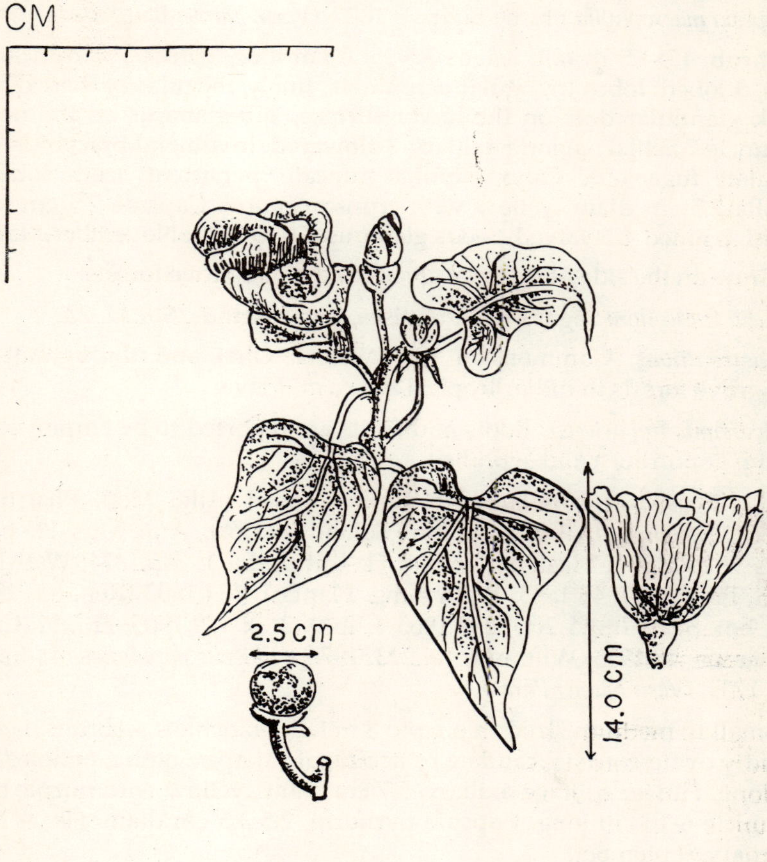

Fig.5. THESPESIA POPULNEA (L.) Soland ex Correa

A. Flowering twig, B. Single flower, C. Single fruit.

Prain Beng. Plants 1: 184.1903 (Rep.ed. 1963) Borss, Blumea 14: 116. 1966; Wlh. Ind. 10: 223.1976. *Hibiscus lampus* Cav. Diss. 3: 154.t.56, f.2.1787. *Thespesia macrophylla* Blume Bijdr. 73.1825. *(Vern. Name: Ban Kapas).*

Shrub, 1.0-1.5 m tall; leaves 7.5-15. 0 cm long, cordate or truncate at base, 3 lobed; lobes triangular acuminate, finely, reticulately veined with black glandular dots on the lower surface, sub-glabrous on the upper. Peduncle 7.5-10.0 cm long axillary, 3-flowered. Involucral bracts 5, small, subulate fugacious. Calyx cupular truncate, persistent, teeth subulate corolla 7.5 cm diam. yellow with crimson centre. Capsule 2.5 cm long, ovoid, pointed 4-5 valved. Seeds glabrous. *Fls. & Frts.:* November-March.

Grow on the side of the tidal creeks of the mangrove forests.

Field Collection: Sagar Mela Premises, Sagar Islands, *S.K.M.-23.*

Distributions: Common on the Western Ghat and throughout the mangrove forests in India; Tropical Africa and Java.

Economic Importance: Roots and fruits are reported to be employed for testing gonorrhoea and syphilis.

12.2. ***Thespesia populneoides*** (Roxb.) Kostel, Allg. Med. Pharm. Fl. 5. 1861. *T. populnea* (auct. *non* Linn.) Soland, ex Corr., Ann. Mus. Hist. nat. Paris 9: 290.1807; Masters, Hook.f. Fl. Brit. India 1: 345.1874; Watt Dict. Econ. Prcd. 6(4): 45.1893; Prain Beng. Plants 1: 184.1903 (Rep. ed. 1963), Rec. bot. Surv. India 2(4): 289.1903 & Ibid. 3(2): 179.1905; Hu, Fl. China Malvaceae 18.1955; Wlh. Ind. 10: 223. 1976. *Hibiscus populaneus* L. Sp. Pl. 694. 1753. *(Vern Name: Paras).*

Small to medium sized tree, upto 8 m tall; branchlets scabrous. Leaves broadly ovate-cordate, caudate or acuminate at apex, entire; petioles 5-10 cm long. Flower solitary, axillary, 5-7 cm diam., yellow with purple base; peduncle 6-10 cm long. Capsule pyriform, 2.5-3.5 cm diam. *Fls. & Frts.:* February-September.

A quick growing tree with dense foliages and spreading branches, which make it suitable as a shade tree. Therefore, very much liked as an avenue tree. Very common throughout the mangrove reclaimed areas of the Sundarbans.

Field Collection: Naraharipur/Sagar Islands, *K.N.- 446.*

Distributions: Almost throughout India. In the Tropics.

Economic Importances: Flowers and fruits produce som(yellow dye and the seed oil has the illuminant properties; bark and wood also produce a red colouring matter. Roots used as tonic, bark astringent and cure dysentery; heart wood used for biliousness and colic. All parts used for healing the skin diseases.

Notes: Thespesia populnea (L.) Soland. ex Corr. is another small tree distributed throughout the coast lines in India and Andaman & Nicobar Islands. (Fig. 5).

13. *URENA* L.

13.1. ***Urena lobata*** L. Sp. Pl. 692.1753; Masters, Hook.f. Fl. Brit. India 1: 329.1874; Watt Dict. Econ. Prod. 6(4): 212.1893; Prain Beng. Plants 1: 178.1903 (Rep. ed. 1963) & Rec. bot. Surv. India 3(2): 178.1905; Hu, Fl. China Malvaceae 73.1955; Borss, Blumea 14: 138.1966. *U. palmata* Roxb. Fl. Ind. 3: 182.1832. *U. sinuta* L. Sp. Pl. 692.1753. *(Vern. Name: Ban Okra).*

Undershrub with rigid stellate hairs. Leaves petioled cordate, angled, even divided beyond the middle, 5-7 lobed. Flowers clustered pink; bracteoles 5; peduncle short. *Fls. & Frts.:* June- October.

Gregariously grow on the waste lands, hedges and village thickets.

Field Collections: Canning, *K.N.-195;* Sarbaria, *K.N.-214.*

Distributions: Throughout India. Tropics of the both hemispheres.

Economic Importance: The stem yield a good fibre; a decoction of the stem and roots is used in Brazil as a remedy in severe windy colic. A poultice prepared from the roots and leaves is used as an emollient.

14. *WISSADULA* Medik.

14.1. ***Wissadula periplocifolia*** (L.) Presl ex Thw. En. Pl. Zeyl. 27.1858; Masters, Hook.f. Fl. Brit. India 1: 325. 1874; Prain Beng. Plants 1: 177. 1903 (Rep. ed. 1963); Hu, Fl. China 153. 1955. *Sida periplocifolia* L. Sp. Pl. 684.1753. (Fig. 6).

Small shrub; stem woody with soft hairs. Leaves palmately lobed, with soft hairs. Inflorescence lax panicles, bracteoles absent. Sepals 5 connate below; petals 5; stamens numerous, connate in a tube below; carpels usually 5, 1-seriate. Fruits 5- many seeded beaked. *Fls. & Frts.:* August-December.

Occasionally cultivated, I have observed this taxon only in cultivation on a village garden at Canning.

Field Collection: Malirdhar/Canning, *K.N.-123.*

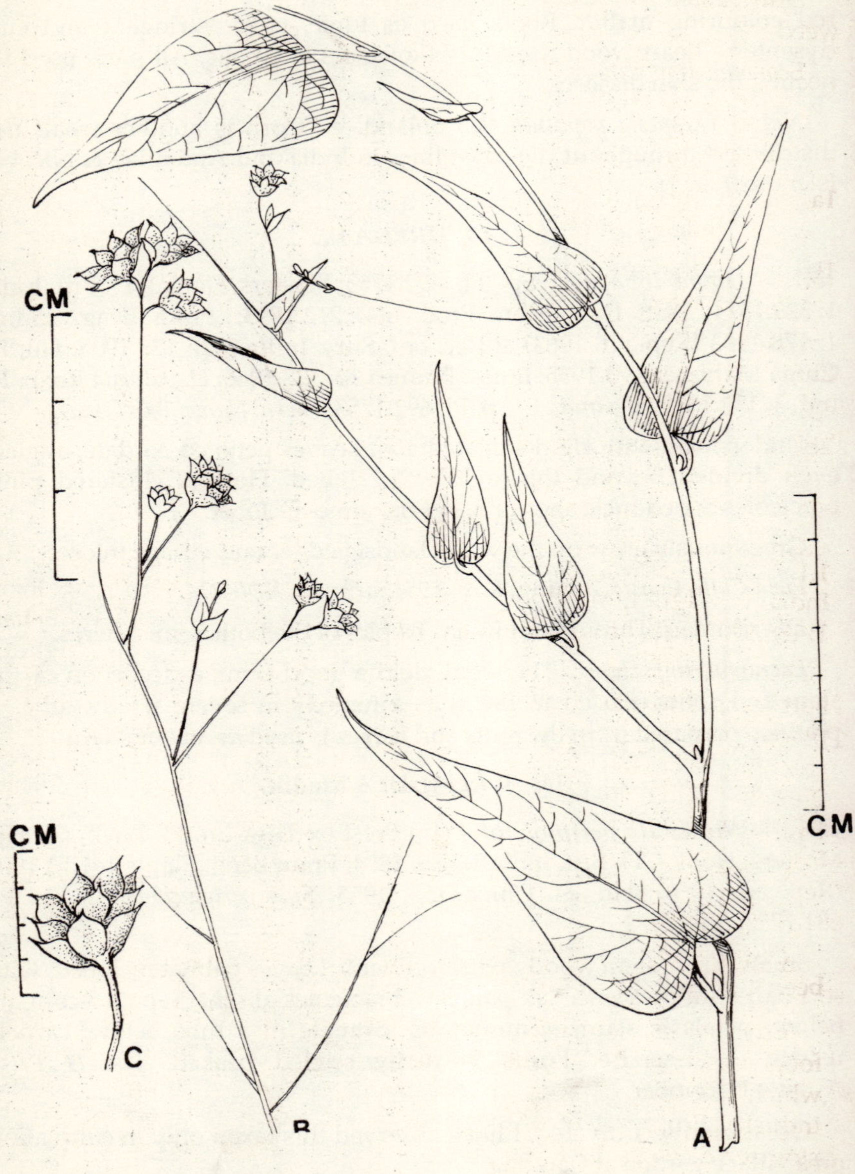

Fig.6. WISSADULA PERIPLOCIFOLIA (L.) Presl ex Thw.

A. Flowering twig, B. Flowering branch, C. Fruits

Distributions: Almost throughout India as cultivated species. Pantropic weed.

Economic Importance: Stem bark is an important fibre for ropes.

XXIV. BOMBACACEAE

Key to the Genera

1a	Calyx 5-10 cleft; fruit oblong, woody, indehiscent; seed not packed in cotton	1. *ADANSONIA* L.
1b	Calyx truncate or irregularly toothed, fruit ovoid, dehiscent, seed packed with cotton	(2)
2a (1)	Flowers red; stamens many; valves of capsule woody	2. *BOMBAX* L.
2b (1)	Flowers white; stamens few; valves of capsule not woody	3. *CEIBA* Medik.

1. *ADANSONIA* L.

1.1. ***Adansonia digitata*** L. Sp. Pl.1190.1753; Masters, Hook.f. Fl. Brit. India 1: 348.1874; Watt Dict. Econ. Prod. 1: 105.1889; Prain Beng. Plants 1: 185.1903 (Rep. ed. 1963); Wlh. India 1: 30.1948. *(Vern. Name: Gorakh-amli).*

Moderate sized tree with swollen trunk, bark smooth; branches spreading. Leaves deciduous, glabrous; leaflets 5-12.5 cm long, obovate oblong, acute or obtuse; petioles upto 15 cm long, villous. Pedicels reaching 20 cm long, softly hairy. Flowers 15- 20 cm diam., white. Ovary densely villous 5-10 celled. Fruits large, gourd like. *Fls. & Frts.:* April-October.

Occasionally planted as avenue tree or in the parks.

Field Collection: Ghoshpur, *K.N.-2124.*

Distributions: Cultivated almost throughout in India; supposed to have been introduced from Africa by the Arab traders. Native of Tropical Africa.

Economic Importances: Fruit pulps make cooling drink and also used for food seasoner; seed kernels are edible. Inner bark yields strong fibre, which used for paper pulp and cordage. Wood is suitable for match industry. Fruit pulp is also used for the treatment of dysentery and other stomach diseases. Bark is used in the treatment of malaria as substitute of quinine in Africa.

2. *BOMBAX* L.

2.1. ***Bombax ceiba*** L. Sp. Pl. 511. 1753 (excl. syn. Baulin); Robyns, Taxon 10: 160.1961. *B. malabaricum* DC. Prodr. 1: 479.1824; Masters, Hook. f. Fl. Brit. India 1: 349.1874; Watt Dict. Econ. Prod. 1: 487.1889; Prain Beng. Plants 1: 185.1903 (Rep. ed.1963) & Rec. bot. Surv. India 3(2): 180.1905. *Salmalia malbaricum* (DC.) Schott. & Endl. Melet. 35. 1832. *(Vern. Name: Simul).*

Spreading tree; trunk buttressed. Leaves upto 16 cm long, stipules small, leaflets 3-7, 7.5-15.0 cm long; petioles 1.5- 2.5 cm; elliptic oblong, lanceolate, ovate lanceolate or ovate oblong, with a cuneate on acute base, acuminate. Flowers many fascicled. Capsules 9-12 cm long, ellipsoid ovoid, hairy, 5- valved, seeds numerous, obovoid, enveloped in dense white cotton. *Fls. & Frts.:* December-March.

Fairly common, wild as well as planted in the village areas.

Field Collection: Malancha/Minakhan, *K.N.-1349.*

Distributions: Common throughout the hotter parts in India; Sri Lanka, Burma, Java and Sumatra. Native of Malaya.

Economic Importance: The wood is soft and is used in the manufacturing of packing boxes. Cotton, surrounding the seeds is extensively used for pillows and beddings. Seeds are used as fodder; the root has stimulant and tonic properties and the flowers are employed to cure skin diseases.

3. *CEIBA* Medik.

3.1. ***Ceiba pentandra*** (L.) Gaertn. Fruct. 2: 244. t.133.1791; Masters, Hook.f. Fl. Brit. India 1: 350.1874; Benthall Trees. Calc. 41.1946. *Eriodendron anfractuosum* DC. Prodr. 1: 479.1824; Masters, Hook.f. Fl. Brit. India 1: 350.1874; Prain Beng. Plants 1: 185. 1903 (Rep.ed. 1963). *E. pentandrum* Kurz., J. As. Soc. Beng. 2: 113.1874 *(Vern. Name: Swet Simul).*

Tall, 15-30 m height, deciduous, glabrous tree with straight trunk; branches horizontal in whorls; young trees armed with prickles; leaves digitate, leaflets 5-8,usually 7, elliptic, acute 7.5-10.0 cm long, subsessile petiole longer than the leaflets. Flowers numerous, axillary fascicles near the ends of branches; calyx green, campanulate with 5 acute teeth; petals 5, oblong concave at the base, 2.5-4.0 cm long, fleshy creamy white; stamens 5 each with 2 or 3 anthers. Capsules 7.5 -12.0 cm long, 5- nerved. *Fls. & Frts.:* January-May.

Occasionally planted in the village areas and roadsides.

Field Collection: Minakhan, *K.N.-2133.*

Distributions: Most of the hotter parts of India; Malaya and tropical America.

Economic Importances: White floss is much used for cushion. Young fruits are cooked as vegetables; seeds are edible; root juice cure diabetes. Gum cures bowel complaints; bark is used as febrifuge.

XXV. STERCULIACEAE

Key to the Genera

1a	Flowers unisexual or polygamous, petals–o	(2)
1b	Flowers harmaphrodite, petals present	(4)
2a (1)	Anthers 10 or more, ripe carpels dehiscent	(3)
2b (1)	Anthers 5 in a ring at the top of column; ripe carpels indehiscent	5. *HERITIERA* Dry. ex. Aiton.
3a (2)	Seeds winged along their upper ends, numerous; carpels globose woody	10. *PTERYGOTA* Schott. & Endl.
3b (2)	Seeds without wing, few in number	11. *STERCULIA* L.
4a (1)	Large to medium sized tree	(5)
4b (1)	Herbs or evergreen shrub	(9)
5a (4)	Trunk erect, tall, branches spreading	(6)
5b (4)	Small to medium tree, branches not spreading	(7)
6a (5)	Leaves glabrescent, roundish palmately leaves; calyx tubular, club-shaped, covered with orange coloured, down, follicles stipitate, glabrous	3. *FIRMIANA* Mar.
6b (5)	Leaves broadly ovate, often cordate; flowers rose coloured in large pendulous panicles; petals 5, shorter than the sepals; capsule membranous	6. *KLEINHOVIA* L.

7a (5)	Leaves roundish, cordate, acuminate, irregularly crenate-dentate	(8)
7b (5)	Leaves oblong lanceolate, obliquely cordate acuminate; flowers yellow; calyx bell-shaped, stellate hairy; fruit woody	4. *GUAZUMA* Plum.
8a (7)	Leaves dotted with small types of stellate hairs above; cymes many flowered	2. *ERIOLAENA* DC.
8b (7)	Leaves covered with whitish or cream colourd tomentum; peduncles axillary generally with white flowers	9. PTEROSPERMUM Sch.
9a (4)	Petals concave at the base, appendages at the tip; androecium tubular; capsule 5-winged and 5-angled	1. *ABROMA* Jacq.
9b (4)	Petals flat; persistent	(10)
10a (9)	Androecium tubular only at base; stamens 5 without staminodes	(11)
10b (9)	Androecium tubular with antheriferous margin; anthers 15, in 5 groups of 3, the groups alternating with 5 staminodes	8. *PENTAPETES* L.
11a (10)	Ovary with 5-celled	7. *MELOCHIA* L.
11b (10)	Ovary with 1-celled	12. *WALTHERIA* L.

1. *ABROMA* Jacq.

1.1 ***Abroma augusta*** L. f. Suppl. 341. 1781 (*Ambroma*); DC. Prodr. 1: 485.1824; Masters, Hook. f. Fl. Brit. India 1: 375. 1874; Watt Dict. Econ. Prod. 1: 8.1883; Prain Beng. Plants 1: 191.1903 (Rep.ed. 1963) & Rec. bot Surv. India 3(2): 181. 1905.*Theobroma augusta* L. *(Vern. Name: Ulat Kambal).*

Evergreen shrubs, branches downy; leaves membranous, finely acuminate; the upper ovate or lanceolate, cordate, sometimes entire. Flowers in terminal of leaf opposed, few flowered cymes; sepals pale green, lanceolate, acute; petals below concave, whitish and dark purple, hairs; anthers in groups of 2 to 4 between each pair of staminodes. Fruits membranous, 5-angled capsule. Seeds envelop in light cottony wools. *Fls. & Frts.:* June-November.

Occasionally planted but frequently found as a weed plant in dry waste lands, roadsides and railway tracts.

Field Collection: Joynagar, *K.N.-1621*.

Distributions: Widely spreading throughout the hotter parts of India. Native of Malaya and possibly of India also; though it is generally thought to be an escape from cultivation in the various parts of the plains, where it is now established in scrub jungles and thickets.Java, Philippines, China and E. Asia.

Economic Importances: The bark contain a strong silky fibre,which has been used for cordage and the plant has been recommended as source of the substitute for hemp. Root bark emmenagogue; leaves are most effective uterine tonic. It regulates irregularities of menstruation and relieves menstrual pain.

2. *ERIOLAENA* DC.

2.1 ***Eriolaena hookariana*** Wight. & Arn. Prodr. 70.1834; Watt Dict. Econ. Prod. 3: 265.1890; Masters, Hook.f.Fl. Brit. India 1: 370.1874; Prain Beng. Plants 1: 189. 1903 (Rep.ed. 1963). *(Vern. Name: Gona-goli* or *Gua-Kashi).*

Small to medium tree. Leaves 10-12 cm, as broad as long, roundish, cordate, acuminate, irregularly crenate, dotted with small tufts of stellate hairs above, finely pubescent beneath, nerves prominent in the underside. Peduncle axillary 8-15 cm long, stellately pubescent when young, afterwards glabrous. Cymes many flowered towards the extremities of the branches. Petals obovate, claw densely pubescent within. Style pubescent. Stigma 8-10 lobed. Capsule ovoid-oblong pointed, 10 valved, valves pubescent. *Fls. & Frts.:* March-May.

Often planted as an ornamental plants in the gardens.

Field Collection: Nimichi/Bhebe, *K.N.-726*.

Distributions: Chota Nagpur, Bihar, the hilly tracts and dry forests of India.

3. *FIRMIANA* Marsili

3.1. ***Firmiana colorata*** Roxb. R. Br., Benn. & Brown. Pl. Jav. Rar. 235.1844; Kostermans, Reinwardtia 5: 386. 1960. *Sterculia colorata* Roxb. Pl. Icor. 1: 26.1795; Masters, Hook.f. Fl. Brit. India 1: 350.1874; Prain Beng. Plants 1: 187.1903 (Rep. ed. 1963) & Rec. bot. Surv. India 3(2): 180.1905. *Erythropsis colorata* Burkill, Coard. Bull. Straits Settl. 5: 231.1931. *(Vern.Name: Sammari).*

Large tree, trunk erect; branches spreading. Leaves glabrescent, roundish palmately lobed, lobe acuminate. Panicle terminal coral like, many flowered. Calyx tubular, club shaped covered with orange coloured down. Follicles stipitate, glabrous, membranous. Seeds ovoid, smooth wrinkled. *Fls. & Frts.:* February-May.

Sometime grow wild in the villages and occasionally planted.

Field Collection: Brace Bridge, *K.N.-2235.*

Distributions: Native of India, Chota Nagpur, Western Peninsula and Tripura. Sri Lanka and Baluchisthan.

Economic Importances: The bark yield strong fibre; leaves are used as fodder.

4. *GUAZUMA* Plum.

4.1. ***Guazuma ulmifolia*** Lamk. var. ***tomentosa*** (Kunth.) Schum., Fl. Jourl. 1: 408.1963. *G. tomentosa* H.B.K. Nov. Gen. Pl. 5: 320. 1821; Masters, Hook.f. Fl. Brit. India 1: 375.1874; Watt Dict. Econ. Prod. 4: 184.1890; Prain Beng. Plants 1: 191. 1903 (Rep. ed. 1963) & Rec. Bot. Surv. India 3(2): 181.1905. *(Vern. Name: Nipal Tunth).*

Medium sized tree. Leaves oblong lanceolate, obliquely cordate, acuminate. Flowers numerous, yellow in terminal and axial panicles. Flower buds globose. Calyx bell shaped, stellate hairy. Sepals ultimately reflexed. Petals exceeding the calyx. Fruit woody, oblong, tubercled capsule. Seed numerous; albumen copious. *Fls. & Frts.:* April-October.

Commonly planted in the shrubberies as an avenue tree, though its growth in Bengal is stunted. It also occurs spontaneously in the thickets and jungles.

Field Collection: Brace Bride, *K.N.-119.*

Distributions: Generally distributed and frequently cultivated in the warmer parts of India. Sri Lanka, Java and very plentiful in tropical

America. The tree is a native of tropical America but is now established in many hotter countries.

Economic Importances: The fruit contains a sweet, gummy substance, which is edible. The leaves are excellent cattle fodder. A glutinous decoction of the inner bark is used in West Indies for clarifying sugar. The young branches yield a strong fibre which is occasionally made into ropes. The bark is employed as a tonic. The inner bark is employed as a remedy for elephantiasis and old bark is used to cure skin troubles and diseases of the chest.

5. *HERITIERA* Dryander ex Aiton.

Key to the Species

1a	Fruits smaller, somewhat corrugated furrowed on the outer side, wingless	1. *H. fomes* Buch. Ham.
1b	Fruits large, smooth, winged, keeled on the inner side	2. *H. littoralis* Dryander

5.1. ***Heritiera fomes*** Buch. Ham. Symes Kingdom Ava, (ed 2) 3: 319- 320. 1800; Masters, Hook.f. Fl. Brit. India 1: 363.1874; Watt Dict. Econ. Prod. 4: 223. 1890; Wlh. India 5: 36.1959; Naskar & Guha Bakshi, Mangr. Swams Sundarbans 167.1987. *H. minor* Roxb. (Hort. Beng. 50.1814) & Fl. Ind. 3: 142. 1832; Prain Beng. Plants 1: 188.1903 (Rep. ed. 1963), Rec. bot. Surv. India 2(4): 287- 290.1903 & Ibid 3(2): 180.1905. *(Vern. Name: Sundari)*.(Fig.7).

Tall tree; trunk straight, often deeply grooved; base of the trunk with buttresses; pneumatophores occasionally present, pointed or as blind rootsuckers. Petiole 1-2 cm, spirally arranged, simple, oblong or ovate elliptic, usually 12-16 cm X 8-10 cm, with cordate at base; margin entire; dark green shining above pale white beneath. Panicles axillary, half the length of the leaves, crowded, much-branched, rusty pubescent. Flowers small 2.5-3.5 mm diam. 3.5-4 mm long with slender short pedicel; orange coloured, bell shaped, unisexual, calyx cup shaped. Fruits a cluster of woody indehiscent keeled or winged ripe carpels, knobby with a ventral ridge together with a transverse ridge. Seeds solitary without vivipary but can float on the tidal water. *Fls. & Frts.:* January-April.

It is believed that the name of the Ganga delta was derived out of the vernacular name of the *H. fomes* 'Sundari'; as it was the common species in the area during the early period. But at the present time it is very scarce in the Sundarbans due to changes of the ecological conditions of the area and

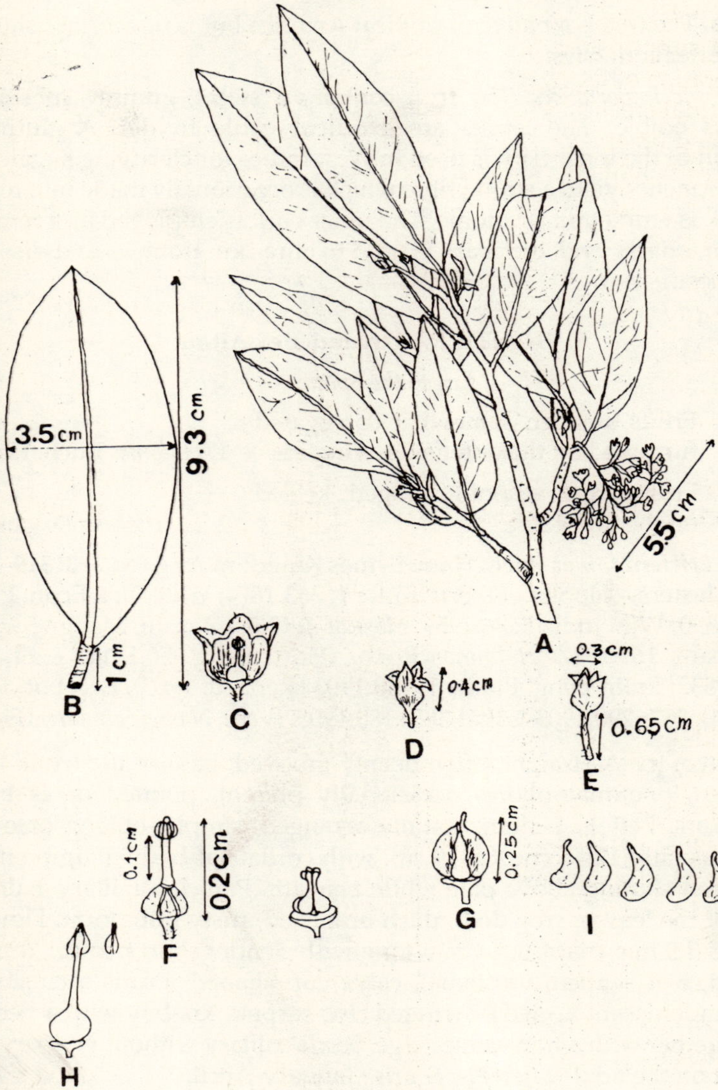

Fig.7. HERITIERA FOMES Buch. Ham.

A. Flowering twig, B. Single leaf, C. Single flower, D. Gynoecium, E. Ovary, F. Anthers, G. Single fruit.

lack of supply upstream fresh water, which is essential for the natural regeneration of the species. It prefers to grow in the area-down to the edge of the tidal creeks and swampy islands, frequently flooded with tidal sea water and upstream fresh water, as well. Exploitation of the plant is much more for valuable timber. All these factors caused the diminising for *H. fomes*.

Field Collections: Satzelia, *K.N. 1554;* Netidhopani, *K.N.-2290.*

Distributions: Common in Sundarbans of Bangladesh, India and Irrawaddy Delta of Burma; occasionally to other tropical coastal areas.

Economic Importances: Timber is much superior than teak and used for boat building and other domestic furnitures; check soil erosion in the tidal forests and loosely consolidated silted up soil.

5.2. ***Heritiera littoralis*** Dryand, Ait., Hort. Kew (ed. 1), 3: 546. 1789; Masters, Hook.f. Fl. Brit. India 1: 363.1874; Watt Dict. Econ. Prod. 4: 223.1890; Naskar & Guha Bakshi, Mang. Swams Sundarbans 169.1987. *(Vern. Name: Sundari).*

Tree up to 25 m tall with buttressed trunk, evergreen, bark fissured, dark. Leaves coriaceous, elliptic oblong, white beneath with flat scales, petiolate; stipules lanceolate, caducous. Flowers monoecious; clustered in much branched axillary; calyx bell shaped; corolla pinkish. Fruit oblong, woody, smooth or tubercled. Capsule membranous, inflated, pyriform, locucidally 3-valved. Seeds 1-2 in each, black. *Fls. & Frts.:* February-August.

A member of the back mangals and occasionally on the fringe area of the shore lines; this species cannot stand strong salinity and poorly drained soils. Occasionally planted in the non-mangrove area.

Field Collections: Ranigarh, *K.N. 1263;* Surjyapur, *K.N.-1405.*

Distributions: Mangrove forests in Sundarbans of India, Andamans, Bangla Desh, Africa, Madagascar, Australia, Qeensland, and Cape York.

Economic Importances: Useful timber and medicinal treatment for diarrhoea.

6. *KLEINHOVIA* L.

6.1. ***Kleinhovia hospita*** L. Sp. Pl. (ed.2) 2: 1365.1763; Masters, Hook.f. Fl. Brit. India 1: 364.1874; Watt Dict. Econ. Prod. 4: 566. 1890; Prain Beng. Plants 2:962. 1903 (Rep. ed. 1963), Rec. bot. Surv. India 2(4): 290.1903 & Ibid. 3(2): 180.1905: Wlh. Ind. 5: 321.1959.

Trees with straight trunk, bark smooth, branches spreading. Leaves broadly ovate, often cordate, smooth, 7.5-30.0 cm long, petiole as long as broad. Flowers bisexual, rose coloured in large pendulous panicle. Petals 5, shorter than the lanceolate sepals. Seeds tubercled black. Capsule membranous, inflated, pyriform, loculicidally 5-valved. *Fls. & Frts.:* May-September

Fairly common, often planted in the roadsides as avenue tree.

Field Collection: Gosaba, *K.N.-1532.*

Distributions: It was introduced in Calcutta from Maluccas in 1788 and now widely planted in India as an avenue tree, especially in the south. Native of Tropical America, Malaya and Australia.

Economic Importance: In Philippines the shoots and young leaves are eaten as vegetable; decoction of the leaves is used in Cochin, China to cure skin diseases.

7. *MELOCHIA* L.

7.1. ***Melochia corchorifolia*** L. Sp. Pl. 675.1753; Masters, Hook.f. Fl. Brit. India 1: 374. 1874; Watt Dict. Econ. Prod. 5: 225. 1891; Prain Beng. Plants 1: 190.1903 (Rep. ed. 1963) & Rec. bot. Surv. India 3(2): 181.1905.

Erect branching herb or much branched undershrub; thinly stellate hairy. Leaves oblong ovate with 5-veins from the base; serrate pointed, base rounded or cordate. Flowers yellow in terminal or axillary heads with bracteoles in terminal. Sepals 5, persistent. Petals 5, obovate white or pink. Capsule globose. *Fls. & Frts.:* August-September.

Occurs near the banks of streams and in shady areas, fairly abundant during monsoon and post monsoon time.

Field Collections: Dakshineswar, *K.N.-685;* Sonarpur, *K.N.-1000.*

Distributions: Generally distributed in the hotter part of India; widely diffuse Tropical weed.

Economic Importances: The stem bark yields a valuable fibre. Stem and leaves decoction with oil are useful for preventing bad consequences from the snake bites. Leaves occasionally edible.

8. *PENTAPETES* L.

8.1. ***Pentapetes phoenicea*** L. Sp. Pl. 698.1753; Masters, Hook.f.Fl. Brit. India 1: 371.1874; Prain Beng. Plants 1: 190.1903 (Rep. ed. 1963) & Rec. bot. Surv. India 3(2): 181. 1905. *(Vern. Name: Katlata).*

Branched, annual glabrous herb. Leaves hastate-lanceolate, crenate-serrate. Stipules subulate. Peduncles simple, axillary. Flowers scarlet, solitary or in pairs, on very short peduncles; fertile stamens 15. Capsule sub-glabrous, shorter than the persistent sepals. Seeds angular, rough but not winged. *Fls. & Frts.:* July-October.

A common weed in the rice field and other marshy lands.

Field Collection: Gobardanga, *K.N.-536.*

Distributions: Throughout the hotter parts of India and Western Peninsula. In the Tropics.

Economic Importances: Medicinally the capsule is used in bowel diseases; root is astringent and leaves are used as tea substitute.

9. *PTEROSPERMUM* Sch. *nom. cons.*

Key to the Species

1a	Leaves roundish or oblong, simple or lobed, entire, palmately nerved, base cordate, often peltate peduncle shorter than the petiole	1. *P. acerifolium* Willd.
1b	Leaves lanceolate and nerve one only	(2)
2a (1)	Shoot slender, virgate; leaves glabrescent above, covered with whitish or cream coloured tomentose, beneath nerve stellate; flowers sweet scented white	2. *P. lanceaefolium* Roxb.
2b (1)	Small tree; leaves coriaceous, glabrous above; above nerve stellate hairy not so prominant; flowering buds oblong; sweet scented, white, seed with a broad terminal wing	3. *P. suberifolium* Lamk.

9.1. ***Pterospermum acerifolium*** Willd. Sp. Pl. 3: 729.1801; Masters, Hook.f. Fl. Brit. India 1: 368. 1874; Watt Dict. Econ. Prod. 6(1): 362.1893; Prain Beng. Plants 1: 189.1903 (Rep. ed. 1963) & Rec. bot. Surv. India 3(2): 181.1903. *P. aceroides* Wall. Cat. 1171; Kurz., J. As. Soc. Bengal. 2: 62.1873. *(Vern. Name: Kanak Champa).*

Large evergreen tree; bark smooth. Leaves roundish or oblong, simple or lobed, entire, palmately nerved, base cordate, often peltate; peduncles shorter than the petioles, flowers fragrant, white, axillary solitary or in pairs. Capsule stipitate oblong, 5-angled sulcate glabrescent. *Fls. & Frts.:* February-August.

Planted in the gardens for its sweet scented flowers.

Field Collection: Baruipur, *K.N. 2260.*

Distributions: North West Himalayas. Native of either Bengal or the W. Peninsula; the foot hills of the Himalaya; Chittagong and Burma.

Economic Importance: The leaves are used as plates, for packing tobacco, and as fodder for cattle. The calyx is used as a cure for glandular swelling of neck and ears. Hindu makes use of the flower for religious purposes, but they do not offer this to Vishnu. The flowers are much valued by the Bengalees as a disinfectant and as a means of keeping insects away from clothes. Used medicinally as a general tonic and occasionally as a cure for inflammation, blood troubles, ulcers, leprosy and tumour. The bark and leaves are applied in small pox.

9.2. ***Pterospermum lanceaefolium*** Roxb. Fl. Ind. 3:163.1832; Masters, Hook.f.Fl. Brit. India 1: 368.1874; Benthall Trees Calc. 59.1946. *(Vern. Name: Bankalla).*

Tree, shoots slender, virgate. Leaves lanceolate and 1-nerved, acuminate, base rounded, glabrescent above, covered with whitish or cream coloured tomentum; beneath stellate hairy. Peduncle axillary. Flower buds oblong striated. Flowers sweet scented, white. Capsule ovoid acute, obscurely 5-angled. *Fls. & Frts.:* May-August.

Commonly cultivated as ornamental plants.

Field Collection: Sodepur, *K.N.-1488.*

Distributions: North West Himalaya and Khasi Hills. Throughout Chittagong and Burma.

Economic Importance: In Assam, the leaves are chewed to redden the lips. Wood is fairly hard and used for paking boxes and furnitures.

9.3. ***Pterospermum suberifolium*** Lamk. Tab. Encyc. et Method. 3: 136.t.576.1823 (not of Roxb.); Masters, Hook.f. Fl. Brit. India 1: 367. 1874; Watt Dict. Econ. Prod. 6(1): 362. 1893. *P. canescens* Roxb. Hort. Beng. 50.1814 & Fl. Ind. 3: 1162.1832. *(Vern. Name: Muskunda).*

Small tree. Leaves coriaceous, glabrous above, nerves stellate hairy, not prominant. Peduncles axillary, twice as long as the petiole. Bracteoles very caducous. Flowering buds oblong. Flowers white, fragrant. Sepals oblong, revolute. *Fls. & Frts.:* March-June.

Rarely cultivated; a few plants were seen in the Alipore Zoological Gardens; also found in the public garden at Shyamnagar.

Field Collections: Shyamnagar, *K.N.-2099.*

Distributions: W. Peninsula, Madras and Pondicherry. Sri Lanka and South West Asia.

Economic Importances: Bark and flowers are used to cure small pox eruptions. Fruits are sometime edible.

10. *PTERYGOTA* Schott & Endl.

10.1. ***Pterygota alata*** (Roxb.) R. Br. Benn. Pl. Jav. Rar. 234.; Datta & Majumder, Bull. bot. Soc. Bengal 20(2): 85.1966; Benthall, Trees Calc. 50.1946. *Sterculia alata* Roxb. Fl. Ind. 3: 152.1832; Masters, Hook.f. Fl. Brit. India 1: 360.1874; Prain Beng. Plants 1: 274. 1903 (Rep. ed. 1963) & Rec. bot. Surv. India 3 (2): 180. 1903. *(Vern. Name: Buddha-Narikel).*

Large tree, very tall. Bark ash coloured. Leaves clustered about the ends of the branches, cordate, ovate, acute, glabrous. Panicles from the leaflets axils, few flowered, shorter than the leaves. Bracteoles caducous. Calyx campanulate, densely tomentose. Follicles globose, depressed, woody, clothed. Seeds many, oblong, compressed; cotyledons not separable from the albumen. *Fls. & Frts.:* January-April.

Commonly cultivated for its nut, as well as ornamental plants.

Field Collection: Dewanganj, *K.N. 2261.*

Distributions: Native of South India, Assam and Andamans. Chittagong, Burma and Sikkim.

Economic Importances: Seeds are edible, it is sold as a cheap substitute for opium, though the tree does not seem to be otherwise recognised as having any narcotic properties.

11. *STERCULIA* L.

Key to the Species

1a	Leaves digitate, branches whorled horizontal	1. *S. foetida* L.
1b	Leaves palmately lobed	(2)

2a (1) Follicles armed with sharp stringing hairs ... 2. *S.urens* Roxb.

2b (1) Follicles rusty villous, without stinging hairs ... 3. *S. villosa* Roxb.

11.1. ***Sterculia foetida*** L. Sp. Pl. 1008.1753; Masters, Hook.f. Fl. Brit. India 1: 354. 1874; Prain Beng. Plants 1: 187.1903 (Rep. ed. 1963) & Rec. bot. Surv. India 3 (2): 1905 *(Vern. Name: Jangli Badam).*

Tall, deciduous tree. Branches whorled, horizontal. Leaves crowded at the ends of the branches, leaflets 7-9, elliptic lanceolate, adult glabrous beneath; stipules ensiform, caducous. Panicles erect, branches glabrous, ultimate pedicels shorter than the flower, jointed in the middle. Bracteoles minute. Calyx deeply 5 partied. Follicles woody, oblong, boat shaped shortly beaked. *Fls. & Frts.:* February-June.

Sometime planted in the village gardens. Foliages falls at the end of the cold season and the flower appear in great profusion in the latter part of February and in March, closely followed or sometime just proceeded by the new leaves.

Field Collection: Malatipur/Deganga, *K.N.-1590.*

Distributions: Almost throughout India. Western Peninsula, a native of South East Asia, East Tropical Africa, Burma, Sri Lanka, Malaya and Australia.

Economic Importances: The seeds are roasted and eaten as chestnuts. Oil from the seed is said as a laxative and carmative; the fruit is astringent.

11.2. ***Sterculia urens*** Roxb. Cor. Pl. 1: 25,t.24.1795; Masters, Hook.f. Fl. Brit. India 1: 355. 1874; Watt Dict. Econ. Prod. 6(3): 364.1893; Prain Beng. Plants 1: 187.1903 (Rep. ed. 1963). *(Vern. Name: Keonji).*

Large tall tree with erect and straight trunk; bark white, smooth papery. Leaves 20-30 cm diam., crowded at the ends of the branches, palmately lobed, glabrous or nearly so above, velvety beneath; lobes 5, caudate acuminate, base cordate. Flowers yellow, numerous, small. Calyx campanulate, hairy on both surfaces, lobes oblong, acute. Male flowers-staminal column short; filaments 10. Female flowers-carpels usually 5, on a short stout gynophores. Follicles 4-6, ovoid oblong. Seeds 3-6, oblong, black. *Fls. & Frts.:* December-April.

Occasionally planted as avenue tree in the gardens and roadsides.

Field Collection: Madarhat/Baruipur, *K.N.-1242.*

Distributions: Almost throughout India. Sri Lanka and Burma.

Economic Importances: Seeds are edible. Gum is used in cosmetics, lotions, sprays, pastes, laxatives, in paper industry as pulp binder; in leather industry as ingredient of dressing compositions. Gum also used in food, baking and dairy industries in dressings and spreads. Bark yields fibre, which is used for ropes and coarse cloth and seeds are roasted and eaten. Wood pulp is suitable for wrapping paper.

11.3. ***Sterculia villosa*** Roxb. Fl. Ind. 3: 153.1832; Masters, Hook.f. Fl. Brit. India 1: 355.1874; Prain Beng. Plants 1: 187.1903 (Rep. ed. 1963): *(Vern. Name: Udal).*

Deciduous tree with white bark. Branches few, spreading with large scars. Leaves 5-7 lobed velvetry beneath; and with spreading. Bracts filiforms, caducous panicles pendulous, follicles, downy. Flower buds globose. Seeds oblong, black. Cotyledons thin. *Fls. & Frts.:* December-March.

Occasionally planted in the gardens and roadsides.

Field Collection: Brace Bridge, *K.N.-2229.*

Distributions: North West India, Bengal and Malabar coast.

Economic Importances: The inner bark of this tree yields a coarse fibre, which is much used in many parts of India for ropes, cordage and bag-making. The root of the tree is occasionally eaten and the trunk yields a transparent gum; gum is used as a substitute for gum tragacanth.

12. *WALTHERIA* L.

12.1. ***Waltheria indica*** L. Sp. Pl. 673.1753; Masters, Hook.f. Fl. Brit. India 1: 374.1874; Prain Beng. Plants 1: 190.1903 (Rep. ed. 1963) & Rec. bot. Surv. India 3(2): 181.1905. *(Vern. Name: Khardundhi).*

Perennial erect herb; stem terete, softly canopubescent. Leaves ovate or elliptic, obtuse, rounded or subcordate at the base, toothed or serrate, softly pubescent on both surfaces, strongly nerved, the nerves impressed on the upper surface. Flowers yellow, sessile, in dense, axillary heads. Calyx campanulats; teeth triangular, very acute. Capsule very small; enclosed in the calyx membrans. Seed solitary. *Fls. & Frts.:* June-August.

Fairly common weed in the cultivated fields.

Field Collections: Bhasila/Barasat, K.N.-1575

Distributions: Throughout India and throughout the warmer parts of the Old World.

Note: Dombya mastersii Hook.f., a horticultural variety is common.

XXVI. TILIACEAE

Key to the Genera

1a	Sepals connate below in a cup; anther cells ultimately confluent at the top; carpels distinct, 2-valved.	1. *BROWNLOWIA* Roxb.
1b	Sepals distinct	(2)
2a (1)	Petals without a basal scale; capsule opening loculicidally, many seeded	2. *CORCHORUS* L.
2b (1)	Petals with a more or less adnate scale	(3)
3a (2)	Fruits not prickly, drupaceous	3. *GREWIA* L.
3b (2)	Fruits prickly, small, globose, indehiscent	4. *TRIUMFETTA* L.

1. BROWNLOWIA Roxb.

1.2. ***Brownlowia lanceolata*** Benth.; Masters, Hook. f. Fl. Brit. India 1: 381. 1874; Watt Dict. Econ. Prod. 1: 469. 1889; Prain Beng. Plants 1: 193. 1903 (Rep. ed. 1963) & Rec. bot. Surv. India 3 (2): 181. 1905. *(Vern. Name: Bhola Sundri).*(Fig. 8).

Medium tree; leaves lanceolate, acuminate, entire, pinnately 3-5 veined, pubescent, scaly beneath. Flowers very small, in large terminal or axillary panicles. Calyx campanulate; petals 5; filaments stout; staminodes petaloid; ovary 5 locular. Fruits a group with 2-valves ripe carpels. Seeds solitary in each carpels; cotyledon thick fleshy. *Fls. & Frts.:* July-November.

Occasionally on the mangrove swam forests in the Sundarbans.

Field Collection: Satzelia/Sundarbans, *K.N.-1551.*

Distributions: Mangrove forests in India and other tropical mangals of the Old World.

Notes: The other mangrove species under the genus *Brownlowia* Roxb. are *B. argentata* Kurz, J. Asiat. Soc. Bengal 39(2): 67.1870 and *B. tersa* (L.) Kosterm. Reinwardtia 4(4): 731.1959.

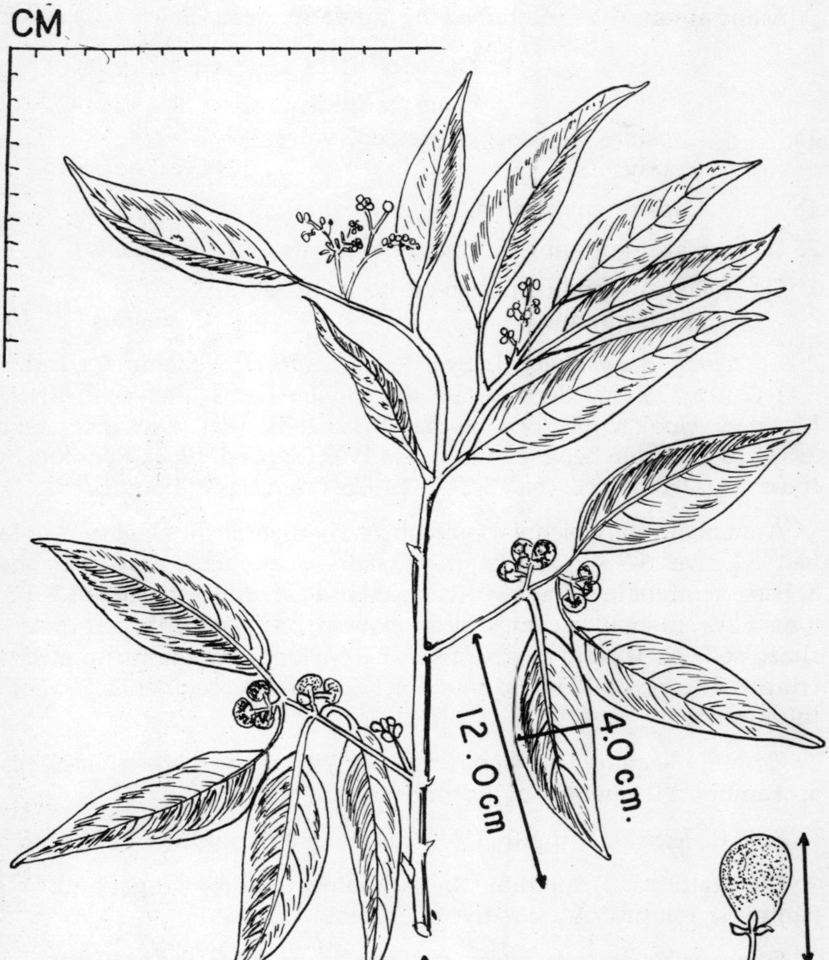

Fig.8. BROWNLOWIA LANCEOLATA Benth.
A. Flowering twig, B. Fruit (Single)

Both these species are distributed in Malay Peninsula through Indonesia to New Guinea and the Solomon Islands.

Many authors have described the Indian species as *B. tersa* (L.) Kosterm.

2. *CORCHORUS* L.

Key to the Species

1a	Capsules globose, depressed, valves woody	2. *C. capsularis* L.	
1b	Capsule enlongated, valves thin		(2
2a (1)	Capsule ending in 3 radiating points	1. *C. aestuans* L.	
2b (1)	Capsule ending in a single beak, lobed at the summit	3. *C. olitorius* L.	

2.1. ***Corchorus aestuans*** L. Syst. (ed.10) 1079.1759; Majumder, Ind. Agr 6 (1 & 2) : 126.1962 (Rep.ed.). *C. acutangulus* Lamk. Encycl. 2: 104.1786 Masters, Hook.f. Fl. Brit. India 1: 398.1874; Watt Dict. Econ. Prod 2: 534.1889; Prain Beng. Plants 1 : 196.1903 (Rep. ed. 1963), Rec. bot,. Surv India 2(4): 290. 1903 & Ibid. 3(2): 182.1905. *(Vern.Name: Tita pat).*

Annual or rarely biennial, suffruticose herb; branches glabrous or laxly hairy. Leaves 6-9 X 2.2-3.5 cm thicky hairy, ovate, acute or oblong, obtuse at base, acuminate serrate with awned basal, serratures; petiole 1.2-1.8 cm long. Flowers small in cymes; buds obovoid; bracts filiform; peduncle very short; pedicels 1-2 mm long. Capsule 2-3 cm long, erect, linear, angled with winged angles; beak 3-5 fid into notched or retuse segments. *Fls. & Frts.* July-October.

Commonly cultivated; frequently met with as an escape in waste places and amongst the hedges of the gardens.

Field Collections: Nimpith, *K.N.-1614;* Budge Budge, *K.N.-1625.*

Distributions: Throughout India, mainly Assam, Bengal and Orissa Pantropic, but probably a native of Tropical Asia.

Economic Importances: Fibres are very important in the industry, as it is the raw material for sac and different types of fine threads for cloth making.

2.2. ***Corchorus capsularis*** L. Sp. Pl. 529.1753; Masters, Hook. f. Fl. Brit. India 1: 397.1874; Watt Dict. Econ. Prod. 2 : 535.1889; Prain Beng Plants 1: 196.1903 (Rep. ed. 1963) & Rec. bot. Surv. India 3 (2) : 182.1905. *(Vern. Name: Pat).*

Erect annual shrub, 140-175 cm tall. Leaves 7.5 X 3.5 cm ovate to ovate lanceolate, rounded at base, acute or shortly acuminate at apex, glabrous, with awned basal serratures; petiole 5-14 mm long, pubescent. Flowers in cymes; bud obovoid; pedicels very short. Capsule upto 1 cm diam., globose ridged, not beaked, 5-valved. Seeds upto 2 mm long, smooth. *Fls. & Frts.:* July-October.

Widely cultivated in moist places throughout the lower Ganga delta; often meet with as an escape in the wastelands and roadsides.

Field Collection: Naihati, *K.N.-1542.*

Distributions: Native of India and China. Tropical countries of the world.

Economic Importances: The most important fibre obtained from the secondary phloem of the plant. The dried leaves are used medicinally as a remedy for dysentery. Fresh leaves are edible as vegetable.

2.3. ***Corchorus olitorius*** L. Sp. Pl. 529.1753; Masters, Hook. f. Fl. Brit. India 1: 397. 1874; Watt Dict. Econ. Prod. 2: 540.1889; Prain Beng. Plants 1: 197. 1903 (Rep. ed. 1963) & Rec. bot. Surv. India 3 (2) : 182.1905. *(Vern. Name: Pat).*

Erect herb or undershrub with a woody base, 150-190 cm high. Leaves 8 X 3.5 cm, glabrous except thinly hairy nerves beneath, elliptic to elliptic oblong, obtuse or rounded at base, acute or acuminate serrate, with awaned basal serratures. Flower buds obovoid, beaked; bracts lanceolate; peduncles and pedicles very short. Capsule 3-6 cm long, glabrous, linear 10-ribbed, beaked, 5-valved. Seeds black, trigonous. *Fls. & Frts.* July to September.

Commonly planted; very often found as garden escape.

Field Collection: Habra, K.N.-1442.

Distributions: Common in the Estern parts of India. Native of the Old World; now Pantropic.

Economic Importances: Most common fibre yielding plants. An important commercial articles of Bengal.

3. *GREWIA* L.

3.1. ***Grewia subinaequalis*** DC. Prodr. 1 : 511.1824; Narayanaswami and Rao, J. Ind. bot. Soc. 29.178.1950. *G. asiatica* L. Mant. 122. 1767; Masters, Hook.f. Fl. Brit. India 1 : 386.1874; Watt Dict. Econ. Prod. 4 : 177.1890;

Prain Beng. Plants 1: 194. 1903 (Rep. ed. 1963) & Rec. bot. Surv. India 3 (2): 181.1905. *(Vern. Name : Phalsa).*

Small tree, young branches pubescent. Leaves obliquely cordate, irregularly serrate, sub acuminate; stipules lanceolate. Flowers 1-3 in axillary umbelliform cymes; peduncles 1-4 together, little longer than the petioles, pedicels shorter than the peduncle tomentose. Fruits red or purple, globose, slightly lobed; *Fls. & Frts.:* April-June.

Sometime cultivated but often met with as an escape in the waste lands; village thickets and shrubberies.

Field Collection: Dakshineswar, *K.N.-695.*

Distributions: Cultivated throughout India; native of North-West India; Pakisthan, Baluchisthan and Afganisthan.

Economic Importances: Fruits are very acidic and used for making a pleasant cooling drink. Used as tonic and as a cure for heart and chest troubles. Rootstock and root barks are also used to cure rheumatism and the leaves are applied as a remedy for certain skin diseases.

Notes: The other species of *Grewia* L. is *G. tiliaefolia* Vahl also found in Bengal.

4. *TRIUMFETTA* L.

4.1. *Triumfetta rhomboidea* Jacq. Enum. Pl. Garib. 22.1760; Masters, Hook. f. Fl. Brit. India 1: 395. 1874; Watt Dict. Econ. Prod. 6(4): 202.1893; Prain Beng. Plants 1: 196. 1903 (Rep. ed. 1963) & Rec. bot. Surv. India 3(2): 182.1905. *Bartramia indica* L. Sp. Pl. 378.1753. *Triumfetta bartramia* L. Syst. Nat. (ed. 10) 1044.1759 *(non. ambig.). (Vern. Name : Ban Okra).*

Suffruticose herb or undershrub. Branches glabrous or scabrous. Leaves 13 X 8 cm rhomboid ovate to lanceolate, entire to 3 lobed, cordate at base acute or acuminate at apex, crenate, 3-7 nerved from the base. Flowers in dense cymes in the axils of upper leaves, upto 7 mm across, yellow. Fruit 4-5 mm diam., armed with glabrous hooked spines. *Fls. & Frts.:* October- January.

Very common during rainy season, on the waste lands and railway tracks.

Field Collection: Sasan Road/Baruipur, K.N.-28.

Distributions: Throughout India. In the Tropics.

Economic Importances: Stem produces soft glossy fibre.

Remarks: It is a polymorphic and highly variable in habitat, size and shape of leaves and degree or density of indumentum.

XXVII . ELAEOCARPACEAE

Key to the Genera

1a	Fruit drupe, with 5-seeds	1. *ELAEOCARPUS* L.
1b	Fruit globose berry; witn many seeds in juicy pulp	2. *MUTINGIA* L.

1. *ELAEOCARPUS* L.

1.1. *Elaeocarpus ganitrus* Roxb. Fl. Ind. 2:592.1824; Masters, Hook.f. Fl. Brit. India 1: 400. 1874; Watt Dict. Econ. Prod. 3: 205.1890; Prain Beng. Plants 1: 197.1903 (Rep. ed. 1963). *E. sphaericus* (Gaertn.) Schum. *(Vern.Name: Rudrakia).*

Medium tree. Leaves 10-15 cm X 2.5-4.0 cm, oblong lanceolate, acute, irregularly crenate serrate, serrate or sub entire, glabrous. Racemes 5.0-7.5 cm long; buds ovoid conical, pointed. Flowers small sepals linear oblong, acuminate pubescent outside; petals oblong, laciniate about half way down, ciliolate; stamens about 40 in groups, opposite each petals; filaments very short; anther linear. Ovary pilose, 5-celled. Drupes globose or ovoid, purple, stone tubercled, 5-celled, 5-seeded. *Fls. Frts.:* April-June.

Planted in the gardens or in the roadsides, occasionally.

Field Collection: Alipore, *K.N.2086.*

Distributions: Mainly on the Westrn Ghats. In the Tropics.

Economic Importances: Fruits are edible. Hard tubercled nuts are made into rosaries, bracelets and buttons. Rosaries and bracelets are worn by Brahmins, Holymen and modern Hippies.

Notes: The other species found in Bengal is *E. floribundus* Blume; the seeds are also used for rosaries and said to have a very valuable medicinal effects.

2. *MUTINGIA* L.

2.2. ***Mutingia calabura*** L. Sp. Pl. 509.1753; Benthall Trees Calc. 66.1946. Datta & Majumdar, Bull. bot. Soc. Bengal 20 (2) : 81.1966. *(Vern. Name : Suji Phal).*

Small spreading tree; branches drooping, pubescent. Leaves distichous, obliquely lanceolate, acute, serrate, sub-glabrous above, silvery tomentose below, about 7.5 cm long. Flowers white, solitary on the extra axillary peduncles; sepals 5, linear lanceolate; petals obcordate; stamens many, stigma 5-lobed; ovary ovoid, glabrous. Berry globose or obovoid. Seeds many in juicy pulp. *Fls. & Frts.* Throughout the year.

Sometimes planted in the garden as ornamental plants; also observed to grow wild in the railway tracks, in the building debrises and protected waste lands.

Field Collection: Ghoshpur/Kulti, *K.N.- 2126.*

Distributions: It has been very recently introduced in Bengal. Native of Central and Tropical South America, but it is cultivated in most parts of the Tropical Asia and elsewhere.

Economic Importance: Fruits are edible; in West Indies, its leaves infusion is used as tea substitute. A quick growing plant and produces fuel and timbers of medium qualities.

XXVIII. LINACEAE

1. *LINUM* L.

1.1. ***Linum usitatissimum*** L. Sp. Pl. 277. 1753; Hook. f. Fl. Brit. India 1: 410.1874; Watt Dict. Econ. Prod. 5:2.1891; Prain Beng. Plants 1: 199. 1903 (Rep.ed. 1963) & Rec. bot. Surv. India 3 (2): 183.1905; Wlh. India 6: 119. 1962. *(Vern. Name: Tisi).*

Annual herb; stem cylindric, erect simple below. Leaves linear, lanceolate. Flowers blue in corymbose panicles. Style free; capsule 5-celled, globose. Seeds ovoid oblong, dark brown, compressed. *Fls. & Frts.:* March-April.

Mostly cultivated as crop plant, also found as an escape near the cultivated fields and waste lands.

Field Collection: Dhopagachhi/Kalyanpur, *K.N.- 2135.*

Distributions: Cultivated throughout India and West Asia.

Economic Importance: The commercial linseed oil is extracted from the mature seeds; seed oil used for paint, varnish, soft soap, printers's ink and also edible. Oilcake used as cattle feed and manure. It is also laxative oil for cattle and horses. Seed mucilage used for cosmetic and pharmaceutical industries as demulcent. Crushed seed is applied for poultice in inflamations, ulcers and boils. Stem fibre is useful for textile manufacture and paper pulp.

XXIX. MALPIGHIACEAE

Key to the Genera

1a.	Style 3, carpels winged all round	1. *ASPIDOPTERYS* Juss.
1b.	Style solitary; carpels 3-winged	2. *HIPTAGE* Gaertner

1. *ASPIDOPTERYS* A. Juss.

1.1. ***Aspidopterys rotundifolia*** A. Juss., Ann. Soc. Nat. Bot. (ser.2) 13: 267.1840; Hook.f. Fl. Brit. India 1: 421.1874; Prain Beng. Plants 1: 200. 1903 (Rep. ed. 1963).

Climbing shrub. Leaves opposite, entire; stipules absent. Flowers small, white in axillary panicles, peduncles bracteate; calyx short, 5-partite, eglandular; petals 5, stamens 10. Ovary 3 locular. Fruit 1-3 samaras. *Fls.& Frts.:* August - November.

Occasionally found as the climber on the village thickets.

Field Collection: Palta/Barrackpore, *K.N.-1836.*

Distributions: Scanty in distribution; Bangladesh, Malaya and Sri Lanka.

2. *HIPTAGE* Gaertner

2.1. ***Hiptage benghalensia*** (L.) Kurz. Journ. An. Soc. Bengal 43 (2): 136.1874; Jacobs in Steenis, Fl. Malesiana Ser I, 5: 132.1955. *Banisteria benghalensis* L. Sp. Pl. 427.1753. *Hiptage madablota* Gaertn. Fruct. 2: 169.t.116.1791; Hook.f. Fl. Brit.India 1: 418.1874; Watt Dict. Econ. Prod. 4: 252. 1890; Prain Beng. Plants 1: 200. 1903 (Rep. ed. 1963) & Rec. bot. Surv. India 3 (2): 183. 1905. *(Vern.Name: Basanti).*

Climbing shrub; leaves opposite, coriaceous, elliptic oblong. Flowers white, pinkish, fragrant, petals silky, clawed. Calyx glandular, persistent, densely pubescence outside. Petals longer than sepals, orbicular, seeds solitary, globose. *Fls. & Frts:* June-October.

Fairly common in the gardens and occasionally as escape.

Field Collection: Baguihati, *K.N.-1884.*

Distributions: Abundantly found throughout India. Sri Lanka, China and Java.

Economic Importance: Ornamental garden plants.

XXX. ZYGOPHYLLACEAE

Key to the Genera

1a.	Branches upto 1 m long, terete; leaves 2-3 jugate, obovate, subacute at apex; fruit ovoid, beaked and tuberculate	1. *KALLSTROEMIA* L.
1b.	Branches upto 40 cm long, young parts hairy; leaves 5-8 jugate; oblong, mucronate; fruits globose with 2 sharp rigid spines and 2 shorter	2. *TRIBULUS* L.

1. *KALLSTROEMIA* L.

1.1. ***Kallstroemia pubescens*** (G. Don) Dandy, Kew Bull. 1 : 138.1955; Bennet, Indian For . 91: 282.f.1-6.1965. *Tribulus pubescens* G. Don, Gen. Syst. 1: 769. 1831.

Deep-rooted, diffused branched, trailing, pubescent annual herb; branches upto 1 m tall, fistular, terete, slightly swollen at the noses with pale reddish. Leaves paripinnate, 2-3 jugate, usually opposite, anisophyllous, obovate, green above, pale below, subacute at apex, obtuse at base. Flowers white to brownish white, solitary axillary. Fruits upto 5mm across, ovoid, beaked ribbed and tuberculate. Seeds white pubescent. *Fls. & Frts.:* June-December.

Occasionally found along the railway tracts and sometime on the canal sides

Field Collections: Shyamnagar, *K.N.-2094;* Garia, *R.N.B-s.n.*(Cal).

Distributions: Throughout India; a native of Tropical America; introduced in West Africa.

Notes: Prain (1903a, 1903b and 1905) did not mention this taxon from Bengal.

2. *TRIBULUS* L.

2.1. ***Tribulus terrestris*** L. Sp. Pl. 387. 1753; Edgeworth & Hook.f. Fl. Brit. India 1: 423.1874; Prain Beng. Plants 1: 201.1903 (Rep. ed. 1963) & Rec. Bot. Surv. India 3(2) : 183.1905. *T. lanuginosus* L. Sp. Pl. 387. 1753. *(Vern. Name : Gokhure).*

Small, procumbent annual herb; branches upto 40 cm long; young parts hairy. Leaves opposite upto 1 cm long in equal pairs, leaflets 5-8 pairs, oblong, mucronate. Flowers solitary, axillary or leaf -opposed. Fruit 5-angled, globose, hairy, with 2 very sharp rigid spines and 2 shorter, one on each cocci. *Fls. & Frts.:* April-October.

Rarely found on the waste places, near river banks, paddy fields and dry waste lands.

Field Collection: Kulpi, *K.N.-1791.*

Distributions: Throughout India; hotter parts of the world; native of the Mediterranean regions.

Economic Importances: Leaves are said to be used in genito-urinary diseases. The entire plants, especially the fruits and leaves are locally used in various kidney diseases. Leaves paste is used for the treatment of stones in gallbladder. Roots aperient and has tonic properties.

XXXI. OXALIDACEAE

(*nom. alt.* Geraniaceae)

Key to the Genera

1a.	Trees; fruits a berry	1. *AVERRHOA* L.
1b.	Herbs; furits a capsule	(2)
2a.	Leaves pinnate	2. *BIOPHYTUM* DC.
2b.	Leaves 3-foliate, palmate	3. *OXALIS* L.

1. *AVERRHOA* L.

1.1. ***Averrhoa carambola*** L. Sp. Pl. 428. 1753; Edgeworth & Hook. f. Fl. Brit. India 1: 439.1874; Watt Dict. Econ. Prod. 1 : 359.1889; Prain Beng. Plants 1: 204.1903 (Rep. ed. 1963) & Rec. bot., Surv. India 3(2) : 184.1905; Wlh. Ind. 1 : 139.1948. *(Vern. Name : Kamranga).*

Small to medium tree with close drooping branches. Leaves alternate; imparipinnate; leaflets 5-11, ovate-oblong, acuminate, entire; base oblique. Flowers pinkish in short racemes; stamens 10, 3 or 4 shorter and without anthers. Fruits oblong, acutely angled, yellow; seeds arillate. *Fls. & Frts.:* July- December.

Often planted in the gardens.

Field collection: Bireshwarpur, near Kulpi, *K.N.-1761.*

Distributions: Cultivated almost throughout India. Duthie (1903) suggested that the species was introduced in India from America by the Portuguese in the 16th Century. A tropical world species; now widely spread in the hotter regions. Hill (1952) considered it to be a native of S.E. Asia. Hayes (1957) believed it to be a native of Moluccas or of Malayan regions.

Economic Importance: Ripe fruits edible as jam and as a substitute of tamarind in cooking; it is the good source of iron and Vitamin B, Vitamin C.

Notes: The genus *Averrhoa* L. is now separated by many Botanists as a distinct family–Averrhoaceae.

2. *BIOPHYTUM* DC.

2.1. ***Biophytum sensitivum*** (L.) DC. Prodr. 1: 690.1824; Edgeworth & Hook. f. Fl. Brit. India 1: 436.1874; Prain Beng. Plants 1: 204. 1903 (Rep. ed. 1963) & Rec. bot. Surv. India 3 (2): 183.1905. *Oxalis sensitiva* L. Sp. Pl. 434.1753. *(Vern.Name: Bannanga* or *Jhalai).*

Erect, annual herb; stems hispidly pubescent; leaves 3.5-12.0 cm long, crowded into a rossette at the top of stem; leaflets 6-15 pairs, variable in size, oblong, obliquely rounded, apiculate at the apex. Flowers at the apex. Flowers yellow, peduncles very variable, sometime swollen at the tip. Capsules elliptic, shining; seeds minute. *Fls. & Frts.:* May - September.

A common weed during rainy season in the waste lands.

Field Collections: Dewanganj, *K.N.-2263;* Behala, *Dutta-17.*

Distributions: Throughout India; Tropical Asia, Africa and America.

Economic Importance: Leaves are said to be diuretic

3. *OXALIS* L.

Key to the Species

1a.	Wild species	1. *O. corniculata* L.
1b.	Cultivated species	2. *O. latifolia* H.B. & K.

3.1. ***Oxalis corniculata*** L. Sp. Pl. 435. 1753; Edgeworth & Hook.f. Fl. Brit. India 1: 436.1774; Watt Dict. Econ. Prod. 5: 658.1891; Prain Beng. Plants 1: 203.1903 (Rep. ed. 1963) & Rec. bot. Surv. India 3 (2): 183.1905; Eiten, Taxon 4: 99.1955. *(Vern. Name: Amrul).*

Small procumbent herb; rooting from the nodes; leaves digitately 3 foliolate, with slender petioles; leaflets 3-5 X 0.7-1.0 cm, subsessile, obcordate, cuneate at the base. Flowers yellow, about 1 cm diam. in axillary, solitary, long peduncled , umbellate cymes. Stamens 10, alternately long and short. Capsules 2 cm long, linear brown. *Fls. & Frts.:* Throughout the year, particularly in monsoon. Very common on the shady and waste places, during rainy season.

Field Collections: Brace Bridge, *K.N.-2230;* Behala, *Dutta-16.*

Distributions: Throughout India; a cosmopolitan species.

Economic Importances: Occasionally used as pot-herb; edible as vegetable—a good source of vitamin C; plant is astringent and has antiseptic properties. Cures dyspepsia, piles, anaemia and tympanitis. Also used for treatment of cornea.

3.2. ***Oxalis latifolia*** H.B. & K., Nov. Gen. & Sp. 5: 237.t.467.1821; Kunth, Pfreich. 95: 273.1930; Wlh. India 7: 199.1966; Guha Bakshi, Bull. bot. Soc. Bengal 23 (1): 32.1969. *(Vern. Name: Khatmitthi).*

Stemless, pubescent, perennial herbs. Leaves radical; leaflets with divergent, more or less oval lobes. Flowers pale pink in cymes, peduncles axillary, radical. Fruits loculicidal capsule. *Fls. & Frts.:* December -June.

Rarely grow on the waste, marshy lands.

Field Collection: Santoshpur, *K.N. 1957.*

Distributions: Naturalized in North West Himalaya, Chota Nagpur, Assam, Palmi Hills and Khandala. Native of Mexico.

Economic Importances: Efficient soil cover in tea estates in Sri Lanka during rainy seasons. The plant is used by the Oraons as a soporific. Leaves and barks used as green manure.

XXXII. BALSAMINACEAE

Key to the Genera

1a.	Lateral petals free, imbricate; fruit drupaceous, fleshy, indehiscent	1. *HYDROCERA* Bl.
1b.	Lateral petals connate in pairs; fruit capsular, elastically, dehiscent	2. *IMPATIENS* L.

1. *HYDROCERA* BL.

1.1. ***Hydrocera triflora*** Wt. & Arn. Prod. 140.1834; Hook.f. Fl. Brit. India 1: 484.1874; Prain Beng. Plants 1: 205. 1903 (Rep, ed. 1963) & Rec. bot. Surv. India 3 (2): 184.1905; Wlh.India 5:146.1959. (*Vern. Name: Domuti*).

Herb or small shrub, 0.3-0.5m high; branches 5-angled. Leaves alternate, sessile, linear lanceolate, serrate. Flowers variegated red, white or yellow; 1-3 on axillary peduncles. Fruits subglobose, succulent. *Fls. & Frts.:* Almost throughout the year, except winter.

Sometime found as an garden escape, mostly cultivated .

Field Collection: Goranbose, *K.N.-1881.*

Distributions: Eastern parts of the Indian plain lands and Tropical Asian countries, elsewhere.

Economic Importance: Ornamental garden plants and the flowers are sometime used for dying finger nails.

2. *IMPATIENS* L.

2.1. ***Impatiens balsamina*** L. Sp. Pl. 938.1753; Hook. f .Fl. Brit. India 1: 453.1874; Watt Dict. Econ. Prod. 4: 334.1890; Prain Beng. Plants 1: 205.1903 (Rep. ed. 1963) & Rec. bot. Surv. India 3 (2): 184. 1905; Wlh. India 5: 168.1959. *(Vern. Name: Dopati).*

Erect, branched, annual herb. Leaves narrow, lanceolate, 4-15 cm long, acuminate, deeply serrate ; petioles glandular. Flowers variable-purple, violet, pink, white, red and spotted, axillary or fascicled; lateral sepals broad, ovate, minute, standard orbicular, retuse wing petals vary, broad, lip small spur short or long. Capsule tomentose, many seeded. *Fls. & Frts.* Throughout the year, except winter.

Commonly planted in the ornamental gardens, but occasionally as an garden escape.

Field Collection: Ichhapur, *K.N.-1796.*

Distributions: Throughout India and a Cosmopolitan species.

Economic Importances: Seeds are edible; flowers extract has antibiotic activity against some fungi and bacteria.

XXXIII. RUTACEAE

Key to the Genera

1a. Tall to medium sixed tree, armed with spines; fruits larger, globose, pericarp woody (2)

1b. Shrub to small tree, armed with spines; fruits generally smaller, round-globose; pericarp not woody (3)

2a (1)	Leaves 3-foliolate, leaflets lanceolate or ovoid; flowers few, axillary cymose; fruit globose, 10-25 cm diam., pulp orange or yellowish brown	1. *AEGLE* Corr.
2b (1)	Leaves alternate and in axils, 4-5 in clusters; leaflets 3-9, ovate; flowers many, panicles; fruits round 7-12 cm diam., pulp whitish to blckish when ripe	3. *FERONIA Corr.*
3a (1)	Petioles not winged; flowers milky white and generally highly scented, berries globose, ovoid or broadly ellipsoid, comparatively smaller in size, not much acidic, 1-2 seeded and without rind	(4)
3b (1)	Petioles generally winged; flowers white to pinkish, fragrant; fruits globose or ovoid, yellow when ripe, rind tightly attached, pulp highly acidic	2. *CITRUS* L.
4a (3)	Leaves 1-5 foliolate, leaflets variable in size and shape, ovate lanceolate; berries globose pinkish or bluish when ripe	4. *GLYCOSMIS* Corr.
4b (3)	Leaflets alternate, elliptic, ovate or elliptic lanceolate; fruits ovoid or broadly ellipsoid blackish or reddish, when fully ripe	5. *MURRAYA* L.

1. *AEGLE* Correa *nom. cons.*

1.1. ***Aegle marmelos*** (L.) Correa, Trans. Linn. Soc. London 5: 223. 1800; Hook.f. Fl. Brit. India 1: 516.1875; Watt Dict. Econ. Prod. 1: 117 .1889; Prain Beng. Plants 1: 212. 1903 (Rep. ed. 1963), Rec. bot. Surv. India 2 (4): 292.1903 & Ibid.3 (2): 185.1905; Wlh. India 1: 33.1949. *Crataeva marmelos* L. Sp. Pl. 444. 1753. *(Vern Name: Bel).*

Medium tree, deciduous, spinous, aromatic tree. Leaves 3-foliolate; leaflets lanceolate or ovate, acuminate or rarely acute at apex, cuneate to obtuse at base, middle one longer than the later two. Flowers few, axillary cymose, panicles combined into a large terminal panicle. Fruit globose 10-25 cm diam. Pulp orange or yelloish brown. *Fls. & Frts.:* November-April.

Occasionally planted but in general as self-sown in the garden edges, village thickets or hedges.

Field Collection: Bhasila/Barasat, *K.N.-1576.*

Distributions: Cultivated throughout India, indigenous in India and Burma.

Economic Importances: Both ripe and green fruits are edible. Leaves, stems barks and roots are used in a number of very useful medicine. A valuable remedy for the cure of dropsy, beriberi, weakness of heart, bilious fever and all other biliary complaints. It is also useful as mild purgative. Unripe fruit is astringent and is believed to be a valuable remedy in dysentery.

2. *CITRUS* L.

Key to the Species

1a	Petioles mostly broad winged; fruits large	2. C. *maxima* Merr.
1b	Petioles naked or with wings seldom more than 1 cm wide; fruits medium to small	(2)
2a (1)	Young shoots usually purplish, petals pink	3. C. *medica* L.
2b (1)	Young shoots green or greenish white; petals white	1. C. *aurantifolia* Swingle

2.1. *Citrus aurantifolia* (Christm.) Swingle, J. Wash. Acad. Sci. 3: 465.1912; Benthal Trees Calc. 83.1946; Wlh. India 3:189.1952. *Limon aurantifolia* Christm., Linn. Pflanzensyst. 1: 618. 1777. *(Vern.Name: Kagzilebu).*

Shrub or small tree. Leaves alternate, elliptic, oblong or ovate; petiole winged. Flowers white; Fruits usually small, globose or ovoid, yellow when ripe, smooth; rind thin, tightly attached, pulp acidic. *Fls. & Frts.:* May to November.

Commonly cultivated in the orchard and on the pond sides, also planted as hedge plant, rarely as an garden escape.

Field Collection: Baruipur, *K.N.-2256.*

Distributions: Almost throughout India; probably a native of India and Tropical countries of the world.

Economic Importances: Fruits contains Vitamin-C and it is edible. Leaves oil is used in cosmetics, confectionery and flavouring foods.

2.2. ***Citrus maxima*** (Burm.) Merr., Interpret. Herb. Amb. 296.1917; Wlh. India 3: 191.1950. *Aurantium maximum* Burm., Auct. Herb. Amb. 16. 1755. *Citrus grandis* Osbeck, Dagbok Ostind. Res. 98.1757. *Citrus decumana* L. Syst. 508. 1754; Hook.f. Fl. Brit. India 1: 516.1874; Watt Dict. Econ. Prod. 2: 348. 1889; Prain Beng. Plants 1: 213.1903 (Rep. ed. 1963) & Rec. bot. Surv. India 3 (2): 185.1905. *(Vern. Name: Batabi Lebu).*

Small tree. Leaves large upto 24 cm long,; leaflets large, ovate- oblong, frequently emarginate, usually downy beneath; petiole broadly winged. Flowers large, white. Fruits globose or pear- shaped, rind thick. *Fls. & Frts.:* April to August.

Commonly cultivated in the orchards for its fruits.

Field Collection: Dewanganj, K.N.-2262.

Distributions: Almost throughout India; introduced in India from Java by Captain Shaddock. Native of Malaysia and Polynesia, but it is now cultivated in most tropical countries.

Economic Importances: Ripe fruits are edible and somewhat delicious. Used in flavouring beverages and confections.

2.3. ***Citrus medica*** L. Sp. Pl. 782.1753 var. ***acida*** Brandis, For. Fl. 52; Hook.f.Fl. Brit. India 1: 515. 1875; Watt Dict. Econ. Prod. 2: 349.1889; Prain Beng. Plants 1: 213.1903 (Rep.ed.1963) & Rec.bot. Surv. India 3(2): 185.1905; Benthall Trees Calc. 85.1946. *C. bergamia* Dalz. & Gibs. Suppl. 13. *(Vern. Name: Patilebu).*

Shrub, young shoots glabrous. Leaves 1-foliolate, eliptic oblong. Racemes short; flowers small; petals usually 4. Fruits small, globose or ovoid; rind thick or thin, pulp very acidic. *Fls. & Frts.:* May-January.

Commonly planted in the orchard of the villages.

Field Collection: Harinbari/Sagar Island, *K.N.- 474.*

Distributions: Throughout India.

Economic Importance: Fruits contain high value of Vitamin C and Citric acid, which is very useful in digestive purposes. Used for flavouring confectionery.

2.4. ***Citrus medica*** L. Sp. Pl. 782.1753 var. ***limonum*** (W. & A.) Hook. f. Fl. Brit. India 1: 515.1875; Prain Beng. Plants 1: 213.1903 (Rep. ed. 1963) & Rec. bot. Surv. India 3 (2): 185.1905. *(Vern. Name: Karna Lebu).*

Shrub usually armed with spines. Leaves alternately, leaflets ovate, petiole distinct. Stipules absent. Calyx imbricate. Flowers pinkish. Fruits medium, oblong or globose. *Fls. & Frts.:* February-May.

Occasionally planted in the orchrd.

Field Collection: Brace Bridge, *K.N.-2234.*

Distributions: Almost throughout India.

Economic Importance: Fruits contain vitamin-C and edible.

3. *FERONIA* Correa

3.1. ***Feronia limonia*** (L.) Swingle, J. Washinton Acad.Sci. 4: 328. 1914; *Schinus limonia* L. Sp. Pl. 389.1753. *Limonia acidissima* L. Sp. Pl. (ed.2) 554. 1762; Airy Shaw, Kew Bull. 293. 1939. *Feronia elephantum* Corr., Trans. Linn. Soc. 5: 225.1800; Hook.f.Fl.Brit.India 1: 516. 1875; Prain Beng. Plants 1: 212.1903 (Rep. ed. 1963) & Rec. bot. Surv. India 3(2): 185.1905. *(Vern.Name: Kathbel).*

Large tree, armed with spines. Bark shallowly furrowed spines strong, straight. Leaves alternate and in axils 2-4 in clusters glabrous; leaflets 3-9, obovate, often retuse at apex. Penicles many flowered. Pericarp woody. *Fls. & Frts.:* November-January.

Generally self-sown, sometime planted in the garden edges and as avenue trees.

Field Collection: Gadkhali, *K.N.-845.*

Distributions: Throughout the drier parts of India and Europe.

Economic Importances: Fruits are edible but a low quality food. Stem and branch gum is used as substitute for gum-arabic. Wood used for fuel and carpentary works.

Notes: The correct nomenclature should be *Limonia acidissima* L. suggested by many authors (Guha Bakshi, 1984).

4. *GLYCOSMIS* Correa

Key to the Species

1a. Leaflets entire; inflorescence short; ovary fusiform, stalked; fruit glandular-pitted ... 1. *G. mauritiana* Tanaka

1b. Leaflets crenate serrate; inflorescence longer than leaf rachis; ovary ovoid, sessile; fruit glandular mammillate ... 2. *G. pentaphylla* Corr.

4.1. ***Glycosmis mauritiana*** (Lamk.) Tanaka, Bull. Soc. Bot. France 75: 708. 1928. *Limonia mauritiana* Lamk., Encycl. 3: 517.1792. *L. pentaphylla* auct. non Retz. 1788; Pl. Cor. 1: 60.t.84.1798. *Glycosmis pentaphylla* DC., Prodr. 1: 538. 1824; Hook.f. Fl. Brit. India 1: 499.1874. *G. pentaphylla* (Retz.) Correa (in correctly given by Narayanswami, Rec. bot. Surv. India 14(2): 12.1941). *G. triphylla* Wight, Bot. Misc. 3: 298, t. 39.1833. *(Vern Name: Sheora).*

Evergreen shrub. Leaves 3-5 foliate; leaflets variable in size and shape, ovate lanceolate, ovate or obovate, obtuse acute or acuminate, entire or obscurely crenate. Flowers small, white, fragrant. Berries globose, pink or bluish when ripe. *Fls. & Frts.:* Throughout the year.

Very common throughout the moist places.

Field Collections: Baruipur, *K.N.-15;* Sarisa, *K.N.-387.*

Distributions: Common throughout the evergreen forest in India; Malaya, Sri Lanka, China, Borneo and Australia.

Economic Importance: The twigs are used as tooth picks and also used for ripening fruits.

4.2. ***Glycosmis pentaphylla*** Correa, Ann. Muss. Par. 6: 386.1905; Hook.f. Fl. Brit. India 1: 499.1874; Watt Dict. Econ. Prod. 3: 512.1890; Prain Beng. Plants 1: 208.1903 (Rep. ed. 1963) & Rec. bot. Surv. India 2(4): 291.1903; Wlh. India 4: 150.1959; Mitra & Subramanyam, J. Arn. Arb. 50: 155. 1969. *Limonia pentaphylla* Retz, Obs. Bot. 5: 24.1788. *L. arborea* Roxb. Pl. Coromand. 1: 60.Pl. 85.1798. *Glycosmis arborea* (Roxb.) DC. Prodr. 1: 60. Pl.85. 1798 (in correctly given by Narayanaswami, Rec. bot. Surv. India 14(2): 20.1940); Brizick, J. Arn. Arb. 43: 90.1962. *(Vern. Name: Arsaora).*

Shrub or small tree. Leaves 1-5 foliate, leaflets sub-opposite or alternate, oblong or elliptic, obtuse or acuminate, petiolate. Flowers in erect pubescent, puberulous, cymose panicles of white slightly fragrant, minute. Berry globose, cream or light or pinkish white. *Fls. & Frts.:* All the year long.

Commonly grow on the waste lands, hedges, thickets, road-sides.

Field Collection: Shyamnagar, *K.N.-2079.*

Distributions: Throughout India; Indo-China and Malaya.

Economic Importances: Branches used as tooth brushes by the villagers. Fresh leaves is most powerful anti-malarial tonic and can also be used with full confidence in case of biliary colic, worm-colic, diarrhoea and dysentery. Fruits are edible and juice extracted from leaves are used in various medicine as for fever, liver troubles, vermifuge. Leaves paste with ginger is said to cure eczema and skin affections.

5. *MURRAYA* L. *nom. cons.*

Key to the Species

1a.	Leaves aromatic; leaflets pubescent; berries black	1. *M. koenigii* (L.) Spreng.
1b.	Leaves not aromatic; leaflets glabrous; berries red	2. *M. paniculata* (L.) Jack.

5.1. ***Murraya koenigii*** (L.) Spreng., Syst. 2: 315.1826; Hook.f. Fl. Brit. India 1:503. 1875; Prain Beng. Plants 1: 209. 1903 (Rep. ed. 1963) & Rec. bot. Surv. India 3 (2): 185.1905. *Bergera koenigii* L. Mant. 1: 565.1767. *(Vern. Name: Barsanga).*

Small tree or shrub. Leaflets alternate, elliptic ovate or elliptic lanceolate or rhomboid, acuminate at apex. Flowers in much branched terminal short peduncled, corymbose cyme. Fruits ovoid or subglobose, apiculate, 2-seeded, puncate, black when fully ripe. *Fls. & Frts.:* February-September.

Mostly cultivated, but sometime as an garden escape.

Field Collection: Panihati/Sodepore, *K.N.-1612.*

Distributions: Native of India, distributed throughout India.

Economic Importance: Leaves are mostly used for flavouring curries; leaves are also useful for dysentery and in skin diseases.

5.2. ***Murraya paniculata*** (L.) Jack. Mal. Misc. 1 (5): 31.1820; Benthall Trees Calc. 73.1946. *Chalcas paniculata* L. Mant. 1: 68.1767. *Murraya exotica* L. Mant. Alt. 2: 563.1771 *(Murraea);* Hook.f.Fl. Brit. India 1: 502.1875; Prain Beng. Plants 1: 209.1903 (Rep. ed. 1963) & Rec. bot. Surv. India 3: 184.1905. *(Vern. Name: Kamini).*

Shrub or small tree; much branched. Leaflets 3-7, ovate or elliptic or elliptic ovate, obtuse or slightly elongated into an obtuse tip, shining above, petiolate. Flowers in terminal corymbose cymes, fragrant, milky white. Fruit berry, red or orange, ellipsoid or broadly ellipsoid, 1-2 seeded. *Fls. & Frts.:* March-July.

Commonly cultivated for its highly scented flowers and as ornamental foliages.

Field Collection: Birati, *K.N.-615.*

Distributions: Almost throughout India; pantropic.

Economic Importances: Ornamental plants for its dense spreading branches and sweet scented flowers; in Burma the bark used as cosmetic making and considered to be cure for snake bite. Leaves are valuable in the treatment of dysentery.

Notes: Atalanta monophylla DC. is another cultivated species under the family-Rutaceae.

XXXIV. SIMAROUBACEAE

1. *AILANTHUS* Desf. *nom. cons.*

1.1. ***Ailanthus excelsa*** Roxb. Cor. Pl. 1: 24, t.23.1795; Bennett, Hook.f. Fl. Brit. India 1: 513.1875; Prain Beng. Plants 1: 224. 1903 (Rep. ed. 1963) & Rec. Bot. Surv. India 3(2): 215.1905; Wlh. India 1: 41.1948. *(Vern. Name: Madala).*

Large, deciduous handsome tree. Leaves large, crowded near the ends of the branches, alternate; leaflets ovate or falcate lanceolate, unequal at base, coarsely toothed, often lobed. Flowers yelloish in large lax often much branched panicles. Samaras about 5 cm long, tapering at each end, twisted at the base, copper-red. *Fls. & Frts.:* February-May.

Fairly common as avenue tree.

Field Collection: Titagarh, *K.N.-2208.*

Distributions: Bihar, Chota Nagpur, Gujarat and the forest of Ganjam; South Asia and Australia.

Economic Importance: The extracts of the boiled leaves is used for washing wounds and skin eruptions.

XXXV. OCHNACEAE

1. *OCHNA* L.

1.1. ***Ochna squarrosa*** L. Sp. Pl. 735.1762; Watt Dict. Econ. Prod. 5: 439.1891; Bennett, Hook.f. Fl. Brit. India 1: 523.1875; Wlh. India 7: 76.1966.

Large shrub or small tree; bark smooth, brown. Leaves elliptic or elliptic lanceolate, parallel nerved. Flowers in umbellate panicles large, yellow, fragrant. Fruits compound 3-6 drupes. *Fls. & Frts.:* February-April.

Frequently planted in parks and fairly common in the hedges and the gardens.

Field Collection: Titagarh, *K.N.-2222.*

Distributions: Almost throughout India; Burma, Sri Lanka and Peru.

Economic Importances: The bark is a digestive tonic; boiled leaves are used as an emollient cataplasm. Root decoction is said to be useful in menstrual complaints and asthma.

XXXVI. BURSERACEAE

1. *GARUGA* Roxb.

1.1. ***Garuga pinnata*** Roxb. Pl. Cor. 3: 5.t. 208.1811; Watt Dict. Econ. Prod. 3: 483.1890; Bennett, Hook.f. Fl. Brit. India 1: 528.1875; Prain Beng. Plants 1: 216.1903 (Rep. ed. 1963) & Rec. bot. Surv. India 3 (2): 186.1905. *(Vern. Name: Nilbhadi).*

Tree, upto 15 m high. Leaflets 5-10 pairs, sparsely pubescent on the nerves beneath, ovate lanceolate or oblong lanceolate, acute to rounded or truncate, often oblique at base. Flowers yellow. Drupe black. Seeds with numerous winged. *Fls. & Frts.:* March-July.

Rarely found on the village jungles.

Field Collection: Dhopagachhi/Baruipur, *K.N.-2137.*

Distributions: Throughout India. Malayan Archipelago and Philippines.

Economic Importances: Leaves possess cooling and digestive properties; said to be used in asthma.

XXXVII. MELIACEAE

Key to the Genera

1a	Ovule 1 or 2 in each cell, seed not winged	(2)
1b	Ovule numerous, seed winged	(7)
2a (1)	Staminal tube complete, oblong or linear	(3)
2b (1)	Staminal tube complete, globose	(5)
3a (2)	Leaves pinnate	(4)
3b (2)	Leaves bi- or tri-pinnate, leaflets serrate, 2.5 - 5.0 cm broad	5. *MELIA* L.
4a (3)	Leaflets mostly serrate, 1.0-2.5 cm long and 2 cm broad	3. *AZADIRACHTA* Juss.
4b (3)	Leaflets coarsely serrate, 4.0-7.5 cm long and 2.0-3.5 cm broad	4. *CIPADESSA* Blume
5a (2)	Littoral trees	(6)
5b (2)	Non littoral; flowers polygamous	2. *APHANAMIXIS* Blume
6a (5)	Leaflets 3-7; flowers bisexual	8. *XYLOCARPUS* Koen.
6b (5)	Leaflets 5-9; flowers unisexual	1. *AGLAIA* Pellegrin
7a (1)	Staminal tube complete	6. *SWIETENIA* L.
7b (1)	Staminal tube distinct, ovary 4-6; 6 celled	7. *TOONA* Roem.

1. *AGLAIA* Pellegrin

1.1. ***Aglaia cucullata*** (Roxb.) Pellegrin. *Amoora cucullata* Roxb. Corr. Pl. 3: 54, t.258.1819; Hiern. Hook.f. Fl. Brit. India 1: 560.1875; Watt Dict. Econ. Prod. 1: 224.1889; Prain Beng. Plants 1: 221.1903 (Rep. ed. 1963) & Rec. bot. Surv. India 2(4) 292.1903, Naskar & Guha Bakshi, Mang. Swam. Sundarbans 136.1987. *(Vern. Name: Amur).*

Medium tree 10-15 m high. Leaves imparipinate, 30-40 cm long; leaflets 2-4 pairs and an odd one, opposite or subopposite, 7.5- 15.0 cm X 4.0-7.0 cm, oblong-elliptic, sub-falcate, sub-acute, glabrous on both surfaces. Male flowers in axillary branched, peduncles long, pedicels short. Calyx 3-lobed, lobes rounded, ciliolate; petals-3, longer than the calyx. Staminal tube scarcely shorter than the petals, obovoid; rudimentary ovary stalked, ovoid-oblong. *Fls. & Frts.:* October-January.

Sometime on the mangrove forests of the Sundarbans, along the side of the tidal canals and creeks; a swamp forest species with aerial pneumatophores, the characteristic features of the mangrove flora.

Field Collection: Henry Island/Bakkhali, *K.N.- 1916.*

Distributions: Tidal influenced forests in Sundarbans, Burma and Sri Lanka.

Economic Importance: Produces fire-woods and timbers.

2. *APHANAMIXIS* Blume

2.1. ***Aphanamixis polystachya*** (Wall.) Parker., Indian For. 57: 486.1931. *Aglaia polystachya* Wall., Roxb. Fl. Ind. 2: 429.1824. *Andersonia rohituka* Roxb. (Hort. Beng. 87.1814), *nom. nud.* & Fl. Ind. 2: 213.1832. *Amoora rohituka* (Roxb.) Wt. & Arn. Prodr. 1: 119. 1834; Hiern, Hook. f. Fl. Brit. India 1: 559.1875; Watt Dict. Econ. Prod. 1: 224.1889; Prain Beng. Plants 1: 220. 1903 (Rep. ed. 1963) & Rec. bot. Surv. India 3: 186.1905. *(Vern. Name: Tikta Raj).*

Small trees, 6-10 m tall. Leaves 30-60 cm long; leaflets opposite, 4-8 pairs, accrescent upwards, glabrous, acuminate at apex, 5-8 mm. Male flowers numerous in axillary or supra axillary panicled spikes, globular, upto 4 mm long, bracteate sessile or subsessile. Female flowers larger than the male, in axillary or supra-axillary, solitary spikes. Capsule 3-valved. Seeds oblong, with a scarlet aril. *Fls. & Frts.:* September-December.

Occasionally cultivated, very casually found as an garden escape; sometimes in the thickets and village shrubberies.

Field Collection: Roynagar/Barasat, K.N. 2890.

Distributions: Western Peninsula, Assam and Oude; Sri Lanka, Malaya and Philippines.

Economic Importance: Seed oil is used as liniment in rheumatism, bark is used in enlarged glands, livers and spleen.

3. *AZADIRACHTA* A. Juss.

3.1. ***Azadirachta indica*** A.Juss., Mem. Mus. Hist. Nat. Paris 19: 221, t. 2, f.5.1830; Wlh. India 1: 140.1948. *Melia azadirachta* L. Sp. Pl. 385. 1753; Hiern. Hook.f. Fl. Brit. India 1: 544.1875; Watt Dict. Econ. Prod. 5: 221.1892;

Prain Beng. Plants 1: 218.1903 (Rep. ed. 1963) & Rec. bot. Surv. India 3 (2): 186.1905. *(Vern. Name: Neem).*

Large tree. Leaves crowded near the ends of branches, imparipinnate; leaflets serrate, fluctely lanceolate, accuminate, glabrous, dark green above, pale beneath. Flowers white, scented in numerous large panicles; anthers 10; ovary 3-celled. Drupes ovoid-oblong, smooth, yellow when ripe. *Fls. & Frts.:* March-June.

In the road sides and village shrubberies, every where.

Distributions: Throughout India, it is considered as a species of India. Tropical zones of the World and many others thought that a native of Burma.

Economic Importances: A medicinal oil is extracted from the seeds; which is valuable as an anitseptic and a reputed medicine for skin affection and possesses parasitotropic properties. Its anti-spirochaetal properties have been tried with beneficial results in syphilis-primary and secondary. The juice from the leaves has marked anthelmintic properties. Neem oil is commonly utilized in the manufacture of medicine soap; used as powerful anti-malarial medicine and in rheumatic pains in various parts of the body; the neem oil cake is also used as insecticides. Neem leaves paste is also applied on the body for curing skin diseases.

4. *CIPADESSA* Blume

4.1. ***Cipadessa fruticosa*** Blume, Bijdr 162.1825; Hiern, Hook. f. Fl. Brit. India 1: 545. 1875; Prain Beng. Plants 1: 219.1903 (Rep. ed. 1963). *Ekebergia indica* Roxb. Fl. Ind. 2: 302. 1824. *Mallea rothii* A. Juss. Mem. Mus. Par. 19: 222, t.13.f.6.1930.

Small, much branched shrub, young parts silky. Leaves imparipinnate, 10-25 cm long, leaflets 4-6 pairs and an odd one, opposite, elliptic lanceolate, acute at both ends, very coarsely serrate, petioles short, slender. Flowers small, white in axillary corymbose panicles; peduncles long, slender, pedicels short. Calyx pubescent outside, 5 toothed short, triangular; petals membranous, oblong, acute, pubescent outside; stamens shorter than the petals; filaments hairy inside. Ovary glabrous, style short; stigma capitate, shortly 5 lobed. Berry globose, 5- lobed, scarlet. Seeds angular. *Fls. & Frts.:* April-August.

Very rare, found in the railway tracks and waste lands.

Field Collection: Baguihati/Kestopur, *K.N.-1885.*

Distributions: Almost throughout India; in Sri Lanka and Java.

Notes: D. Prain (1903a,) mentioned the locality for this taxon from Chota Nagpur and Orissa only.

5. *MELIA* L.

5.1. *Melia azedarch* L. Sp. Pl. 384.1753; Hiern, Hook.f. Fl. Brit. India 1: 544.1875; Watt Dict. Econ. Prod. 5: 211.1892; Prain Beng Plants 1: 218.1903 (Rep. ed. 1963) & Rec. bot. Surv. India 3(2): 186.1905; Wlh. Ind. 6: 323.1962 *(Vern. Name: Maha Neem).*

Medium, deciduous tree with stout trunk and spreading crown. Leaves bipinnate occasionally 3-pinnate, leaflets serrate, ovate or lanceolate, acuminate. Flowers lilac blue and purple, fragrant in lax panicles of cymes; ovary 5-celled. Drupes globose, more or less dry, yellow and wrinkled at maturity. *Fls. & Frts.:* January-May.

Occasionally planted as avenue tree or in the railway stations.

Field Collection: Rabindra Sarabar, *K.N.-2245.*

Distributions: An introduced cultivated plants almost throughout India, now it is naturalized almost everyhere. Native of Persia.

Economic Importance: The leaves, barks and fruits are accredited with insect-repellent properties.

6. *SWIETENIA* L.

Key to the Species

1a.	Leaflets 5.5cm long or more; capsule 15 cm long or more	1. *S. macrophylla* King
1b.	Leaflets 5 cm long or less; capsule 10 cm long or less	2. S. *mahagoni* Jacq.

6.1. ***Swietenia macrophylla*** King, Hook. Icon. Pl. 16.t.1550.1886; Prain Beng. Plants 1: 223 1903 (Rep. ed. 1963) & Rec. bot. Surv. India 3(2): 187.1905; Wlh. India 10 : 81.1976. *(Vern. Name: Bara Mahagoni).*

Lofty 30-40 m high, evergreen tree; trunk straight, cylindrical; bark greyish brown. Leaves paripinnate, upto 0.6 m long; leaflets 3-8 pairs, generally 4, dark green and shining when mature. Flowers greenish, sweet scented in narrow supra axillary panicles. Capsule large, inverted club like. *Fls. & Frts.:* April- May.

Planted as avenue tree and on the railway stations.

Field Collection: Santoshpur/Budge Budge, *K.N. 1957.*

Distributions: Introduced into the Royal Botanic Garden, Calcutta in 1795 from West Indies; now commonly planted almost throughout India. Native of Tropical America and Jamaica.

Economic Importances: Valuable timber yielding plant.

6.2. ***Swietenia mahagoni*** Jack. Enum. Pl. Carib. 20.1760; Watt Dict. Econ. Prod. 6(3): 393.1893; Prain Beng. Plants 1: 222.1903 (Rep. ed. 1963) & Rec. bot. Surv. India 3 (2): 187.1903; Wlh. India 10: 84.1976. *(Vern. Name : Mahagony).*

Large evergreen tree, 25-35 m tall. Leaves paripinnate; leaflets 2.5.-5.0 cm long, 2.5. cm broad, obliquely ovate or lanceolate, entire, glabrous. Flowers greeish yellow, in axillary panicles. *Fls. & Frts.:* April - May.

Largely planted as avenue tree.

Field Collection: Majerhat, *K.N-2246.*

Distributions: Planted almost throughout India; an introduced species and first sown in the Indian Botanic Garden at Calcutta. Central America is its Native land.

Economic Importance: Valuable timber yielding plant.

7. *TOONA* (End.) Roem.

7.1. ***Toona ciliata*** Roem., & Syn. Hesp. 139.1846. *Cedrela toona* Roxb. ex Rottb. & Willd. Ges. Naturf. Fr. Berlin Neue Schr. 2 : 188. 1803; Hiern. Hook. f. Fl. Brit. India 1: 568.1875; Prain Rec. bot. Surv. India 3(2): 187.1905; Wlh. India 2 : 104.1950. *(Vern. Name : Tunt).*

Large 25 m tall, handsome, deciduous tree. Leaves paripinnate; leaflets 5-10 cm long, lanceolate or ovate lanceolate, acuminate, entire, slightly undulated, often oblique at the base. Flowers white, scented in short terminal panicles; stamens 5; ovary 5 celled. Capsule oblong, smooth; seeds reddish brown, winged at both ends. *Fls. & Frts.:* March-July.

Planted everywhere and often self-sown in the village shrubberies.

Field Collection: Chandpara/Gobardanga, *K.N.-1670.*

Distributions: Throughout India, native of India. Burma, Malaya and Australia.

Economic Importance: The bark is astringent and antiperiodic; it is used for chronic infantile dysentery; also used in as an external application for ulcers.

8. *XYLOCARPUS* Koen.

Key to the Species

1a. Medium tree; leaflets ovate or broadly or with pointed apex; fruit upto 12 cm diameter ... 2. *X.mekongensis* Pierre

1b. Small- medium tree; leaflets obovate to elliptic oblong with rounded apex; fruit upto 25 cm diameter ... 1. *X. granatum* Koeng.

8.1. ***Xylocarpus granatum*** Koeng. Naturforscher 20:2.1784; Naskar & Guha Bakshi, Mang. Swms. Sundarbans 137.1987. *Carapa obovata* Blume Bijdr. 179.1825; Hiern, Hook.f. Fl. Brit. India 1: 567. 1875; Prain Beng. Plants 1. 222.1903 (Rep.ed. 1963) , Rec. bot. Surv. India 2 (4): 293.1903 & Ibid. 3 (2): 192.1905. *Xylocarpus obovatus* A. Juss. Mem. Mus. Par 19: 244. 1830. *(Vern. Name: Dhundul)*. (Fig. 9).

Medium tree, trunk generally single, bark smooth, pale green or yellowish brown. Buttress formed from the base of the trunks; undulated ribbon like surface roots . First leaves are simple but the mature leaves are paripinnate with short corky petiole; leaflets vary in shape and size, 6.0 x 12 cm; apex rounded or blunt and occasionally slightly emarginate. Panicle 4-7 cm long, stout, few flowered, spreading and rather irregularly branched, infloresence axis. Flowers are unisexual, male flowers with more slender ovary, stamen with functional pollen; the female flowers have a more swollen ovary with empty stamens. Ovules few and larger in female flowers. Fruit up to 25 cm in diameter with 8-10 seeds. Seeds angular with thick corky testa; embryo with plumule and radicle. Seed germination starts during its floating condition in the tidal water. *Fls. & Frts.:* July- December.

Very common on the side of the tidal creeks, river sides and intertidal mangrove forests of the Sundarbans.

Field Collection: Satzelia/Sundarbans, *K.N.-1555.*

Distributions: In the tidal forests of the Sundarbans, Andaman and Nicobar Islands and other mangrove forests. In the coastal sandy, rocky or typically mangrove environment of the Old World Tropics, from East Africa to Australia and Polynesia.

Economic Importances: Timber is reddish brown, hard and having anti-ants properties; therefore it is much used for carpentary works and boat building purposes. Bark has tannin properties.

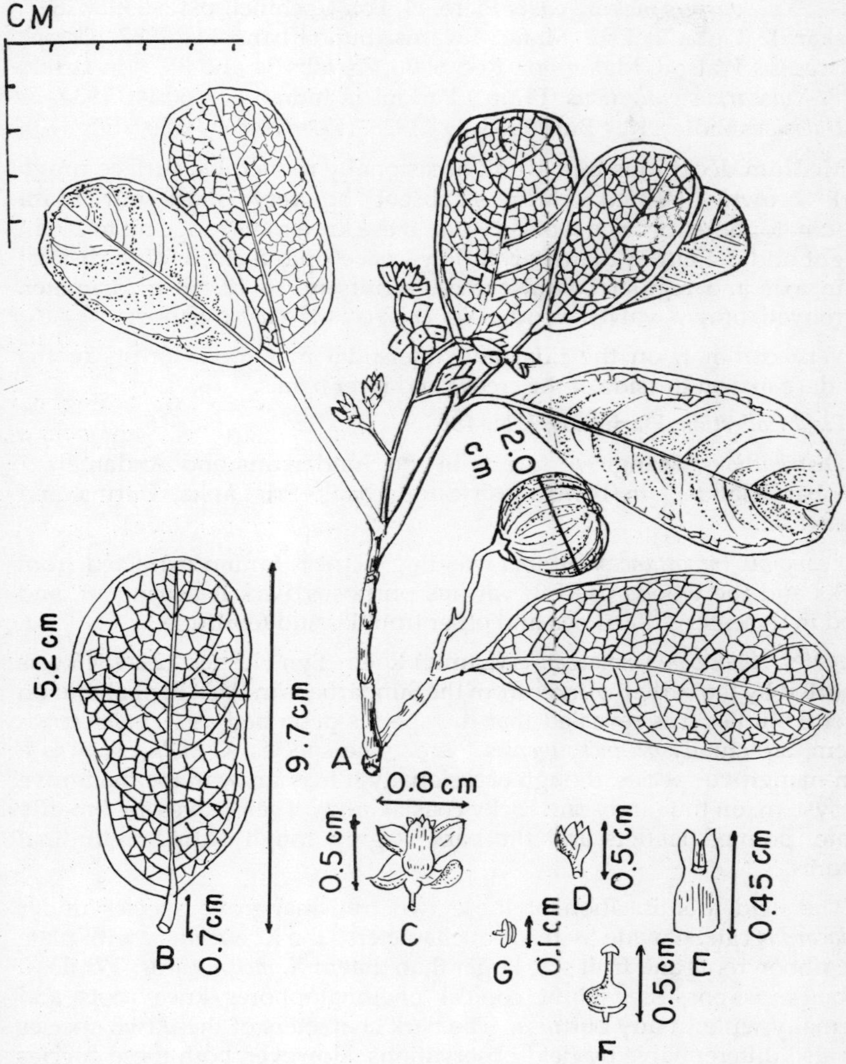

Fig.9. XYLOCARPUS GRANATUM Koenig.
A. Flowering twig, B. Single leaf, C. Single flower,
D. Calyx tybe, E. Petal with anther lobes, F. Gynoecium.

8.2. ***Xylocarpus mekongensis*** Piere, Fl. For. Cochinch plate 359B. 1897; Naskar & Guha Bakshi Mang. Swams. Sundarbans 71. 1987. *Carapa moluccensis* Watson, Malay For. Rec. 6: 70, 75, tabs 34 and 35, non Lamk. 1928. *Xylocarpus gangeticus* (Prain) Parkinson Indian For. 60:140.1934. *X. australasicus* Ridley, Key Bull. 1938: 191.1938. *(Vern. Name: Pasur)* (Fig. 10).

Medium deciduous tree, trunk occasionally numerous, surface rough dark brown, fissured. Buttress absent or very short but blunt pneumatophores or knee roots present; these knee roots range up to 20 cm height and 3-4 cm in diameter. Inflorescence upto 8 cm, with a distinct main axis and regular lateral cymes. Fruits up to 12 cm in diameter. 4-grooved, brown, valves woody. *Fls. & Frts.:* March-November.

Very common on the tidal and inter-tidal mangrove forests of the Sundarbans; on the side of the creeks and river banks.

Field Collection: Harinbari, *K.N.-486.*

Distributions: Mangrove forests in the Sundarbans and Andaman & Nicobar Islands; in Bangladesh Sundarbans, Sri Lanka, Burma and Malayasia.

Economic Importances: Timber yielding plants; tannin extracted from barks and the woods used in various purposes; bark is astringent, and used in dysentery, diarrhoea and other troubles and febrifuge.

Notes: Xylocarpus moluccensis (Lamk.) Roem. Syn Hesper. 124. 1846 was reported by the earlier worker from the Sundarbans in West Bengal. But on careful studies it is revealed that the species described as *X. moluccensis* Roem. is actually *X. mekongensis* Piere. The species *X. moluccensis* is a non-mangrove species, though occasionally it may grow in the mangrove ecosystem, on the sandy and rocky coastal lands. Their leaves are broadly ovate, pointed leaflets and the bark is grey, rough with longitudinal fissures.

The field identification of these two true mangrove species under *Xylocarpus* can be made by the root characters; the *X. granatum* with plate like ribbon roots and fruit size larger than that of *X. mekongensis.* While *X. mekongensis* possesses blunt conical pneumatophores/knee roots and normally without any buttress. The bark characters of these two species are also differentiable by field observations. However, both these species grow side by side in the intertidal mangrove forests and it is very difficult to differentiate until it is observed carefully.

The Chromosome number of *X. granatum* was reported n=21 (Taxon 27: 229 . 1978).

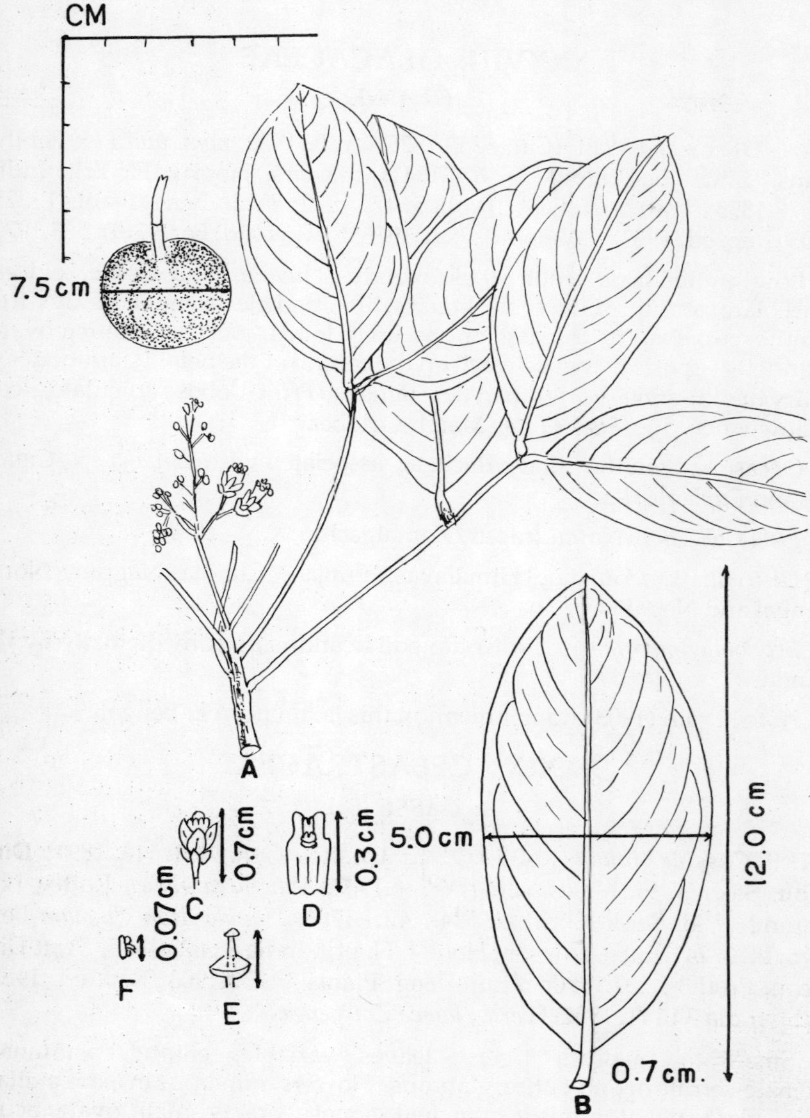

Fig. 10. XYLOCARPUS MEKONGENSIS Pierre.
A. Flowering twig, B. Single leaf, C. Single flower,
D. Petal with anther lobes, E. Ovary with gynoecium,
F. Anther lobes, G. Single fruit.

XXXVIII. OLACACEAE

1. *OLAX* L.

1.1. ***Olax nana*** Wall. (Cat. 6783.1828; A. B. 1832, *nom. nud.)* ex Benth., Trans. Linn. Soc. Lond. 18: 678.1841; Masters, Hook.f. Fl. Brit. India 1: 577.1875; Watt Dict. Econ. Prod. 5: 479.1892; Prain Beng. Plants 1. 227. 1903 (Rep. ed. 1963); Wlh. Ind. 7: 91.1966; Biswas Ind. For. 97(9): 508.1971.

Low undershrub, root stock woody. Leaves sub-sessile, oblong lanceolate, strong. Flowers solitary white, pedicles slender. Calyx free minute, cup shaped, truncate or obscurely lobed; petals 3, oblong ovate rounded at apex; stamens 3, each on the central of the petal; staminodes 6 ovary ovoid; style shorter than the stamens. Fruit globose apiculate, deep yellow when ripe. *Fls. & Frts.:* March-October.

Occasonally in the village thickets; associated with *Lantana* sp., *Croton* sp. and *Euphorbia* sp.

Field Collection: Narendrapur/Kamalgachhi, *K.N.- 1984.*

Distributions: Gujarat, Himalayas, Punjab, Chhota Nagpur, North Bengal and Nepal.

Economic Importances: Fruits are edible and is used medicinally by the Santals.

Notes: Prain (1903)did not mention this taxon from L. Bengal.

XXXIX. CELASTRACEAE

1. *CASSINE* L.

1.1. ***Cassine glauca*** (Rottb.) O. Kuntze, Rev. Gen. Pl. 1: 114. 1891; Din Hou, Steenis, Fl. Males. ser. 1, 6:286.1962. *Mangifera glauca* Rottb., Ny Samml. Vid. Selsk. Skrift. 2: 534.t..4.f.1.1783. *Elaeodendron glaucum* Per Syn. Pl. 1: 241.1805; Lawson, Hook f. Fl. Brit. India 1 : 623.1875; Watt Dic Econ. Prod.3 : 207.1890; Prain Beng. Plants 1: 230.1903 (Rep. ed. 1963 Whl. India 3 : 141. 1952. *(Vern. Name : Chikyeng).*

Small to medium tree; leaves variable, elliptic, acuminat crenate-serrate or sub-entire, glabrous. Flowers numerous in extra axillar paniculate dichotomously branched cymes; bracts small, ovate, acut Calyx deeply divided; lobes unequal, orbicular. Petals 0.5 cm long oblon obtuse distinct. Stamens shorter than the petals, anthers roundish. Drup obovoid. *Fls. & Frts.:* March-August.

Occasionally planted as a ornamental tree.

Field Collections: Bhangar, *K.N.610;* Kulti, *K.N.-2125.*

Distributions: Native of most of the hotter parts of India and Malaya.

Economic Importance: Timber is very hard and close grained; root is believed to be a remedy for snake bite and also in swellings.

XL. HIPPOCRATEACEAE

1. *SALACIA* L.

1.1. ***Salacia chinensis*** L. Mant. 3: 293.1763; Wlh. Ind. 9: 168. 1972. *S. prinoides* DC. Prodr. 1: 571.1824; Lawson, Hook. f. Fl. Brit. India 1: 626. 1875; Prain Beng. Plants 1: 232.1903 (Rep. ed. 1963) & Rec. bot. Surv. India 2 : 293.1903. *S. podopetala* Kurz. Bull. Soc. Nat. Mosc. 1: 581.1863. *(Vern. Name : Madhu Phal).*

Large straggling or climbing shrub; branches divaricate; young twigs angular, glabrous. Leaves thinly coriaceous, obtusely acuminate, distinctly serrate. Flowers yellow, from small tubercles in the axils of the leaves or from the smaller branches below the leaves; pedicels slender. Calyx cupular, persistent, 5-lobed, rounded; petals larger than calyx lobes. Disk convex, fleshy; stamens 3 from the face of the disk; ovary conical trigonous. Fruit sub-globose, minutely apiculate, fleshy, red when ripe, 1-celled, 1-seeded. *Fls. & Frts.:* November-March.

Sometimes on the river banks of the Sundarbans; climbing on the river side hedges or shrubby on small forest belt.

Field Collection: Herobhanga/Dhosha, *K.N.-1858.*

Distributions: Eastern and Western Peninsula; Bangladesh, Java and Philippines.

Economic Importance: Fruits edible and roots applied as an antidiabetic medicine.

XLI. RHAMNACEAE

Key to the Genera

1a.	Usually climber, unarmed with pinniveined leaves; fruits samaroid; 1-celled	1. *VENTILAGO* Gaertn.
1b.	Usually shrubby or small tree, armed with stipular prickles, with 3-veined leaves; fruit drupaceous; stone 1-4 celled	2. *ZIZIPHUS* Miller.

1. *VENTILAGO* Gaertn.

1.1. ***Ventilago madraspatana*** Gaertn., Fruct. 1: 223.t.49.f. 2.1788; Lawson, Hook.f.Fl. Brit. India 1: 631.1875; Watt Dict. Econ. Prod. 6(4): 227.1893; Prain Beng. Plants 1: 235.1903 (Rep. ed. 1963) & Rec. bot. Surv. India 3(2): 188.1905. *V. denticulata* Willd. Ges. Naturf. Fr. New. Schr. 3: 417.1801. *V. calyculata* Tulasne, Arn. Soc. Nat. 8: 124. 1857. *(Vern. Name: Rakta pita).*

Large woody climber with long sarmentose branches. Leaves bifarious, 4-8 X 2-4 cm, elliptic or oblong lancelolate, nearly entire, acute at the base; petiole 5-8 cm long; stipules small, subulate. Flowers small, fascicled in groups on large leafy pubescent panicles. Nut globose, 4 cm diam., yellow, wing 3-4 cm long, glabrous, shining, round at apex. *Fls. & Frts.:* December-April.

Very rare, occasionally climbing on the village thickets.

Field Collection: Dakkhin Barasat, *K.N.-1652.*

Distributions: Throughout India ; Sri Lanka and Malaya.

Economic Importances: Powdered root bark carminative, stomachic, stimulant, useful in chronic dyspepsia, debility and fever. Powder stem bark used for skin diseases and itches.

2. *ZIZIPHUS* Miller.

1a.	Small to medium sized tree; leaves glabrous above, woody beneath; flowers yellow or red; drupe 1.5-2.0 cm or longer	1. *Z.mauritiana* Lamk.
1b	Straggling to climbing shrub; leaves softly pubescent above, silky beneath; flowers smaller, black; drupe upto 7 mm long	2. Z. *oenoplia* (L.) Miller

2.1. ***Ziziphus mauritiana*** Lamk. Encycl. 3: 319.1789; Suessenguth, Pflanzenfam. (ed.2.) 20.d: 124.1953. *Z.jujuba* Lamk. Encycl. 3: 318.1789; Laswson, Hook.f. Fl. Brit. India 1: 632. 1875; Watt Dict. Econ. Prod. 6(4): 367.1893; Prain Beng. Plants 1: 234.1903 (Rep. ed. 1963) & Rec. bot. Surv. India 3: 188.1905. *(Vern. Name: Kul)*

Small to medium tree with spreading crown;young twigs and undersurfaces of the leaves densely rusty or grey tomentose. Leaves variable, ovate, ovate-oblong or sub-orbicular, obtuse, entire. Flowers greenish yellow in short axillary subsessile cymes. Drupes oblong-globose or ovoid. *Fls. & Frts.:* August- February.

Fairly abundant; rarely planted but generally it grows wild.

Field Collection: Ghoshpur/Kulti, *K.N.-1122.*

Distributions: Throughout India, mainly in the hotter parts. Indigenous in many parts of Asia and Australia.

Economic Importances: Acid fruits are commonly eaten by the villagers and sold in the markets. Fruits are said to be useful in fever and for wounds and ulcers. Barks used for tannin purposes and also used as a cure for diarrhoea. The branches are largely used for garden fencing and fodder. The root and stem bark have tonic properties and are used as a remedy for diarrhoea; leaves are valuable for the treatment of asthma, wounds and liver complaints.

2.2. ***Ziziphus oenoplia*** (L.) Miller, Gard. Dict. (ed. 8) No. 3.1768; Lawson, Hook. f. Fl. Brit. India 1: 634.1875; Watt Dict. Econ. Prod. 6(4): 371.1893; Prain Beng. Plants 1: 294.1903 (Rep. ed. 1963) & Rec. bot., Surv. India 3: 188. 1905. *Rhamnus oenoplia* L. Sp. Pl. 194.1753. *(Vern. Name: Siakul* or *Bitkul).*

Straggling or climbing shrub; prickles paired. Leaves obliquely ovate-lanceolate or ovate, acute, entire, pubescent above, brown silky beneath. Drupes globose or obovoid, black, shining. *Fls. & Frts.:* April-September.

Comon in the village shrubberies, roadside thickets, railway tracks, dry waste lands.

Field Collections: Masjidbati, *K.N.-849;* Garia, *G. Sengupta*-2.

Distributions: Throughout the hoter parts of India; Tropical Asia and Australia.

Economic Importances: The fruits are eaten by the poor villagers. The branches are largely used for fencing the village gardens. Decoction or root barks used to heal fresh wounds. Fruits as an ingredient of stomach-ache pills.

XLII. VITACEAE

(*nom. alt.* Ampelidaceae)

Key to the Genera

1a.	Tendrils arising from the peduncles	1. *AMPELOCISSUS* Pl.
1b.	Tendrils arising directly from the stem	(2)

2a. (1)	Leaves compound	2. *CAYRATIA* Juss.
2b. (1)	Leaves simple	(3)
3a.(2)	Berry globose, usually one seeded or few	3. *CISSUS* L.
3b. (2)	Berry rounded, usually many seeded	4. *VITIS* L.

1. *AMPELOCISSUS* Planch. *nom. cons.*

1.1. ***Ampelocissus latifolia*** (Roxb.) Planch. Journ. Vigne Amer. 374. 1884 & DC. Monogr. Plan 5(2): 370.1887; Suessenguth, Pfamilien 20D: 300.1953; Lawson, Hook.f. Fl. Brit. India 1: 652.1875; Watt Dict. Econ. Prod. 6(4): 255. 1893; Prain Beng. Plants 1: 337.1903 (Rep. ed. 1963) & Rec. bot. Surv. India 3 (2): 188.1905; Wlh. India 10:527. 1976. *Vitis latifolia* Roxb. Pl. Ind. 1: 661.1820. *(Vern. Name: Govila).*

Extensive climber. Stem hollow, striate. Leaves long petioled 8-10 cm long, orbicular cordate, 3-7 angled or lobed; lobes dentate, basal nerves 5. Peduncles long, bifurcate, one of the branches consisting usually of a slender forked tendrils. Flowers brown red, in small compact cymes. Berries succulent, black, 2-seeded. *Fls. & Frts.:* June-October.

In the open lands, hedges and shrubberies.

Field Collection: Hasanabad, *G.Sen Gupta-129* (Cal).

Distributions: Major parts of India; most of the Tropical areas.

Economic Importances: The berries are edible; the roots are applied to wounds. Plants are having insecticidal properties.

2. *CAYRATIA* A. Juss. *nom. cons.*

Key to the Species

1a.	Leaves pedately 5-foliate, glabrous	1. *C. japonica* (Thunb.) Gagn.
1b.	Leaves pedately 7-9 foliate	2. C.pedata (Lamk.) Juss.

2.1. ***Cayratia japonica*** (Thunb.) Gagn., Lecomte, Notul. Syst. (Paris) 1: 349.1911; Saldanha, Fl. Hassan 358.1976. *Vitis japonica* Thunb. Fl. Jap. 104.1784; Prain Beng. Plants 1: 238. 1903 (Rep. ed. 1963). *V.mollis* Wall. Cat. 6025; Lawson, Hook.f.Fl. Brit. India 1: 660. 1875.

Slender herb. Leaves 5-foliate; leaflets elliptic, serrate, glabrous, membranous. Flowers white. *Fls. & Frts.:* June- October.

Occassionally found on the waste lands and village shrubberies.

Field Collection: Sagar Islands, *A.K.Mukherjee- 4321* (CAL).

Distributions: Throughout India ; Bangladesh and Malaysia.

2.2. ***Cayratia pedata*** (Lamk.) Juss. ex Gagnep., Lecomte, Notual. Syst. 1: 346. 1910. *Cissus pedata* Lamk. Encycl. 1 : 31.1783. *Columella pedata* Lowr. Fl. Cochinch. 1: 85.1790. *Vitis pedata* (Lamk.) Wall. ex Wt & Arn. List. No. 6027.1831.1832; Lawson, Hook.f.Fl. Brit. India 1 : 661.1875; Prain Beng. Plants 1 : 238. 1903 (Rep. ed. 1963), Rec. bot. Surv. India 3(2) : 189.1905. *(Vern. Name : Goali Lata).*

Large, weak climber; pubescent or puberulous, tendril long, branching at the ends, softly hairy. Leaves compound 7-11 foliolate, petiole 10-15 cm long, leaflets 3.5 cm -20 cm X 2 cm-8 cm, central ones longer obliquely, elliptic-oblong, variable in shape, crenate or shallowly serrate; mid-nerve and lateral nerves prominent, petioles short. Flowers in pedunculate, spreading, sub-corymbose cymes, white. Berry 6-10 mm diam., sub-globose, often 4-lobed, cream coloured when ripe. Seeds usually 4, sometimes 3 or 2, smooth, hemispheric, flat. *Fls. & Frts.:* August-October.

Fairly common in the village shrubberies, hedges and thickets.

Field Collection: South Garia/Champahati, *K.N.- 831.*

Distributions: Major parts in India; Sri Lanka, Burma and Malaya.

Economic Importance: A good number of medicinal properties are attributed in the decoction of the plants.

3. *CISSUS* L.

Key to the Species

1a.	Stem quadrangular	2 C. *quadrangula* L.
1b.	Stem terete, not angular	(2)
2b (1)	Large climber, berry ovoid or globose	(3)
2b (1)	Scandent; berry pyriform, mucronate, 1-seeded	3. C. *repanda* Vahl
3a (2)	Berry ovoid, angular, pitted; 1 or 2 seeded	1. C. *adnata* Roxb.
3b (2)	Berry globose; 3-4 seeded	4. C. *trifolia* L.

3.1. ***Cissus adnata*** Roxb. Fl. Ind. 1: 423.1820. *Vitis adnata* (Roxb.) Wall. (Cat. 5998.1828) ex.Lawson, Hook.f. Fl. Brit. India 1: 649.1775; Watt Dict.

Econ. Prod. 6 (4): 251.1893; Prain Beng. Plants 1:235. 1903 (Rep. ed. 1963) & Rec. bot. Surv. India 3 (2): 188. 1905.

Large climber, young shoots brown tomentose; tendrils tomentose, branched. Leaves 10-20 cm X 8-11 cm wide, ovate, abruptly acuminate or rarely obtuse or rounded at apex, toothed, cordate at base, glabrous above, pubescent beneath in older leaves, tomentose peduncle; pedicels short; bracteoles small, early caducous. Berry obovoid, small on ripening changes from reddish brown to black. Seed 1 or rarely 2, angular, pitted. *Fls. & Frts* :May-August.

Rare in the shrubberies and hedges.

Field Collections: Garia, *K.N.323;* Saktinagar, *K.N.-546.*

Distributions: Fairly common in the semi-evergreen forests. Indomalayasia, Sri Lanka and Java.

3.2. ***Cissus quadrangula*** (*quadrangularis*) L. Mant. 1: 39.1767; Wlh. India 2: 184.1950; Backer & Brink, Fl. Java 2: 90 1965. *Vitis quadrangularis* (L.) Wall. ex Wt. & Arn. Prodr. 125. 1834; Lawson, Hook. f. Fl. Brit. India 1: 645.1875; Watt Dict. Econ. Prod. 6 (4): 256.1893; Prain Beng. Plants 1: 237. 1903 (Rep. ed. 1963) & Rec. bot. Surv. India 3 (2): 188.1905. *(Vern. Name: Harjora).*

Fleshy cactus like, jointed climber with 4 angled stem, internodes and a tendril at some of the nodes. Leaves short-petioled, cordate, broadly ovate or reniform, crenate-serrate, glabrous. Flowers greenish white in short glabrous umbellate cymes. Berries red, 1 seeded. *Fls. & Frts.:* July-November.

Fairly common in the hedges and thickets. Occasionally planted in the ornamental gardens. Leaves formed during monsoon and post monsoon months only.

Field Collection: Canning, *K.N.-169.*

Distributions: Almost throughout India; Java, Africa and in the Tropics.

Economic Importance: Leaves are used in stomach and digestive troubels. Stem juice is given internally and applied topically for fracture of bones.

3.3. ***Cissus repanda*** Vahl, Symb. 3: 18. 1874; Wlh.India 2 : 186. 1950. *Vitis rependa* Wt. & Arn. Prodr. 125. 1834; Lawson, Hook.f. Fl. Brit. India 1: 648.1875.

Scandent; branches terete, covered with tomentum when young. Leaves 5-14 cm X 7-16 cm, broadly ovate, cordate, acuminate, coarsely crenate-dentate. Peduncles leaf-opposed, solitary. Calyx truncate, hairy outside; petals 4, pink, triangular oblong. Berry pyriform, mucronate, 1-seeded. Seed obovoid, smooth. *Fls. & Frts.:* March-August.

Very rare on the waste lands and village shrubberies.

Field Collection: Shyamnagar, *K.N.-2096.*

Distributions: Throughout India and Sri Lanka, Bangladesh.

Notes: D. Prain (1903 and 1905) did not mention this taxon from the Lower Ganga Delta.

3.4. ***Cissus trifolia*** L. Sp. Pl. (ed. 2) 170.1763. *Cayratia trifolia* (L.) Domi, Biblioth. Bot. 89: 371.1827. *Vitis trifolia* L. Sp. Pl. 1: 203.1753; Prain Beng. Plants 1: 238.1903 (Rep. ed. 1963) & Rec. bot. Surv. India 3(2): 189.1905. *Vitis carnosa* Wall. Cat.6018.1832, *nom. nud. (Vern. Name: Amal Lata).*

Large creeping or twining shrub with angular branches; young parts usually clothed with grey pubescence. Leaves trifoliolate, leaflets variable, broadly ovate, ovate lanceolate or obovate, obtuse or acute, crenate-serrate; branched. Flowers greenish white, in umbellate cymes. Berries globose, 3-4 seeded. Seeds trigonous, rugose. *Fls. & Frts.:* July-October.

Field Collection: Diamond Harbour, *K.N.-396.*

Common in the thickets, hedges and forest areas, everywhere.

Distributions: Throughout the hotter parts of India; Sri Lanka, Burma, Malaya, Malacca and in the Tropics.

4. *VITIS* L.

4.4. ***Vitis vinifera*** L. Sp. Pl. 202 . 1753; Lawson, Hook.f. Fl. Brit. India 1: 652.1875; Watt Dict. Econ. Prod. 6(4): 4.258.1893; Wlh. India 10: 527. 1976. *(Vern. Name: Angur).*

Large, woody climber with bifid tendrils. Leaves glabrous, clothed beneath with grey deciduous tomentum, sub-orbicular, cordate more or less deeply 5-lobed, margins unequally out into acute tooth. Flowers greenish on slender pedicels, arrangd in umbel like cymes, which form pyramidal compound panicles. *Fls. & Frts.:* March - October.

Occasionally cultivated for its delicious fruits.

Field Collection: Nilgunge/Barrackpore, *K.N.-2109.*

Distributions: Common in the Deccan Peninsula. Most of the Tropical Countries.

Economic Importance: Fruits are highly delicious and nutritious.

Notes:: Prain (1903) has not mentioned this taxon from Bengal Proper.

XLIII. LEEACEAE

1. *LEEA* L.

Key to the Species

1a	Erect herb, stem 0.3-1.0 m high; leaves simple, broadly ovate or cordate; acute or acuminate; flowers white; berries globose	3. *L. indica* Merr.
1b	Erect shrub, stem 1-4 m high; leaves pinnately compound; flowers white or greenish white	(2)
2a (1)	Stem 1-4 m high, young shoot pubescent or terete; leaves mono-or tri-pinnate, irregularly serrate	(3)
2b (1)	Shrub; leaves 2-3 pinnate, glabrous; flowers greenish white in much branched terminal puberous and corymbose cymes	4. *L. macrophylla* Roxb.
3a (2)	Inflorescence short compact pubescent cymes; flowers white	1. *L.aquata* L.
3b (2)	Inflorescence much branched trichotomous, pubescent cymes flowers white	2. *L. crispa* van Royen

1.1. ***Leea aquata*** L. Mant. 1: 124.1767; Prain Beng. Plants 1: 239.1903 (Rep. ed. 1903) ; Wlh. India 4: 56.1962. *Leea hirta* Roxb. ex. Hornem Enum. Pl. Hoit. Hafin. 1: 131.1807; Lawson, Hook.f. Fl. Brit. India 1: 668. 1875; Watt Dict. Econ. Prod. 4: 617.1890. *(Vern. Name: Kak-jhanga).*

Shrub 1-4 m high; young shoots pubescent. Leaves bi-pinnate, pubescent; leaflets 2.5-4.0 cm X 7.5-16.0 cm, oblong, or ovate-oblong, acuminate, irregularly serrate. Flowers white, very shortly pedicelled, in short compact pubescent cymes; bracts large, broadly ovate, obtuse, caducous. Berry 0.6 cm diam., depressed globular. *Fls. & Frts.:* April-September.

Rarely found in the waste lands.

Field Collection: Salipur/Baruipur, *K.N.-2148.*

Distributions: Throughout India. Sikkim, Himalaya and Bangladesh.

Economic Importances: The whole plant possesses anti-tubercular properties. Leaves and young twigs are antiseptic and are applied to wounds.

Notes: D. Prain (1903 & 1905) did not mention this taxon from the lower Ganga Delta.

1.2. ***Leea crispa*** van Royen ex L. Syst. Nat. (ed.2) 627. 1767 & Mant. 124.1767. *Leea herpacea* Hamilton, Trans. Linn. Soc. 14: 228.1823. *L. aspera* Edgew., Trans. Linn. Soc. 20: 36.1846 (not of Wall.); Lawson Hook.f. Fl. Brit. India 1: 665. 1875; Clarke, Trim. J. Bot. 136.1881; Watt. Dict. Econ. Prod. 4: 616.1890; Prain Beng. Plants 1: 239.1903 (Rep. ed. 1963). *Leea edgeworthii* Santapau, Fl. Khandala 54.1953.

Shrub, 2-3 m high; stem terete. Leaves simple pinnate; leaflets 5-7, elliptic, more or less acuminate, more or less glabrous finely pubescent beneath, irregularly serrate, cordate at the base; main nerve 12-18 pairs. Flowers white, in much branched trichotomous pubescent cymes; bracts, linear subulate. Berry depressed globular, smooth, black when ripe. *Fls. & Frts.:* July-November.

Rarely on the waste lands of the village shrubberies.

Field Collection: Champadalia / Barasat, *K.N.-902.*

Distributions: Tropical Himalaya and South India.

Economic Importances: Berries are edible; root tubers are used for guineaworm and leaves applied to wounds. It is the best plant of lac insect.

Notes:: D. Prain (1903a, 1903b and 1905) did not mention this taxon from the lower Ganga Delta.

1.3. ***Leea indica*** (Burm.) Merrill, Philip. J. Soc. 14: 245.1919; Wlh. India 6: 56.1962. *Staphylea indica* Burm. Fl. Ind. 75,t.23.f.2.1768. *Leea sambucina* Will. Sp. Pl. 1: 1177.1798, *nom. illegit.;* Lawson, Hook.f. Fl. Brit. India 1: 666.1875; Watt Dict. Econ. Prod.4: 618.1892; Prains Beng. Plants 1: 239.1903 (Rep.ed. 1963), Rec. bot. Surv. India 2: 295.1903 & Ibid. 3(2): 188. 1905; Baker & Bakh. f. Fl. Java 2: 94.1965. *(Vern. Name: Kukur Jhiwa).*

Shrub, leaves compound 2-3 pinnate, glabrous, ovate, oblong-lanceolate, acute-acuminate, truncate cordate. Flower greenish white, stamens united. Berry 0.6 cm diam. *Fls. & Frts.:* August-December.

Occasionally on the village shrubberies.

Field Collection: Santoshpur, *K.N.-1960.*

Distributions: Throughout India, from Garhwal eastwards to Assam and Southwards to Kerala, also in Andaman & Nicobar Islands. Bangladesh and in the Tropics.

Economic Importances: Tender shoots are used as vegetable, fruits are also edible; leaves are used as green manure.

1.4. ***Leea macrophylla*** Roxb. ex Hornem. Hort. Hofn. 1: 231.1813; Lawson, Hook.f. Fl. Brit. India 1: 664.1875 (excl. Syn. *L. latifolia);* Watt Dict. Econ. Prod.4: 617.1890; Prain Beng. Plants 1: 240. 1903 (Rep. ed. 1963) & Rec. bot. Surv. India 3(2): 189.1905; Wlh. India 6:56.1962. *(Vern. Name.: Dhol Sumudra).*

Erect herb 0.3-1.0 m high; root tuberous, perennial. Leaves simple, broadly ovate, cordate, acute or acuminate, dark green and glabrous above, nerve 8-10 pairs, petioles 5-10 cm long deeply striate glabrous. Flowers white, in terminal much branched puberulous corymbose cymes; buds oblong; peduncles deeply grooved. Petals oblong; staminal tube deeply divided. Berry 0.6 -0.8 cm diam., black, 3-6 celled, depressed globose , usually 3-6 lobed *Fls. & Frts.:* June - October.

Common on the dry waste lands, roadsides and railway tracks.

Field Collection: Jugdia/Benipur, *K.N.-220.*

Distributions: Throughout the hotter parts of India; in the Tropics.

Economic Importances: Root is said to be used in native medicines, for the cure of guineaworm. Leaves and furits are edible.

XLIV. SAPINDACEAE

Key to the Genera

1a	Flowers petalous	(2)
1b	Flowers apetalous, very rarely petalous	(4)
2a (1)	Herbs or shrubs	(3)
2b (1)	Erect tree	6. *SAPINDUS* L.

3a (2)	Climbing herb	2. *CARDIOSPERMUM* L.
3b (2)	Glabrous or pubescent shrub	2. *ALLOPHYLUS* L.
4a (1)	Leaves simple	4. *DODONAEA* L.
4b (2)	Leaves compound	(5)
5a (4)	Calyx 4-6 lobes; fruits smooth or prickly	7. *SCHLEICHERA* Willd.
5b (4)	Calyx nearly truncate or shallowly toothed; fruits tubercled	(6)
6a (5)	Petals O; fruits covered with prominant tubercles	5. *LITCHI* Sonner
6b (5)	Petals 4-6; fruits with indistinct flattened tubercles	3. *DIMOCARPUS* L.

1. *ALLOPHYLUS* L.

Key to the Species

1a	Plants pubescent; leaf margin serrate	1. *A. serratus* Kurz
1b	Plants glabrous; leaf margin obscurely crenulate	2. *A. triphyllus* Merr.

1.1. ***Allophylus serratus*** (Roxb.) Kurz, As. Soc. Bengal 44 (2): 185.1876; Radlkofer, Pfamilien. 3(5): 313.1895; Bennet, Fl. Howrah 224.1979. *Ornitrophe serrata* Roxb. Pl. Corom. 1: 44.t.61.1795. *Allophylus cobbe* sensu Hiern, Hook.f.Fl.Brit. India 1: 674.1875 quoad from *serratus*, non. Blume; Prain Beng. Plants 1: 241. 1903 (Rep. ed. 1963) & Rec. bot. Surv. India 3 (2): 190.1905.

Pubescent shrub, leaflets 5-10 X 3-6 cm, ovate or broadly ovate, acuminate, margin distinctly serrate. Inflorescence densely brownish, pubescent. *Fls. & Frts.:* May-August .

Sometime in the hedges and village thickets, rarely epiphytic.

Field Collection: Doodhoi, *M.K.Ghosh-969* (Cal).

Distributions: Assam, Bihar and Bengal; Bangladesh, Sri Lanka and North Australia.

Economic Importances: Root decoction is astringent and occasionally applied to check diarrhoea.

1.2. ***Allophyllus triphyllus*** (Burm. f.) Merr. Philip. J. Sci.19: 363. 1921; Bennet Fl. Howrah 224. 1979. *Usubis triphyllus* Burm.f.Fl. Ind.89.t. 32,f.1.1768. *Ornitrophe glabra* Roxb. Fl. Ind. 267.1832, *non. illeg. Allophylus cobbe* non Bl. sensu Hiern, Hook.f. Fl. Brit. India 1: 674.1875 quoad form *glaber;* Prain Beng. Plants 1: 242.1903 (Rep. ed. 1963) & Rec.bot.Surv. India 3(2): 295.1905. *A. glaber* (Roxb.) Badlk., Pfamilion 3 (5): 313.1895 & Pfreich. 98B: 566.1932.

Glabrous shrub, 1-2 m high. Leaves usually 3 -foliolate; petiole 2.5-4.0 cm long; leaflets ovate, elliptic or obovate, apex acute, base cuneate to rounded. Flowers in fascicles on simple raceme, 1.0 -1.5 mm diam., white, shortly pedicelled. Drupe orange, globose. *Fls. & Frts.:* May-August.

Common on the village shrubberies, especially under the shade.

Field Collections: Magrahat, *K.N.-273;* Taki, *K.N.-1012.*

Distributions: Throughout India , especially on the Peninsular India; Bangladesh, Burma, Sri Lanka and north Australia.

2. *CARDIOSPERMUM* L.

2.1. ***Cardiospermum halicacabum*** L. Sp. Pl. 366. 1753; Hiern, Hook.f.Fl. Brit. India 1: 670.1875; Watt Dict. Econ. Prod. 2: 113.1890; Prain Beng. Plants 1: 241.1903 (Rep.ed. 1963) & Rec. bot. Surv.India 3(2): 189.1905; Whl. India 2: 75.1950. *(Vern. Name: Shibjhul).*

Climbing herb, stem furrowed. Leaflets lanceolate, deeply cut, acuminate, glabrous, thin and flaccid. Flowers white polygamous, in small umbellate cymes on axillary, slender, striate peduncle; peduncle 4-6 cm long with a pair of opposite, closely twisted spring like tendrils below the cymes; stamens 8; ovary 3-celled. Capsule trigonous, bladder like, winged at three angled. Seeds globose, smooth black. *Fls. & Frts.:* February-July.

Common in waste and shady places, climbing on bushes and hedges.

Field Collection: Bhebe Choumatha/Basirhat, *K.N.-717.*

Distributions: Throughout India; Sri Lanka, Malacca and most of the Tropical and Sub-tropical countries.

Economic Importances: Used in rheumatism, stiffness of limbs, lumbago and nervous diseases. Leaf juice used for cure of earache.

3. *DIMOCARPUS* L.

3.1. ***Dimocarpus longan*** Lour Fl. Cochinch. 233. 1790; Guha Bakshi, Fl. Murshidabad 342.1984. *Euphoria longan* (Lour.) Steud. Nomencl. 328. 1821; Merrill, Fl. Cochin. 248.1935; Benthall Tree Calc. 122. 1946; Wlh. India 3:230.1952. *Nephelium longana* Camb., Mem. Mus. Hist. Nat. Paris 18: 30.1829; Hiern, Hook.f. Fl. Brit. India 1: 688. 1775; Watt Dict. Econ. Prod. 5: 348.1892; Prain Beng. Plants 1: 244. 1903 (Rep. ed. 1963) & Rec. bot. Surv. India 3 (2): 190.1905. *(Vern. Name: Ashphal)*

Tall tree 10-15 m high. Leaves 20-40 cm long; leaflets usually 3-6 pairs, 7-16 cm X 2.5-4.0 cm, elliptic-oblong, linear-oblong. Flowers numerous in large rusty tomentose terminal or upper axillary panicles. Fruit supported on the persistent calyx, globose, reddish. seed-1, black. *Fls. & Frts.:* January -June.

Sometimes planted in the orchards.

Field Collection: Subuddhipur/Baruipur, *K.N.-360.*

Distributions: Native of S.W. India; in Bangladesh, Sri Lanka.

Economic Importances: Low quality edible fruits; timber is hard and durable; medicinally the fruits are reputed to be nutrient, stomachic and anthelmintic. In Indo-China, the dired fruits are used as a tonic and brain stimulant.

4. *DODONAEA* L.

4.1. ***Dodonaea viscosa*** (L.) Jacq. Enum. Syst. Pl. Ca. 10. 1760; Hiern, Hook.f. Fl. Brit. India 1: 697.1875; Watt Dict. Econ. Prod. 3: 172. 1890; Prain Beng. Plants 1: 244. 1903 (Rep. ed. 1963); Wlh. India 3: 97.1952. *Ptelea viscosa* (N.Jacq.) L. Sp. Pl. (ed.2): 173. 1962, non Linn. 1753. *(Vern. Name: Belati-mehedi).*

Very common in the hedges, found wild.

Field Collection: Dhablat/Sagar Islands, *A.K.Mukherjee-5944* (Cal).

Distributions: Throughout India; Bangladesh, Sri Lanka, Malacca and all worm countries.

Economic Importances: Leaves are used in the treatment of wounds, swellings and burns.

Notes: Prain (1903a) has reported this taxon from Tirhat, Chhota Nagpur, N. Bengal and Chittagong; no mention has been made so far from the lower Ganga Delta.

5. *LITCHI* Sonn.

5.1. ***Litchi chinensis*** Sonner., Voy. Ind. 2: 230. Pl.129.1782; Benthall Trees Calc. 124.1946; Wlh. India 6: 154.1962. *Dimocarpus litchi* Lour. Fl. Cochin 1: 233. 1790. *Nephelium litchi* (Lour.) Gamb. Mem. Mus. Hist. Nat. Paris 18: 30.1829; Hiern, Hook.f. Fl. Brit. India 1: 687.1875; Watt Dict. Econ. Prod. 5: 346.1893; Prain Beng. Plants 1: 244.1903 (Rep. ed. 1963). *(Vern. Name: Litchi).*

Spreading, evergreen tree. Leaves alternate, leaflets 2-5 pairs, 5-16 cm long, oblong elliptic, lanceolate, acute or acuminate, coriaceous smooth and shining above, glaucous beneath. Flowers small, in pubescent terminal panicles; petals absent. Fruit globose covered with pointed tubercles, red when ripe; seeds 1; aril large, fleshy. *Fls. & Frts:* April-June.

Commonly cultivated in and around the village orchards.

Field Collection: Dhanyakuria, *K.N.-564.*

Distributions: Eastern parts of India, reported to have been introduced into India towards the ends of 18th Century from S. China. A chinese species.

Economic Importances: Fruits are very delicious and juicy; ripe fruits are edible. In Indo-China the seeds are used for intestinal troubles.

6. *SAPINDUS* L.

Key to the Species

1a	Handsome tree; leaves not crowded near the ends of branches; leaflets oblong acuminate; flowers white; fruits lobed	1. *S. laurifolia* Vahl
1b	Medium sized tree; leaves crowded near the ends of branches; leaflets lanceolate oblong, acuminate or obtuse; flowers white or purple; fruits sub-globose	2. *S. mukorossi* Gaertn.

6.1. ***Sapindus laurifolia*** Vahl Symb. 3: 54.1794; Roxb. Fl. India 2: 278.1824. *S. trifoliatus* L. Sp. Pl. 367.1753; Hiern, Hook.f. Fl. Brit. India 1: 682.1875; Watt Dict. Econ. Prod. 6(2): 468. 1893; Prain Beng. Plants 1: 242.1903 (Rep. ed. 1963); Wlh. India 9: 227.1972. *S. emarginata* Vahl, Symb.3: 54.1794; Roxb. Fl. Ind. 2: 279.1824. *(Vern. Name: Bara Rita).*

Handsome tree. Leaves pinnately compound; leaflets oblong acuminate. Flowers white, hairy; sepals 5, elliptic, obtuse. Fruit fleshy, 2-3 lobed, saponaceous. *Fls. & Frts.:* March-May.

Barely cultivated.

Field Collection: Harinbari/Sagar Islands, *K.N.- 477.*

Distributions: Common in S. India, may be a native of S. India.

Economic Importances: Fruits are used as emetic tonic, astringent and anthelmintic properties; also used for treatment of asthma colic due to indigestion, diarrhoea, cholera and tubercular glands.

6.2. ***Sapindus mukorossi*** Gaertn., Fruct. 1: 342.1788; Hiern, Hook. f. Fl. Brit. India 1: 683.1885; Watt Dict. Econ. Prod. 6(2): 468. 1893; Prain Beng. Plants 1: 242.1903 (Rep. ed. 1963) & Rec. bot. Surv. India 3(2): 190.1905; Wlh. India 9: 225.1972. *(Vern. Name: Etaphal).*

Medium sized tree. Leaves pinnate, crowded near the ends of branches; leaflets 10-16, shortly stalked, 5-15 cm long, lanceolate oblong, alternate or sub-opposite, acuminate or obtuse, glabrous, base oblique, rachis sometimes margined. Flowers white or purple, in terminal and lateral panicles. Fruits sub-globose, fleshy, saponaceous. *Fls. & Frts.:* May-February.

Planted in the gardens, but very rare.

Field Collection: Kestopur, *K.N.-1891.*

Distributions: Found in the Himalayas from Himachal Pradesh and towards Assam. Native of China and Japan.

Economic Importances: The fruits are largely used as a substitute for soap, for washing woolen clothes.

7. SCHLEICHERA Willd.

7.1. ***Schleichera oleosa*** (Lour.) Oken. Allg. Naturgesch. 3(2): 1341. 1841; Wlh. India 9: 250.1972. *Pistacia oleosa* Lour. Fl. Cochinch. 2: 615.1790. *Schleichera trijuga* Willd. Sp. Pl. 4(2): 1096.1806; Hook.f. Fl. Brit. India 1: 681.1875; Watt Dict. Econ. Prod. 6(2): 487.1893; Prain Beng. Plants 1: 243.1903 (Rep. ed. 1963). *(Vern. Name: Kusum).*

Large, handsome, deciduous tree. Leaves paripinnate; leaflets 4- 8, opposite, 5-23 cm long, oblong, elliptic-oblong, obtuse, acute or short acuminate, entire. Flowers yellowish green, in pedicellate clusters, on interrupted slender drooping racemes. Fruits pointed, smooth or echinate. *Fls. & Frts.:* March-November.

Planted in gardens and as road side tree; also found as wild.

Field Collection: Namkhana, *K.N.-1921.*

Distributions: Native of low hills of Himalayas, Central India and the Western Peninsula. Burma and Sri Lanka.

Economic Importances: Kusum oil has been used for hair dressing. It is used for external massage in rheumatism.

Notes: Prain (1903a & 1905) has mentioned this taxon from Bengal proper. Another species *Erioglossum rubiginosum* (Roxb.), Bl. Rhumphia 3: 118.1847 is also distributed in the lower parts of Bengal.

XLV. ANACARDIACEAE

Key to the Genera

1a	Leaves simple	(2)
1b	Leaves pinnately compound	(4)
2a (1)	Ovules pendulous from a basal funicle	(3)
2b (1)	Ovules pendulous from the top or above the middle of the cell; stamens 5-6; styles 3; petals imbricate; drupe superior	4. *SEMECARPUS* Linn. f.
3a(2)	Flowers polygamous; stamens 1-5; fruit a fleshy drupe	3. *MANGIFERA* L.
3b (2)	Flowes hermaphrodites; stamens 10; fruit a drupe, not as fleshy as mangifera	1. *ANACARDIUM* L.
4a (1)	Drupe 1-celled, 1-seeded	2. *LANNEA* A.Rich.
4b (1)	Drupe 2-5 celled, 2-5 seeded	5. *SPONDIAS* L.

1. *ANACARDIUM* L.

1.1. ***Anacardium occidentale*** L. Sp. Pl. 383.1753; Watt Dict. Econ. Pord. 1: 232.1889; Hook.f.Fl. Brit. India 2: 20.1876; Prain Beng. Plants 1: 250.1903 (Rep. ed. 1963) & Rec. bot. Surv. India 3 (2): 191.1905; Wlh. India 1:70.1948. *(Vern. Name: Kajubadam).*

Exotic, cultivated small tree, 7-20 m high. Leaves small, alternate, ovate, petioled, entire; stipules absent, coriaceous. Panicles terminal, longer than the leaves, branches cymose. Sepals 1-1.5 cm lanceolate, minutely puberulous outside. Stamens 9, one longer than the others. Ovary glabrescent. Fruit reniform 2-2.5 cm long. *Fls. & Frts.:* March-May.

Rarely cultivated in the orchards; grow best in the saline soils of the coastal regions.

Field Collection: Harinbari/Sagar Islands, *K.N.-492.*

Distributions: Throughout the coastal regions in India. A Native of West Indies, Tropical America and Brazil; appears to have been under cultivation throughout the Tropical America before the voyage of Columbus to the New World. It seems to have been introduced into India from Brazil by the Portuguese in the 16th century and is now naturalized in the forests of Chittagonge and all over the coastal regions of India.

Economic Importances: Fruits are eaten after frying, a very nutritive and delicious food. It is useful in case of mental derangement, sexual debility, nervous prostration following seminal emission. Useful for morning sickness of pregnency.

2. *LANNEA* A. Rich. *nom. cons.*

2.1. ***Lannea coromandelica*** (Houtt.) Merrill., Journ. Arb. 19: 353. 1939; Wlh. Ind. 6: 27. 1962. *Dialium coromandelicum* Houtt., Nat. Hist. 11(2): 39.t.5, f.2.1774. *Odina wodier* Roxb, (Hort Beng. 29. 1814) & Fl. Ind. 2:293.1832; Hook.f. Fl. Brit. India2: 29.1876; Prain Beng. Plants 1: 251.1903 (Rep. ed. 1963) & Rec. bot. Surv. India 3(2): 191.1905. *Heberlia grandis* Dennst. Schl. Hort. Mal. 30. 1814. *Lannea grandis* (Dennst.) Engl., Benthall Trees Calc. 129.1946. *(Vern. Name: Jiyal).*

Small to medium tree, deciduous. Leaves imparipinnate, crowded at the end of the branches; leaflets 5-11 opposite, ovate lanceolate, entire, acuminate, base often unequal. Flowers small greenish yellow, fascicled on the branches. Drupe oblong, laterally somewhat compressed, red when ripe. *Fls. & Frts.:* February-June.

Fairly common near villages, shrubberies and scrub jungles.

Field Collection: Dewanganj, *K.N.-2264.*

Distributions: Throughout India, indigenous to the hotter parts of India. Burma and Sri Lanka.

Economic Importances: It yields an useful gum, which is used in clothes printing. The leaves and branches are commonly used as fodder of goat, yields fire wood.

3. *MANGIFERA* L.

3.1. ***Mangifera indica*** L. Sp. Pl. 200. 1753; Watt Dict. Econ. Prod. 5: 146. 1891; Hook.f.Fl.Brit. India 2: 13.1876;Prain Beng. Plants 1: 248.1903 (Rep. ed. 1963) & Rec. bot. Surv. India 3 (2): 190.1905; Benthall Trees Calc. 134. 1946; Wlh. India 6:265.1962. *(Vern. Name: Aam).*

Large evergreen tree. Leaves simple. Flowers yellowish green in terminal panicles; stamen only one; edible drupe from its solitary carpel. Drupe ovoid, greenish yellowish or reddish when ripe. *Fls. & Frts.:* January-June.

Commonly planted in the gardens for its delicious fruits. There are several varieties of mango depending upon the shape, taste and time of fruitings.

Field Collection: Padmapukur, *K.N.-332.*

Distributions: Native of India; Burma, Sri Lanka and different parts of the Tropical World.

Economic Importances: Most important and delicious fruits. A number of food items are prepared from green and ripe fruits.

4. *SEMECARPUS* Linn.f.

4.1. ***Semecarpus anacardium*** L. f. Suppl. 182.1781; Roxb. Fl. Ind. 2: 83.1824; Watt Dict. Econ. Prod. 6 (2): 498.1893; Hook.f. Fl. Brit. India 2: 83.1876; Prain Beng. Plants 1: 230.1903 (Rep. ed. 1963); Wlh. Ind. 9: 271.1972. *(Vern. Name: Bhola).*

Moderate tree. Leaves very large, crowded at the ends of branches, alternate, 16-60 cm long, oblong or obovate, entire, rounded at the apex, cordate at base. Flowers greenish, yellow, subsessile in fascicles, arranged on terminal pubescent panicles. Fruits drupe, obliquely ovoid or oblong, smooth, shining black, seated on an orange red fleshy crop. *Fls. & Frts.:* May-September.

Rarely on the village thickets.

Field Collection: Tapsia/Bantala, *K.N.-1743.*

Distributions: Bihar, Chhota Nagpur and Madhya Pradesh; Tropics.

Economic Importances: The fruits contain a corrosive juice, which after mixing with lime used by the washermen as marking ink.

Notes: Prain (1903a) has not mentioned this taxon from L. Bengal.

5. *SPONDIAS* L.

Key to the Species

1a	Leaflets 6-8 jugate, crenate serrate	1. *S. cytherea* Sonn.
1b	Leaflets 4-6 jugate, quite entire	2. *S. pinnate* Kurz.

5.1. ***Spondius cytherea*** Sonn. Voy. Indes. Or. & Chine 3: 242.t.123. 1782; Wlh. India 10: 19 . 1976. *S. dulcis* Frosst.f. Pl. Escul. 33. 1786; Hook.f. Fl. Brit. India 2: 42. 1876; Prain Beng. Plants 1: 252.1903 (Rep. ed. 1963) & Rec. bot. Surv. India 3 (2): 191.1905; Benthall Trees Calc. 133.1946. *(Vern Name: Belati Amra).*

Medium tree; leaves imparipinnate, leaflets 6-8 jugate, crenate serrate. Flowers polygamous, whitish. Ovary 2-5 celled. Fruit ovoid drupe 4-5 cm in length, 2-2.5 cm radius, greenish with prominent nerves. *Fls. & Frts.:* March-November.

Commonly planted in the orchards, almost throughout W. Bengal.

Field Collection: Nabagram, *K.N.-641.*

Distributions: Almost throughout India; native of Polynesia, but is now grown in many hot countries.

Economic Importance: Fruits are edible.

5.2. ***Spondius pinnata*** (Linn.f.) Kurz. Pegu Rep. A . 44. 1875; Wlh. India 10: 20 . 1976. *Mangifera pinnata* Linn. f. Suppl. 156.1781. *Spondias mangifera* Willd. Sp. Pl. 2: 751.1799; Hook.f. Fl. Brit. India 2: 42.1876; Prain Beng. Plants 1: 252.1903 (Rep. ed. 1963) & Rec. bot. Surv. India 3 (2): 191.1905; Benthall Trees Calc. 131.1946. *(Vern. Name: Amra).*

Large, glabrous, deciduous tree. Leaves pinnately compound; leaflets with marginal veins. Flowers greenish white in panicles. Drupe fleshy yellow, 2-6 celled. *Fls. & Frts.:* February-March.

Commonly planted for its acidic edible fruits, which are occasionally noted as self propagated.

Field Collection: Malancha/Minakhan, *K.N.-1358.*

Distributions: Native of India; Burma, Sri Lanka and Bangladesh.

Economic Importances: The pulp of the fruits is useful in dysentery and the fruits are also edible. Bark is used as tann.

XLVI. MORINGACEAE

1. *MORINGA* Adans.

1.1. ***Moringa pterygosperma*** Gaertn. Fruct. 2: 314.1791; Hook.f. Fl. Brit. India 2: 45.1876; Watt Dict. Econ. Prod. 5:276.1891; Prain Beng. Plants 1: 252.1903 (Rep. ed. 1963) & Rec. bot. Surv. India 3 (2): 191. 1905. *M. oleifera* Lamk. Encycl. 1: 398.1785; Benthall Trees Calc. 137.1946; Wlh. India 4: 426.1962. *Guilandina moringa* L. Sp. Pl. (ed. 1) 381.1753. *(Vern. Name: Sajina* or *Nadna).*

Small-medium deciduous tree. Leaves usually 3-pinnate, pinnate 4-6 pairs, leaflets 6-9 pairs, opposite, elliptic, ovate or obovate, pale beneath; glands linear, hairy. Flowers pale white, fragrant. Capsule pendulous, 20-40 cm long, 9-ribbed; seeds trigonous, winged at angles. *Fls. & Frts.:* February-June.

Common everywhere; planted on the embankment of the ponds and vegetable gardens; flowers and fruits appear twice a year in two different varieties of plants.

Field Collection: Joygram/Minakhan, *K.N.-304.*

Distributions: Native of the forests of the Western Himalayas; now cultivated in many tropical countries.

Economic Importances: Both leaves and the capsules are edible as vegetable. Considered as famine food during scarcity. The seeds are useful in ascites resulting from the enlarged lever and spleen. External application of oil extracted from the seeds is found useful in gout and acute rheumatism. The vegetable amorphous base present in it responsible ephedrine and adrenaline in action. It helps the rise of blood pressure and accelarate heart beat.

Remarks: Depending on flowering and fruiting times the local names 'sajina' and 'nadna' are commonly used for this taxon; but the interesting thing is that there different vernacular names applied for the same species.

Since, *M. oleifera* Lamk. is a *nomen ambiguum* as it is composed of *M. oleifera* and *M. myrepsica,* the correct name for this taxon should be *M. pterygosperma* Gaertn., the next oldest binomials.

XLVII. PAPILIONACEAE
(FABACEAE)

Key to the Genera

1a	Plants with hairs on the twigs; leaves and calyx fixed by their centre, connectives of anthers mucronate	20. *INDIGOFERA* L.
1b	Plants with basifixed hairs or glabrous; anthers not mucronate or gland tipped	(2)
2a (1)	Pods indehiscent or rarely opening along the ventral sutures	(3)
2b (1)	Pods dehiscent by both sutures	(15)
3a (2)	Leaves pellucidaly gland-dotted; leaflets-1, toothed; ovule-1	31. *PSORALEA* L.
3b (2)	Leaves not pellucid-dotted	(4)
4a (3)	Pods often 1 or several indehiscent; 1-seeded segments	(5)
4b (3)	Pods generally more than one seeded; non-segmented, always indehiscent	(9)
5a (4)	Leaves exstipulate	(6)
5b (4)	Leaves stipulate	(7)
6a (5)	Leaves even pinnate; joints folded inside the calyx	36. *SMITHIA* Ait.
6b (5)	Leaves pinnate; joints straight, several exserted	2. *AESCHYNOMENE* L.
7a (5)	Pods twisted up	40. *URARIA* Desv.
7b (5)	Pods not twisted up	(8)
8a (7)	Pods twiged often terete	3. *ALYSICARPUS* Desv.
8b (7)	Pods flattened, septate and indented	14. *DESMODIUM* Desv.
9a (4)	Leaves even pinnate, the rachis ending in a bristle	4. *ARACHIS* L.
9b (4)	Leaves odd-pinnate	(10)
10a (9)	Trees or woody climber, leaflets entire	(11)
10b (9)	Herbs; leaflets toothed	(13)

11a (10)	Leaflets opposite	13. *DERRIS* Lour.
11b (10)	Leaflets alternate	(12)
12a (11)	Flowers small, white or pink; pods narrow	12. *DALBERGIA* Linn.f.
12b (11)	Flowers medium, yellow; pods sub-orbicular	32. *PTEROCARPUS* L.
13a (10)	Pods indehiscent	(14)
13b (10)	Pods indehiscent, elongated	39. *TRIGONELLA* L.
14a (13)	Racemes slender; pods sub-globose or very short	24. *MELILOTUS* Miller
14b (13)	Racemes not slender; pods spiral or sickle-shapped	23. *MEDICAGO* L.
15a (2)	Leaf rachis not ending in a tendril; leaves odd pinnate or simple or digitately 3-more foliolate	(16)
15b (2)	Leaf rachis ending in a bristle or tendril; leaflets even pinnate	(37)
16a (15)	Leaves digitately 3-7 foliolate, or simple	(17)
16b (15)	Leaves pinnately compound	(18)
17a (16)	Leaves digitately 3-7 foliolate, not glandular beneath, stamen monadelphous, 10	11. *CROTALARIA* L.
17b (16)	Leaves digitately 3-foliolate with small bracts, glandular beneath; stamens diadelphous, 9+1	17. *FLEMINGIA* Roxb. ex Ait.
18a (16)	Leaves 3-foliolate	(19)
18b (16)	Leaves pinnately 5-many foliolate; pods dehiscing from apex to base	(33)
19a (18)	Pods dehiscing from the apex to base	(20)
19b (18)	Pods dehiscing at the seed bearing apex only, 1-seeded	6. *BUTEA* Roxb. ex Willd.
20a (19)	Leaves glandular beneath, pods compressed	(21)

20b (19)	Leaves not glandular beneath, pods woody or coriaceous	(23)
21a (20)	Ovules 1-2	33. *RHYNCHOSIA* Lour.
21b (20)	Ovules more than 3	(22)
22a (21)	Climbers; pods oblong-obtuse	5. *ATYLOSIA* W. & A.
22b (21)	Woody undershrubs; pods acute	7. *CAJANUS* DC.
23a (20)	Style bearded below th^e stigma	(24)
23b (20)	Style not beared below stigma	(28)
24a (23)	Stigma oblique	(25)
24b (23)	Stigma terminal, not oblique	(27)
25a (24)	Keel spirally twisted	28. *PHASEOLUS* L.
25b (24)	Keel not spiral	(26)
26a (25)	Style flattened upward and circinate	27. *PACHYRHIZUS* Rich.
26b (25)	Style not flattened	42. *VIGNA* Savi.
27a (24)	Pods square, 4-winged	30. *PSOPHOCARPUS* Neck.
27b (24)	Pods flattened , not winged	15. *DOLICHOS* L.
28a (23)	Nodes of racemes not tumid	(29)
28b (23)	Nodes of racemes tumid	(31)
29a (28)	Stamens diadelphous; stipules and bracts persistent; calyx tooth distinct	35. *SHUTERIA* W. & A.
29b (28)	Stamens monadelphous, stipules and bracts caducous	(30)
30a (29)	Anthers all fertile	19. *GLYCINE* L.
30b (29)	Five alternate anthers abortive	38. *TERAMNUS* R. Br.
31a (28)	Petals equal in length; upper lip of calyx projecting; stamens mono- or diadelphous; suffruticose	8. *CANAVALIA* DC.
31b (28)	Petals variable in length	(32)
32a (31)	Climbing shrub; keel exceeding the wings and standard	26. *MUCUNA* Adans
32b (31)	Small tree; standard exceeding keel	16. *ERYTHRINA* L.

33a (18)	Flowers medium to less than 2 cm, standard not longer than the petals	(34)
33b (18)	Flowers large, 3-5 cm long with unequal petals; standard large	10. *CLITORIA* L.
34a (33)	Erect or diffuse herbs or shurbs; pods soon dehiscent	(35)
34b (33)	Medium-small tree; pods woody; tardily dehiscent	(36)
35a (34)	Pods flattened, non-septate	37. *TEPHROSIA* Pers.
35b (34)	Pods cylindric, long, transversely septate between the seed	34. *SESBANIA* Adans
36a (34)	Deciduous tree, leaflets opposite, 12-14 pairs; flowers exserted rosy white	18. *GLIRICIDIA* H.B. & K.
36b (34)	Evergreen tree; leaflets 3-4 pairs; flowers mauve or lilac in pendulous racemes	25. *MILETTIA* W. A.
37a (15)	Small herbs; flowers axillary or racemes; stipules large foliaceous; oblique at base; stamens 9+1	(38)
37b (15)	Twinning shrubs, stem woody. Flowers in terminal racemes; stipules narrow, equal at base; stamens 9	1. *ABRUS* L.
38a (37)	Leaflets toothed; style not bearded; seeds with slender funicle; pods turgid	9. CICER L.
38b (37)	Leaflets smooth, entire; style bearded; seeds with short funicle	(39)
39a (38)	Staminal tube with oblique mouth	(40)
39b (38)	Staminal tube truncate	(41)
40a (39)	Style with dorsal tuft of hairs or bearded round the tip; ovule usually more than two	41. *VICIA* L.
40b (39)	Style longitudinally bearded along inner face; ovule not more than 2	22. *LENS* Mill.

41a (39) Pods compressed; style flat, margin not recurved — 21. *LATHYRUS* L.

41b (39) Pods turgid; style 3-cornored, margin recurved — 29. *PISUM* L.

1. *ABRUS* L.

1.1. ***Abrus precatorius*** L. Syst.(ed.12) 472. 1767; Baker, Hook.f. Fl. Brit. India 2: 175.1876; Watt Dict. Econ. Prod. 1: 10.1889; Prain Beng. Plants 1: 262.1903 (Rep.ed. 1963) & Rec. bot.Surv. India 3 (2): 192. 1905; Wlh. India 1:2.1948. *(Vern. Name: Kunch).*

Twining perennial shrub. Leaves paripinnate, leaflets 10-20, subdeciduous. Stipules minute, persistent. Racemes terminal, fascicled. Flowers pinkish,in dense racemes. Stamens 9, monadelphous. Pods compressed, corrugated, dehiscent by both sutures. Seeds globose, red or pearl white, shinning smooth. *Fls. & Frts.:* July-December.

Common on the dry waste lands, on the bushes or climbing over trees.

Field Collection: Sasan/Baruipur, *K.N. 27.*

Distributions: Almost throughout India. Native of the Tropical Asia; at present cosmopolitan in the Tropics.

Economic Importance: The Goldsmiths used the dry seeds for 'Rathi' to weight gold, as it has a constant weight of 1.75 gms.

2. *AESCHYNOMENE* L.

Key to the Species

1a Stem not woody; calyx hispid; joints of pods echinulate — 1. *A. aspera* L.

1b Stem woody; calyx glabrous; joints of pods smooth — 2. *A. indica* L.

2.1. ***Aeschynomene aspera*** L. Sp. Pl. 713.1753; Baker, Hook.f Fl. Brit. India 2: 152.1876; Watt Dict. Econ. Prod. 1: 125.1889; Prain Beng. Plants 2: 299.1903 (Rep. ed. 1963) & Rec. bot. Surv. India 3(2): 202.1905; Wlh. Ind. 1: 35.1948; Majumdar, Ind. Agr. 6: 138.1962; Subramanyam Aq. Angios. 14.f.9.1962. (*Vern. Name: Bhat Sola* or *Phul Sola).*

Aquatic, erect stout shrub with a few branches, stem nodules numerous. Leaves exstipulate, imparipinnate. Flowers yellow. Stamens in two lateral bundles of 5 each; calyx and the longer corolla hispid. Pods echinulate, straight; separating at the joints. *Fls.& Frts.:* July-November.

Common on the ditches, jhills, roadside canals.

Field Collection: Falta, *K.N.-1291*; Majerhat, *K.N.- 2247.*

Distributions: Almost throughout India, mainly in the eastern and the southern states. Bangladesh, Burma, Malaya and Tropical America.

Economic Importances: Used as fodder, bottle cork, toy manufacturing and decorating purposes; also used as surgical lints and packing materials.

2.2. ***Aeschynomene indica*** L. Sp. Pl. 713.1753; Backer, Hook.f.Fl. Brit. India 2: 151.1876; Watt Dict. Econ. Prod. 1: 126.1889; Prain Beng. Plants 1: 418.1903 (Rep. ed. 1963) & Rec. bot. Surv. India 3(2): 202.1905; Wlh. India 1: 35.1948 *(Vern. Name: Sola).*

Shrub like, annual herb, stem nodule present. Leaves compound, leaflets numerous small, stipules linear lanceloate, acuminate, with acute auricled base. Flowers yellow, in axillary racemes, calyx glabrous. Pods 6-10 joint, dotted with black. *Fls. & Frts.:* September-November.

Commonly found as a weed in the rice field, water logged places and near canals. Base of the stem is often swollen and bears bacterial nodules.

Field Collection: Sarberia/Gocharan, *K.N.-234.*

Distributions: Common in Bengal, Assam and South India. Throughout the Tropics of the Old World.

Economic Importances: Weed effects on the aquatic systems; but the tender leaves are edible and the soft pitch is used for decorating purposes, ornamental purposes and for bottle cork.

3. *ALYSICARPUS* Desv. nom. cons.

Key to the Species

1a	Calyx much longer than the first joint of the pods; pods slightly moniliform, neither veined nor rugose	1. *A. bupleurifolius* DC.
1b	Calyx shorter or very slightly longer than the first joint of the pod	(2)
2a (1)	Flowers pink in close erect pedunculate racemes; pods distinctly moniliform, veinlets	2. *A. monilifer* DC.
2b (1)	Flowers yellow in lax racemes; pods terete, veined	3. *A. vaginalis* DC.

3.1. ***Alysicarpus bupleurifolius*** (L.)DC. Prodr. 2: 352.1825; Backer, Hook.f. Fl. Brit. India 2:158.1876; Prain Beng. Plants 1:307.1903 (Rep. ed. 1963). *Hedysarum bupleurifolium* L. Sp.Pl. 746.1753. *H. graminoum* Retz. Obs. 5: 26.1789. *(Vern. Name: Pan-nata).*

Glabrous, erect slender or diffuse herb. Leaves 1-foliolate, shortly stalked, glabrous; leaflets 2.5-8.-0 cm long, linear-lanceolate or oblong, acute, glabrous. Flowers pink, shortly pedicelled, usually in pairs on spike like loose racemes; calyx much longer than the lowest joint of the pod. Pods stalked, moniliform. *Fls. & Frts.:* August-November.

Commonly found in moist grassy places.

Field Collection: Ichhapur, *K.N.-1797.*

Distributions: Throughout India. Srilanka, Malaya, China, Philippines, Mauritius and Polynesia.

3.2. ***Alysicarpus monilifer*** (L.) DC. Prodr.2: 353.1825; Backer, Hook.f. F., Brit. India 2: 158.1876; Prain Beng. Plants 1: 306. 1903 (Rep. ed. 1963). *Hedysarum moniliferum* L. Mant. 1: 102.1767.

Prostrate, much branched, deep rooted perennial herb. Stem densely tufted. Leaves simple, oblong, obtuse, often cordate. Flowers pink, about 5-10, in close erect pedunculate racemes. Pods distinctly moniliform, jointed densely clothed with hooked hairs. *Fls. & Frts.:* September-January.

Found during rainy season in association with grasses.

Field Collection: Mallakhali/Sundarbans, *K.N.-1842.*

Distributions: Throughout India; and in Abyssinia.

Economic Importance: Good fodder values.

3.3. ***Alysicarpus vaginalis*** (L.) DC., Prodr. 2: 353.1825; Baker, Hook.f. Fl. Brit. India 2: 158.1876; Prain Beng. Plants 1: 306. 1903 (Rep.ed. 1963) & Rec. bot. Surv. India 3 (2): 203.1905; Wlh. India 1: 66.1948. *Hedysarum vaginalis* L. Sp. Pl. 746.1753.

Much branched, prostrate or erect herb, stem ascending. Leaves 1-foliolate; stipules ovate, acute; leaflets variable, both in size and shape. Flowers in lax racemes, yellowish, twinged with pink or purple. Pods terete, 6-8 jointed; faintly pubescent. *Fls. & Frts.:* March-December.

Commonly found in lawns and open pastures, often mixed with grasses.

Field Collection: Fultala/Baruipur, *K.N.-45.*

Distributions: Throughout India, Sri Lanka, Afghanistan, and generally throughout the Tropics of the Old World.

Economic Importances: In Philippines, it is considered as good forage legumes.

4. *ARACHIS* L.

4.1. ***Arachis hypogaea*** L. Sp. Pl. 741.1753; Baker, Hook.f. F. Brit. India 2: 161.1876; Watt Dict. Econ. Prod. 1: 282.1889; Prain Beng. Plants 1: 297.1903 (Rep. ed. 1963) & Rec. bot. Surv. India 3(2): 201.1905; Bailey Man. Cult. Plants 554.1924; Wlh. India 1: 90.1948. *(Vern. Name: Mat-kalai* or *China Badam).*

Prostrate, diffuse or erect annual hearb. Leaves bipinnate, leaflets oval to oval obovate. Flowers yellow, showy, axillary. Pods 1-4 seeded, constricted between the seeds; the pedicles elongated rapidly seen after fertilization. *Fls. & Frts.:* January-August.

Mostly cultivated in the sandy soils for its edible pods.

Field Collection: Barrackpore, *Sen s.n.* (CAL).

Distributions: Introduced into India in the 16th Century; now cultivated almost throughout India. Native of Brazil.

Economic Importances: Seeds edible, kernels are eaten raw and rosted. Seed oil commercially known as 'Badam oil', is edible. Highly proteinous food.

5. *ATYLOSIA* Wt. & Arn.

5.1. ***Atylosia scarabaeoides*** Benth. Miq. Pl. Jung. 243.1851; Baker,Hook.f.Fl. Brit. India 2: 215.1876; Prain Beng. Plants 1: 272. 1903 (Rep. ed. 1963) & Rec. bot. Surv. India 2: 300.1903; Wlh. India 1: 137.1948. *Cantharospermum pauciflorum* Wt. & Arn. Prodr. 255.1831. *Dolichos scarabaeides* Roxb. Fl. Ind. 3: 315.1832. *(Vern. Name: Banar Kalai).*

Copiously branched, pubescent grey green biannual herb with slender trailing or twinning stem. Leaves sub-digitately 3- foliolate, stipules minute, caducous with gland dot on the leaf. Flowers yellow, 2-6 flowered in racemes. Pods small oblong very hairy or silky, 4-6 seeded. *Fls. & Frts.:* July-January.

Common in the waste places and clay soils.

Field Collection: Baruipur, *K.N.-18.*

Distributions: Throughout India; Sri Lanka, Malaya, China, Mauritious and Madagascar.

Economic Importance: Useful as green manure and forage.

6. *BUTEA* Roxb. ex. Willd. *nom. cons.*

6.1. ***Butea monosperma*** (Lamk.) Taub., Engl. Prantt., Pflamzenfam. 3(3): 366.1894; Benthall Trees Calc. 160.1946. *Erythrina monosperma* Lamk. Encyl. 2: 39.1786. *Butea frondosa* Konig. ex Roxb., Ac. Res. 3: 469. 1792& Pl. Cor. 1: 21.1795; Baker, Hook.f. Fl. Brit. India 2: 194. 1876; Watt Dict. Econ. Prod. 1: 548.1889; Prain Beng. Plants 1: 286.1903 (Rep. ed. 1963) & Rec. bot. Surv. India 3(2): 199.1905. *Plaso monosperma* (Lamk.) Kuntze, Rev. Gen. 1: 202.1891. *(Vern. Name: Palas).*

Small to moderate sized, deciduous tree; young parts silky pubescent. Leaves trifoliate; stipule triangular, acute, short, leaflets usually 7-25 cm long, terminal rhomboid or obovate from a narrowed base, lateral leaflets slightly smaller, obliquely ovate or elliptic, obtuse or emerginate at the apex, all coriaceous. Flowers bright orange red. Pods pendulous, silky tomentose, 1-seeded. *Fls. & Frts.:* March-April.

Planted occasionally on the lawns, public gardens and roadsides.

Field Collection: Budge Budge, *K.N.-729.*

Distributions: Throughout the drier parts of India; Indo-China and Sri Lanka.

Economic Importance: Beautiful ornamental plants. It is commonly known as 'Flame of the forest'. A hot poulitce of the leaves is used for cure of boils, pimples and swellings. A beautiful yellow dye is prepared from the flowers and is chiefly used for sprinkling during Holy Festival and Sri Panchami.

7. *CAJANUS* DC. *nom. cons.*

7.1. ***Cajanus cajan*** (L.) Millspaugh, Publ. Field. Columbian Mus. Bot. 2: 53.1900; Wlh. Ind. 2: 6.1950; Santapau, Rec. bot. Surv. India 16(1): 76.1967. *Cytisus cajan* L. Sp. Pl. 739.1753. *Cajanus indicus* Spreng. Syst. 3: 248.1826; Baker, Hook.f. Fl. Brit. India 12: 217.1876; Watt Dict. Econ. Prod. 2: 12.1889; Prain Beng. Plants 1:272.1903 (Rep. ed. 1963) & Rec. bot. Surv. India 3(2): 195.1905 *(Vern. Name: Arhar).*

Erect shrub with many silky branches. Leaves trifoliate, leaflets oblong-lanceolate, acute, entire, densely silky beneath indistinctly gland dotted; stipules minute, soon falling. Flowers yellow in corymbose racemes. Pods finely downy, often clothed with reddish purple streals; seeds varying in colour from yellow and red to brown or black. *Fls. & Frts.:* November-March.

A common field crop, cultivated everywhere in garden and field.

Field Collection: Harinbari/Sagar Island, *K.N.-476.*

Distributions: Cultivated almost throughout India. Native of Africa.

Economic Importances: An extremely nutritious pulse crop. The leaves and the seed coat is a good fodder. It is cultivated as cover crop in plantations. Tall stem used for basket making and the plants is a good host for lac insects.

8. *CANAVALIA* DC.*nom. cons.*

Key to the Species

1a	Pods turgid; stipules large, oblong, obtuse, bracteoles large	1. *C. cathartica* Thour.
1b	Pods not turgid; stipules small, triangular bracteoles minute	(2)
2a (1)	Racemes many flowered; pods many seeded	2. *C. ensiformis* DC.
2b (1)	Racemes few flowered; pods few seeded	3. *C. lineata* DC.

8.1. ***Canavalia cathartica*** Thour, J. Bot. Desv. 1.: 81.1813. *C. stocksii* Dalzell, Dalz. & Gibs. Bombay Fl. 69.1861. *C. ensiformis* (L.) DC. var. *turgida* Baker, Hook.f.Fl., Brit. India 2: 196.1876. *C. obtusifolia* Prain, J. As. Soc. Bengal 66: 417. 1898 & Beng. Plants 1: 281.1903 (Rep. ed. 1963).

Large twiner; stem very long, terete, smooth. Leaves large, petioles 15-20 cm long, oblong, obtuse, nerved; leaflets thinly membranous 12-20 cm long, nearly as broad as long, glabrous, reticulately veined. Flowers several at the apex of an angular naked peduncle; pedicles short; bracteoles elliptic oblong; corolla bluish purple, veined. Pods turgid, almost flat along the dorsal suture. Seeds 4-6. *Fls. & Frts.:* January-March.

Common on the river banks of Sundarbans areas; a littoral species.

Field Collection: Mollakhali/Sundarbans, *K.N.- 1843.*

Distributions: Sea coast and tidal forests of India.

Notes: D. Prain (1898) has been restored this taxon to the rank of a species and not a variety of *C. ensiformis* as Baker has treated it in Hook. f. Fl. Brit. India (1.c.).

8.2. ***Canavalia ensiformis*** DC. Prodr. 2: 404.1825; Baker, Hook.f. Fl. Brit. India 2: 195.1876; Watt Dict. Econ. Prod. 2: 97.1889; Prain Beng. Plants 1: 281.1903 (Rep. ed. 1963) & Rec. bot. Surv. India 3 (2): 197.1905; Wlh. India 2: 55.1950. *C. gladiata* DC. Prodr.2:404.1825. *(Vern. Name: Makhan sim).*

Large climber. Leaves pinnately 3-foliate. Flowers numerous in axillary racemes. Upper lip of calyx projecting . Pods oblong or linear, 8 to 12 seeded, 15-30 cm long. Seed reddish brown or white. *Fls. & Frts.:* October-December.

Commonly cultivated as vegetable crop in village gardens.

Field Collection: Baharu on the public garden, *K.N.-1624*.

Distributions: Throughout India; native of West Indies.

Economic Importances: Pods are highly proteinacious and cooked as vegetable; plant is used as green manure.

Remarks: C. ensiformis DC. var. *viresa* Baker, Hook.f. Fl. Brit. India 2: 196.1876 (Syn. *C. virosa* W. & A) is also commonly cultivated for vegetable crop.

8.3. ***Canavalia lineata*** DC. Prodr. 2: 404.1825; King, J.As. Soc. Beng. 66: 53.1898; Prain, J. As. Soc. Bengal 66: 418.1898, Beng. Plants 1: 281. 1903 (Rep. ed. 1963) & Rec. bot. Surv. India 3 (2): 197.1905. *C. obtusifolia* Baker, Hook.f. Fl. Brit. India 2: 196. (not of DC.) 1876; Watt Dict. Econ. Prod. 2: 98.1890. *Dolichos obcordatus* Roxb. Fl. Ind. 3: 303. 1832.

Perennial, stem spreading prostrate, sometimes twining, glabrous. Leaves 3-foliolate; petioles 8-12 cm long, stipules triangular, leaflets sub-coriaceous. Flowers 4-8 flowered racemes; corolla 2.5 cm long, violet; standard orbicular, emerginate . Pods few, linear oblong, shortly beaked, nearly straight, glabrous. Seeds 4-6, dark brown. *Fls. & Frts.:* January-March.

Seaface of the Sundarbans area; a littoral species.

Field Collection: Sagar Islands, *A. K.M.-7076* (CAL).

Distributions: Throughout the seacoast of India. Malaya, Sri Lanka, China, Japan, Polynesia and Australia.

9. *CICER* L.

9.1. ***Cicer arietinum*** L. Sp. Pl. 738.1753; Baker, Hook.f. Fl. Brit. India 2: 176.1876; Watt Dict. Econ. Prod. 2: 274.1889; Prain Beng. Plants 1: 259.1903 (Rep. ed. 1963) & Rec. bot. Surv. India 3 (2): 191.1905: Wlh. India 154.1950 *(Vern. Name: Chhola).*

Viscid, much branched, annual herb. Leaves usually with a terminal leaflet; stipules small, obliquely ovate, toothed. Flowers pink, blue or white. Pods oblong, pubescent, topped by the persistent base of the style; seeds obovate or sub-globose, beaked, reddish brown, black or white. *Fls. & Frts.:* January-March.

A winter crop, commonly cultivated during November -December.

Field Collection: Diamond Harbour, *K.N.-398.*

Distributions: Cultivated almost throughout India; native of Mediterranean regions.

Economic Importances: The seeds are largely eaten in the form of various types of preparation. It is also fed to horses. The vegetative parts are used as vegetables.

Notes: Cicer soongaricum Step.- a pot herb and occasionally found in the Western Himalaya. Its seeds are edible in Ladakh.

10. *CLITORIA* L.

10.1. ***Clitoria ternatea*** L. Sp. Pl. 753.1753; Baker, Hook.f. Fl. Brit. India 2: 208.1876; Watt Dict. Econ. Prod. 2: 375.1889; Prain Beng. Plants 1: 287.1903 (Rep. ed. 1963) & Rec. bot. Surv. India 3(2): 199.1905; Wlh. India 2: 233.1950. *(Vern. Name: Aprajita).*

Climbing annual herb. Leaflets ovate or oblong, obtuse, sub-coriaceous; stipules small, lanceolate. Flowers bright blue with an orange centre or white, solitary, axillary. Pods flat, 6-10 seeded. *Fls. & Frts.:* November-April.

In gardens everywhere, also self-sown and spontaneous in hedge and thickets.

Field Collections: Jugdia, *K.N.-231;* South Garia, *K.N.-823.*

Distributions: Throughout India; introduced from Malaya. In the Tropics.

Economic Importances: An ornamental plant; seeds purgative and roots as cathartic, diuretic and purgative.

11. *CROTALARIA* L.

Key to the Species

1a	Leaves digitately compound; leaflets always 3, ovate, obtuse	5. *C.pallida* Aiton
1b	Leaves simple	(2)
2a (1)	Stipules persisting as long decurrent wings along the twigs; pods linear oblong, many seeded, stipitate, glabrous	1. *C.alata* Ham. ex Roxb.
2b (1)	Stipules rare but often not decurrent along the twigs	(3)
3a (2)	Erect shrubs or herbs, seeds 10-20	(4)
3b (2)	Diffuse low herbs or rarely stemless shrubs; stipules small or absent	(7)
3c (2)	Prostrate herbs with flexuous, trailing stem	6. *C. prostrata* Roxb.
4a (3)	Shrubs with strict, erect branches; racemes both terminal and lateral; pods pubescent or valvety	(5)
4b(3)	Herbs with stout, unbranched stem, racemes terminal; pods glabrous	(6)
5a (4)	Branches angular; leaves ovate; stipules foliaceous semi-lunar, large; flowers usually blue and white	9. *C. verrucosa* L.
5b (4)	Branches rounded, grooved; leaves lanceolate, obtuse. Flowers yellow; pods sessile	3. *C. juncea* L.
6a (4)	Leaves obtuse; bracts deciduous, stipules subulate; minute	7. *C.retusa* L.
6b (4)	Leaves acute; bracts reflexed persistent; stipules foliaceous, leafy	8. *C. spectabilis* Roth
7a (3)	Leaves linear-oblong, obtuse; bracts lanceolate, foliaceous; stipules linear, persistent; pods sub sessile, glabrous	4. *C.mysorensis* Roth
7b (3)	Leaves linear oblanceolate, obtuse bracts linear, very minute, pods glabrous	2. *C. albida* Hey. ex Roxb.

11.1. ***Crotalaria alata*** Ham. ex Roxb., Don. Prodr. 241. 1825; Baker, Hook.f. Fl. Brit. India 2: 69.1876; Prain Beng. Plants 1:265. 1903 (Rep. ed. 1963 & Rec. bot. Surv. India 3 (2): 193. 1905; Wlh. India 2: 372.1950. *C. bilata* Roxb. Fl. Ind. 3: 274.1832.

Unbranched. Leaves simple, stipules large, persistent. Flowers often showy. Pods linear oblong, many seeded, stipitate, glabrous. *Fls. & Frts.:* October-December.

Common on the railway tracts and roadside hedges.

Field Collection: Baharu/on the railway tracts, *K.N.- 1625.*

Distributions: Almost throughout India; in the Tropical and Sub-Tropical regions of the world.

Economic Importances: Used as fodder in Sumatra and green manure.

11.2. ***Crotalaria albida*** Heyne ex Roth, Nov. Pl. Sp. 333.1821; Baker, Hook.f. Fl. Brit. India 2: 71.1876; Prain Beng. Plants 1: 266.1903 (Rep. ed. 1963). *C. punctata* Grah., Watt Cat. 5401. *C. epunctata* Dalz. Hook. Kew. J. 3: 210.1857. (*Vern.Name: Atasi).*

Much branched undershrub; branches slender, terete, more or less silky pubescent. Leaves linear oblong or oblanceolate, obtuse, apiculate, pellucid-puncate, more or less silky pubescent, base acute. Flowers in terminal standard, sub-orbicular, with a few hairs on the back, wings obovate oblong. Pods glabrous, sessile, oblong cylindric. Seeds 6-12. *Fls. & Frts.:* May-July.

Rare on the shaddy area and on the garden edges.

Field Collections: Dewanganj, *K.N. 2265*; Barrakpore, *Dutta-23.*

Distributions: Throughout the plains of India; Malaya, Sri Lanka, China and Philippines.

Economic Importances: Green manure in the agriculture fields.

11.3. ***Crotalaria juncea*** L. Sp. Pl. 714.1753; Baker, Hook.f. Fl. Brit India 2: 79.1876; Watt Dict. Econ. Prod. 2: 595.1889; Prain Beng. Plants 1: 265.1903 (Rep. ed. 1963) & Rec. bot. Surv. India 3(2): 193.1905; Wlh. India 2: 373.1950. *(Vern. Name: San).*

Annual shrub, 1-3 m high, branches numerous, ascending, slender, terete, striate, silky pubescent. Leaves linear or oblong, obtuse or subacute, apiculate. Flowers large in erect terminal and lateral 12-20 flowered; corolla bright yellow, slightly exserted. Pods sessile clothed with short fulvous silky hairs. *Fls. & Frts.:* December-April.

Cultivated but common as a weed in the roadside hedges.

Field Collections: Beliachandi, *K.N. 213;* Madhyampur, *K.N.-957.*

Distributions: Almost throughout India; probably indigenous in India. Malaya Islands and Australia.

Economic Importances: Stem bark produces valuable fibre for cordages and bags.

11.4. ***Crotalaria mysorensis*** Roth, Nov. Pl. Sp. 338.1821; Baker, Hook.f. Fl. Brit. India 2: 70.1876; Prain Beng. Plants 1: 265.1903 (Rep. ed. 1963).

Erect much branched 0.3-0.6 cm high, branches clothed with dense hairs. Leaves sub-sessile, linear oblong, with rounded apex, apiculate. Flowers yellow in lax racemes. Pods sub-sessile, glabrous when ripe, pale brown. *Fls. & Frts.:* November-March.

Occasionally grow on the waste lands and railway tracts.

Field Collection: Diamond Harbour, *K.N.-1454.*

Distributions: Throughout India; Sri Lanka.

11.5. ***Crotalaria pallida*** Aiton, Hort. Kew 3: 20.1789; Gandhi, Saldhana & Nicolson, Fl. Hassan. Dist. 243.1976. *C. mucronata* Desv. J. Bot. 3: 76.1814. *C. striata* DC. Prodr. 2: 131. 1825; Baker, Hook. f.Fl. Brit. India 2: 84.1876; Watt Dict. Econ. Prod. 2: 614.1889; Wlh. India 2: 382.1950. *C. brownej* Bert. ex. DC. Prod. 2: 130.1825. *C. saltiana* (non Andr. 1811) Prain ex King, J. As. Soc. Bengal 66(2): 41.1897, Beng. Plants 1: 264.1903 (Rep. ed. 1963), Rec. bot. Surv. India 2: 297.1903 & Ibid. 3(2): 193.1905. *(Vern. Name: Ban Can.).*

Undershrub with slender branches. Leaves trifoliate, stipules minute, setaceous, deciduous; leaflets usually sub-obtuse, green and glabrous above, pale and obscurely silky below. Flowers yellow, streaked with purple, in racemes; calyx thinly silky. Pods deflexed rather recurved, 20-30 seeded. *Fls. & Frts.:* February-October.

Wayside, railway tracts and waste lands, every-where.

Field Collection: Sitakundu/Baruipur, *K.N.-54.*

Distributions: Throughout India; Sri Lanka, Malaya, Tropical America and Africa.

Economic Importances: Used as cover plant in tea, rubber and coconut plantations.

11.6. ***Crotalaria prostrata*** Roxb. (Hort. Beng.54.1814) *nom. nud.* Fl. Ind. 3: 270.1832; Baker, Hook. Fl. Brit. India 2: 67.1876; Watt Dict. Econ.

Prod. 2: 613.1889; Prain Beng. Plants 1: 266.1903 (Rep. ed. 1963). *(Vern. Name: Chhota Jhunjhan).*

Small annual herb, 15-40 cm high; branches profuse, ascending, slender, clothed with silky yellowish hairs. Leaves sub-sessile oblong or oblong obovate, apex rounded; stipules absent. Flowers small, pedicelled in 4 flowered racemes, peduncle slender. Calyx densely hairy; corolla yellow, not exerted; standard broadly oblong. Pods inflated glabrous. Seeds 12-13, highly polished. *Fls. & Frts.:* November-December.

Occasionally found on the dry waste lands and roadside thickets.

Field Collection: Subuddhipur/Baruipur, *K.N.-1475.*

Distributions: Drier parts of India; Sri Lanka and Java.

11.7. ***Crotalaria retusa*** L. Sp. Pl. 715.1753; Baker, Hook.f.Fl. Brit. India 2: 75.1876 (excl. syn. Dalz. & Gibs.); Watt Dict. Econ. Prod. 2: 613. 1889; Prain Beng. Plants 1: 265.1903 (Rep. ed. 1963), Rec. bot. Surv. India 2: 297.1903 & Ibid. 3(2): 193.1905; Wlh. Ind. 2: 382.1950. *(Vern. Name: Bhil- jhanjhan).*

Robust undershrub, 1-2m high; branches striate, glabrous or oppressedly pubescent. Leaves oblanceolate oblong, obtuse glabrous above, silky pubescent beneath, base cuneate; petioles very short. Flowers numerous, large and showy yellow with purple tinge in erect terminal racemes. 15-30 cm long. Pod 2.5-4 cm long, stalked, liner-oblong, slightly broader upwards. *Fls.& Frts.:* July-October.

On the railway tracts, roadside thickets and village scrubs.

Field Collection: Ichhapur, *K.N.-1802.*

Distributions: Throughout India; Sri Lanka, China, Malaya, North Australia and Tropical Africa.

Economic Importances: Used as green manure in Java and also used in treatment of scabies and impetigo.

11.8. ***Crotalaria spectabilis*** Roth, Nov. Pl. Sp. 341.1821. *C. sericea* Retz. Obs. Fasc. 5: 26.1789; Baker, Hook.f. Fl. Brit. India 2: 75.1876; Watt Dict. Econ. Prod. 2: 614.1889; Prain Beng. Plants 1: 265.1903 (Rep. ed. 1963); Wlh. India 2: 32.1950. *(Vern. Name: Pipali Jhanjhan).*

Erect herb or undershrub; branches striate subglabrous. Leaves simple, glabrous above and finely silky beneath, oblanceolate to obovate. Flowers in terminal 20-30 cm long, racemes, yellow; standard with dark strains. Pod linear oblong, cylindric, 20-30 seeded. *Fls. & Frts.:* November-March.

Commonly grow on the pasture lands and roadside thickets.

Field Collection: Piyalinagar/Baruipur, *K.N.-102.*

Distributions: Throughout India and Andaman; Malaya.

Economic Importances: Medicinally used in scabies and impetigo; in U.S.A. used as fodder.

11.9. ***Crotalaria verrucosa*** L. Sp. Pl. 715.1753; Baker, Hook.f. Fl. Brit. India 2: 77.1876; Watt Dict. Econ. Prod. 2: 614.1889; Prain Beng. Plants 1: 265.1903 (Rep. ed.1963), Rec. bot. Surv. India 2: 297.1903 & Ibid. 3(2): 193.1905. *(Vern. Name: Bansan* or *Jhunjhun)*

Much branched herb; branches acutely angled. Leaves ovate rhomboid or ovate deltoid, obtuse or occasionally acute at the apex. Flowers in terminal or lateral racemes; blue and white. Calyx membranous, faintly pubescent; tube short, campanulate. Pods densely villous when young but at maturity softly pubescent or valvety. *Fls. & Frts.:* October-March.

Rarely grow on the grasslands and waste places.

Field Collection: Harinbari/Sagar Islands, *K.N.-474.*

Distributions: India generally. Sri Lanka, Malaya Peninsula, Archipelago, China, N. Australia and Tropical America.

Economic Importances: Leaves used as medicine in biliousness, dyspepsia, fever, blood impurities, throat and mouth diseases and heart complaints.

12. *DALBERGIA* L.f.

Key to the Species

1a	Tall to medium sized deciduous tree with glabrous branches; pods thin and flattened except opposite the seeds	(2)
1b	Small tree or erect glabrous shrub	(4)
2a (1)	Stamens in two lateral phalanges of 5-filaments each; calyx teeth as long as tube	1. *D.lanceolaria* L.f.
2a (1)	Stamens 9-10 in one bundle; vexillary stamens united to the edge of sheath or absent	(3)

3a (2)	Leaflets roundish, distinctly cuspidate; panicles small, pedicels short; pods narrow, 6-8 times as long as broad	4. *D. sisoo* Roxb.
3b (2)	Leaflets oblong or orbicular, obtuse or retuse, pods only 2-3 times as long as broad	2. *D.latifolia* Roxb.
4a (1)	Leaflets 9-15, thinly sub-coriaceous; flowers numerous in axillary panicles about equalling the leaves; pods 2.5-7.0cm long, 1-seeded broadly elliptic lanceolate	3. *D. melanoxylon* Guiller & Perr.
4b (1)	Leaflets 7-11, firm, oblong, crowded from the nodes, flowers in congested corymbose axillary panicles; pods 2.5cm long, reniform, flat, glabrous, brown	5. *D. spinosa* Roxb.

12.1. ***Dalbergia lanceolaria*** L. f. Supp. 316.1781; Baker, Hook.f.Fl. Brit. India 2: 235.1876; Watt Dict. Econ. Prod. 3: 6.1890; Prain Beng. Plants 1: 293. 1903 (Rep. ed.1963) & Rec. bot. Surv. India 3(2): 200. 1905. *D. frondosa* Roxb. Fl. Ind. 3: 226.1832; Prain, Ann. Roy. Rot. Gard.(Calcutta) 10(1): 93, t.76.1905. *(Vern. Name: Chakmedia).*

Large, deciduous tree, 20-25 m tall with glabrous branches. Leaves imparipinate. Flowers pinkish white or purplish. Pods narrowed at both the ends, 1-3 seeded. *Fls. & Frts.:* September- January.

Planted on the roadsides, as avenue trees.

Field Collection: Falta, *K.N.-2009.*

Distributions: Bihar and Western Peninsula. Native of Sikkim terai, Sri Lanka.

Economic Importances: Bark contains 14% tannin; bark decoction used in dyspepsia; seed oil is said to be useful for rheumatic affections. The timber is used for fuel and carpentry works.

12.2. ***Dalbergia latifolia*** Roxb. Cor. Pl. 2: 7.t.113.1799; Baker Hook.f. Fl. Brit. India 2: 231.1876; Watt Dict. Econ. Prod. 3: 7.1890; Prain Beng Plants 1: 294.1903, Ann.Roy. Bot. Garden (Calcutta) 10(1): 80.1905 & Rec. bot. Surv. India 3(2): 201.1905. *(Vern. Name: Swet-sal).*

Tall to medium tree. Large imparipinnate leaflets oblong or orbicular. Flowers whitish, long axillary panicles. Pods flat, strap shaped, 2 or 3 times as long as broad. *Fls. & Frts.:* April- August.

Occasionally planted on the avenues or in the roadside gardens.

Field Collection: Majerhat, *K.N.-2248.*

Distributions: Indigenous to Central and South, as well as on the outer ranges of Himalayas; Sikkim and Bangladesh.

Economic Importance: Produce valuable timber, the leaves are used as fodder and green manure.

12.3. ***Dalbergia melanoxylon*** Guiller & Perr., Flor. Seneg. 1: 227. t.53.1830; Prain J. As. Soc. Bengal. 66: 446.1898; Wlh. India 3: 11.1952. *D. stocksii* Benth., J. Linn. Soc. 4: 42.1860; Baker, Hook.f. Fl. Brit. India 2: 234.1876.

Small tree, 5-8 m high; branches more or less armed with stout spines; bark white. Leaves 10-12 cm long, stipules very caducous. Leaflets 9-15, thinly sub-coriaceous. Flowers numerous in axillary panicles about equalling the leaves. Pods 2.5-7.0 cm long, 1-seeded, broad elliptic lanceolate, acute, quite glabrous reticulately veined. *Fls. & Frts.:* November-May.

Planted as avenue tree; very rare.

Field Collection: Champahati, *K.N.-928.*

Distributions: Bengal, Tamil Nadu, Karnataka and Maharashtra. In the Tropics.

Economic Importances: Wood is remarkably hard and useful; this wood was called ebony in ancient Egypt. Wood is useful for musical instruments, handles of surgical instruments, walking sticks and fancy articles.

12.4. ***Dalbergia sissoo*** Roxb. (Hort. Beng. 53.1814) *nom. nud.* & Fl. Ind. 3: 233.1832; Baker, Hook.f. Fl. Brit. India 2: 231. 1876; Watt Dict. Econ. Prod. 3: 13.1890; Prain Beng. Plants 1: 294. 1903 (Rep. ed. 1963) & Rec. bot. Surv. India 3(2): 200.1905; Benthall, Trees Calc. 152.1948; Wlh. India 3: 7.1952. *(Vern. Name: Sisu).*

Large deciduous tree. Leaves imparipinnate, accuminate, glarbous. Flowers yellowish white, recurved on the branches of axillary, pubescent panicles. Pods strap shaped, 1-3 seeded. *Fls. & Frts.: January-February.*

Planted on the roadsides, village gardens and railway stations.

Field Collection: Mallickpore/near rail gate, *K.N.1418*.

Distributions: Temperate regions of India. Native of the outer hills of the Himalayas.

Economic Importances: Very hard and fine grained timber, used for cabinet making, musical instruments, tool handles and different other articles of commerce. Bark is good paper pulp.

12.5. ***Dalbergia spinosa*** Roxb. Fl. Ind. 3: 233.1832; Baker, Hook.f. Fl. Brit. India 2: 238.1876; Watt Dict. Econ. Prod. 3: 15.1890; Prain Beng. Plants 1: 294.1903 (Rep. ed. 1963) & Rec. bot. Surv. India 3(2): 201. 1905. *(Vern Name: Chulia Kanta).*

Stiff, erect glabrous shrub; branches short, ending in a hard spines. Leaves crowded from the nodes of the spinous branchlets; rachis slender, rusty pubescent, leaflets 7-11, firm oblong obtuse, glabrous, petioles very short. Flowers long reniform, flat, glabrous, brown. *Fls. & Frts.:* January-March.

Plentiful in the Sundarbans area, on the dense thickets.

Field Collection: Kakdeep/near river banks, *K.N.-2912.*

Distributions: Throughout the sea-coast in India; Bangla Desh.

Economic Importance: Fuel woods, charcoal and timber.

13. *DERRIS* Lour.

Key to the Species

1a	Small to medium tree; pods thick, not winged	1. *D. indica* (Lamk.) Bennett.
1b	Scandent shrubs; pods thin, winged	(2)
2a (1)	Vexillary stamens united; flowers fascicled on tumid nodes; pods not sinuate	(3)
2b (1)	Vexillary stamens quite free; flowers single in simple thyrsoid panicles; pods winged along upper suture	3. *D. sinuata* Benth.
3a (2)	Pods narrow, pointed at both the ends, seeds more than one	2. *D. scandens* Benth.
3b (2)	Pods sub-orbicular, obtuse, glabrous; seeds solitary	4. *D. trifoliata* Lour.

13.1. ***Derris indica*** (Lamk.) Bennett, J. Bombay nat. Hist. Soc. 65: 302. 1971; Saldhanha & Nicolson Fl. Hassan Dist. 248.1976. *Galedupa india* Lamk. Ency. 2: 594.1888-1889. *Cytisus pinnatus* L. Sp. Pl.1: 741. 1753 (non *D. pinnata* Lour.1790). *Robinia mitis* L. Sp. Pl. 2: 1044. 1763 (*nom. illegit*). *Pongamia glabra* Vent. Jard. Malm. 28.1803; Baker, Hook.f. Fl. Brit. India 2: 240.1876; Watt Dict. Econ. Prod. 6(1): 322. 1893; Prain Beng. Plants 1: 291.1903 (Rep. ed. 1963) & Rec. bot. Surv. India 2: 300.1903. *P. pinnata* (L.) Pierre, Fl. For. Cochinch. sub. t. 385.1899; Benthall Trees Calc. 149.1946. *(Vern. Name: Karanja).*

Moderate sized tree. Leaves compound; leaflets opposite, 5-15 cm long, ovate-oblong or elliptic, acute or shortly acuminate, glabrous; stipules small, oblong, caducous. Flowers pinkish white, in short axillary racemes; calyx purplish truncate or obscurely toothed. Pods 3-5 cm long, oblong or slightly obovoid, compressed, 1-seeded. *Fls. & Frts.:* July-March.

Frequently found near canals, old gardens, waste places and hedges; sometime noticed in the railway stations, common in Sundarbans.

Field Collections: Baruipur, *K.N.118;* Gosaba, *K.N.1993.*

Distributions: Throughout the coastal regions of India and in Tropical Asia and Seychelles.

Economic Importances: The powder of the seed is commonly used in bronchitis and whooping cough. The oil from the seed is used in rheumatism.

13.2. ***Derris scandens*** Benth., J. Linn. Soc. 4: 103. (Suppl.) 1860; Baker, Hook.f. Fl. Brit. India 2: 240.1876; Watt Dict. Econ. Prod. 3: 81.1890; Prain Beng. Plants 1: 292.1903 (Rep. ed. 1963), Rec. bot. Surv. India 2: 301.1903 & Ibid. 3(2): 200.1905; Wlh. India 3: 38. 1952. *Dalbergia scandens* Roxb. Cor. Pl. 2: 192.1805 & Fl. Ind. 3: 232.1932. *Brachypterum scandens* Benth., Ann. Wiea. Muss.2: 101.1840. *(Vern. Name: Noalata).* (Fig. 11).

Large climber; stem smooth, dark green, young part pubescent. Leaves glabrous, stipules small, caducous; leaflets opposite, elliptic-oblong or obovate-oblong, sub-acute. Flowers numerous in short peduncled axillary racemes, arranged in fascicles from the nodes of a puberulous rachis pink coloured. Pods tapering to both ends, pointed, seeds 1-4. *Fls. & Frts.:* February -April.

Very common on the bushes, near the scrub jungles of the river sides in Sundarbans.

Field Collections: Gosaba, *K.N.-179;* Bhangankhali, *K.N.-165.*

Fig. 11. DERRIS SCANDENS Benth. — with pods.

Distributions: Sea coast and river banks throughout India; Sri Lanka, Burma, S.E. Asia to North Australia.

Economic Importances: Seed has been sown on sectional population for plantation in the tidal mangrove forest areas. Seed powder is poisonous for fish as it contains rotenone; it also has insecticidal properties.

13.3. ***Derris sinuata*** Benth. ex Thw. Enum. Pl. Zeyl. 93.1855; Baker, Hook.f. Fl Brit. India 2: 246.1876; Prain Beng. Plant 1: 291.1903 (Rep. ed. 1963) & Rec. bot. Surv. India 2: 301.1903. *(Vern. Name: Sundrilata* or *Mahajanilata).*

Large climbing shrub. Leaves odd-pinnate, alternate; stipules small; leaflets opposite. Flowers single in ample thyrsoid panicles; vexillary stamen quite free. Pods winged along upper suture; sutures sinuate between the seeds. *Fls. & Frts.:* February-March.

Rare on the scrub-jungles of the sea coast of Sundarbans.

Field Collection: Frazergunj, *K.N.-1913.*

Distributions: Coastal regions of India; Burma and Malaya.

Economic Importance: Act as soil binder in tidal area.

13.4. ***Derris trifoliata*** Lour. Fl. Cochinch. 2: 433.1790; Wlh. Ind. 3: 39.1952; Thothatri, Bull. bot. Surv. India 3: 181.1962. D. *uliginosa* Benth., Miq. Pl. Jungh. 1: 252.1852; Baker, Hook.f. Fl. Brit. India 2: 241.1876; Prain Beng. Plants 1: 292.1903 (Rep. ed. 1963), Rec. bot. Surv. India 2: 301.1903 & Ibid. 3(2): 200.1905. *Pongamia uliginosa* DC. Prodr.2: 416. 1824. *Galedupa uliginosa* Roxb. (Hort. Beng. 53.1814) & Fl. Indica 3: 243.1832. *(Vern. Name: Pan-Lata).* (Fig. 12).

Large glabrous scandent shrub; bark dark-grey with white lenticels. Leaves 10-20 cm long; leaflets subcoriaceous, ovate or oblong, acuminate. Flowers in axillary upto 15 cm long recemes; corolla rose coloured. Pods variable in shape, obliquely rounded or orbicular. Seed solitary, sub-orbicular, reddish brown. *Fls. & Frts.:* January-March.

Common on the river banks of Sundarbans.

Field Collection: Diamond Harbour, *K.N.-389.*

Distributions: Muddy and sandy sea coasts and river banks of Indian sub-continent; Sri Lanka, China, North-Australia, Polynesia, Madagascar and Zambesi Islands.

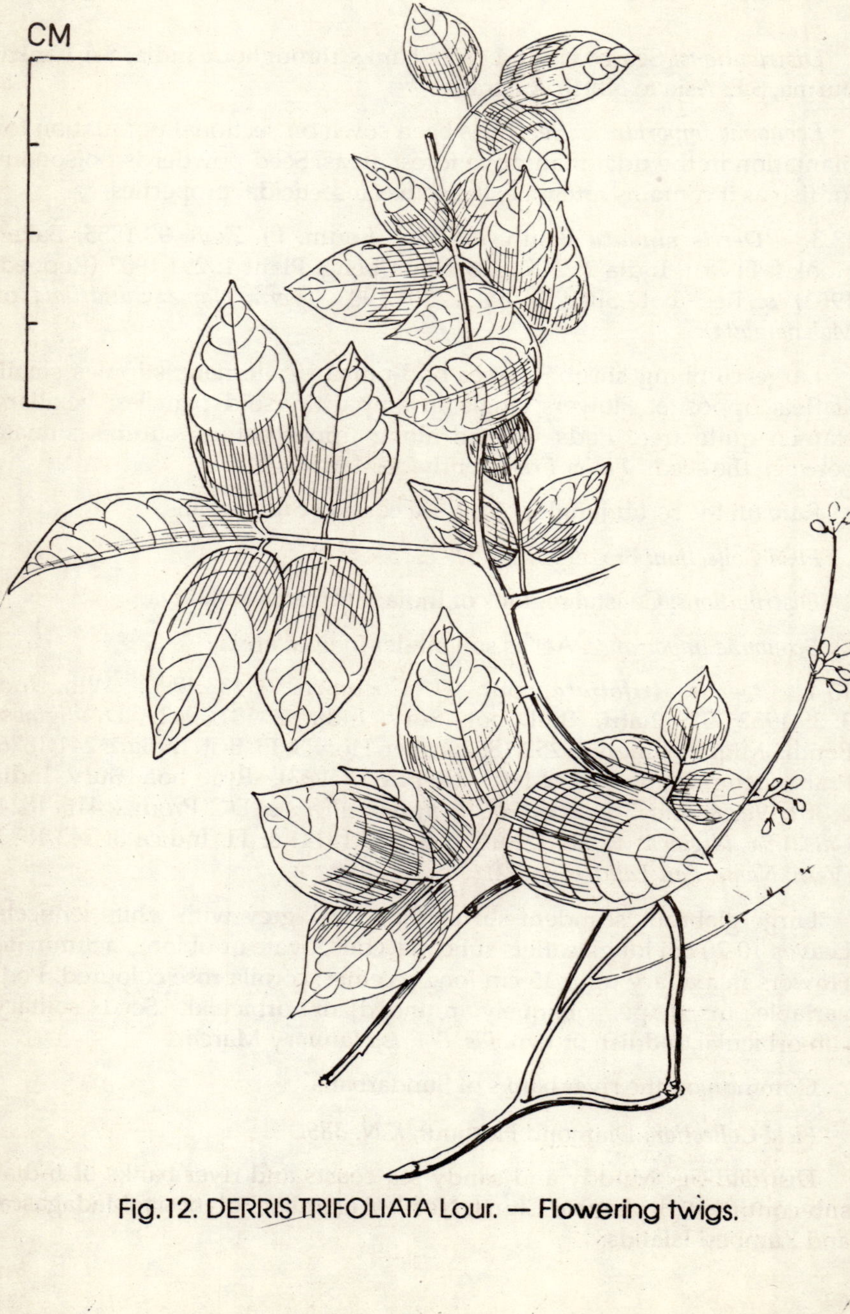

Fig. 12. DERRIS TRIFOLIATA Lour. — Flowering twigs.

Economic Importances: Entire plant is used as stimulant, antispasmodic; bark is used as an alternative in rheumatism; leaves are used as fodder in Guntur and Madras.

14. *DESMODIUM* Desv. *nom. cons.*

Key to the Species

1a	Leaves 1-foliolate	1. *D.gangeticum* (L.) DC.
1b	Leaves 3-foliolate	(2)
2a (1)	Shrubs or undershrubs; branches terete	3. *D. umbellatum* DC.
2b (1)	Erect herb; branches not terete	2. *D. triflorum* (L.) DC.

14.1. ***Desmodium gangeticum*** (L.) DC. Prodr. 2: 327. 1825; Baker, Hook. f. Fl. Brit. India 2: 168.1876 (including var. *maculatum*); Watt Dict. Econ. Prod. 3: 82.1890; Prain Beng. Plants 1: 304.1903 (Rep. ed. 1963); Wlh. Ind. 3: 41.1952. *Hedysarum gangeticum* L. Sp. Pl. 746.1753. *(Vern. Name: Salpani).*

Erect or sub-erect undershrub; stem slightly angular, branches clothed with appressed white hairs. Leaves 1-foliate; leaflets 5-10 cm long ovate elliptic to ovate oblong, glabrous above, grey- puberulous beneath; stipules linear, sub-persistent. Flowers lilac or white in axillary or terminal racemes. Pods deeply indented on lower suture, indehiscent, usually covered with minute hooked hairs. *Fls. & Frts.:* May -February.

Common on the bushes and waste places; orchards, railway tracts.

Field Collections: Baruipur, *K.N.-9;* Canning, *K.N.184.*

Distributions: Throughout India, ascending up to 160 m on hilly areas; Sri Lanka, Tropical Africa, Malaya Islands, Philippines and China.

Economic Importances: Used as green manure. Also useful for the treatment of febrifuge and anticatarrhalic.

14.2. ***Desmodium triflorum*** (L.) DC., Prodr. 2: 334.1824; Baker, Hook.f.Fl. Brit. India 2: 173.1876; Watt Dict. Econ. Prod.3: 84.1890, Prain Beng. Plants 1: 303.1903 (Rep. ed. 1963) & Rec. bot. Surv. India 3(2): 202.1905: Wlh.India 3: 42.1952. *Hedysarum triflorium* L. Sp.Pl.749.1753. *(Vern.Name: Koddalia).*

Prostrate herb with copiously branched trailing stem and trifoliate leaves; rooted at lower nodes. Flowers axillary, white or pinkish. Pods jointed, pubescent. Fruits covered by leaves. *Fls. & Frts.:* June-December.

Common on the grass lands, railway tracts, road-side hedge.

Field Collections: Baruipur, *K.N.-93;* Dandirhat, *K.N.-971.*

Distributions: Throughout the plains of India; cosmopolitan in the Tropics.

Economic Importances: Leaves used stimulating flow of milk, useful as remedy for diarrhoea, dysentery and convulsion. Fresh leaves are applied to wounds and abscesses.

14.3. ***Desmodium umbellatum*** DC. Prodr. 2: 325.1825; Baker, Hook.f. Fl. Brit. india 2: 161.1876; Prain Beng. Plants 1: 304.1903 (Rep. ed. 1963) & Rec. bot. Surv. India 2: 298.1903; Wlh. India 3: 43.1952.

Bushy shrub, 2-5 m high; young branches terete, downy with appressed pubescence. Leaves trifoliolate. Flowers in axillary peduncled 6-12 flowered umbels, white. Pods sparsely silky when young, glabrous when ripe; thick, coriaceous, almost turgid, joints 4-5. *Fls. & Frts.:* November-March.

Rarely on the river banks of Sundarbans.

Field Collection: Kakdwip, *K.N.-1922.*

Distributions: The coastal India; Sri Lanka, Mascarene Islands, Malaya, Philippines and Polynesia.

Economic Importances: Grown as soil binder and for fodder.

15. *DOLICHOS* L.

Key to the Species

1a	Style filiform throughout, penicillate round the stigma; pods linear recurved	1. *D. biflorus* L.
1b	Style clavate thickened upwards; pods oblong, recurved, tapering to apex	2. *D. lablab* L.

15.1. ***Dolichos biflorus*** L. Sp. Pl. 727.1753; Baker, Hook. f. Fl. Brit. India 2: 210.1876; Watt Dict. Econ. Prod. 3: 175.1890; Prain Beng. Plants 1: 278.1903 (Rep. ed. 1963) & Rec. bot. Surv. India 3(2): 197.1905; Wlh. India 3: 101.1952. *D. uniflorus* Lamk. Encyc. Meth. 2: 299.1878. *(Vern. Name: Kuratta Kalai).*

Twining prostrate undershrub. Leaves pinnately 3-foliate; stipules small sub-persistent. Flowers 1-3 axillary, yellow. Pods recurved, 5-6 seeded. *Fls. & Frts.:* September-January.

Cultivated pulse crop.

Field Collection: Santoshpur/near Budge Budge, *K.N.-1961.*

Distributions: Native of India and found throughout; Tropics of the Old World.

Economic Importances: Grains and the leaves are good fodder; seed contain high protein and cooked as pulse.

15.2. ***Dolichos lablab*** L. Sp. Pl. 725.1753; Baker, Hook.f. Fl. Brit. India 2: 209.1876: Watt Dict. Econ. Prod. 3: 183.1890; Prain Beng Plants 1:278.1903 (Rep. ed. 1963) & Rec. bot. Surv. India 3 (2): 197.1905: Wlh. India 3: 104.1952. *(Vern. Name: Sim).*

Glabrous twinning shrub. Leaves 3-foliolate; stipules basifixed ovate, acute, entire, base cuneate or deltoid. Flowers white or pink. Pods flat, beaked. *Fls. & Frts.:* September - January.

Cultivated everywhere and almost in the every vegetable gardens.

Field Collection: Bhangankhali, *K.N.-141.*

Distributions: Cultivated throughout India; Tropics of the Old World.

Economic Importances: Pods are used as vegetable and as fodder also.

Notes.: Pods are highly varibale in shape, size and colours.

16. *ERYTHRINA* L.

Key to the Species

1a	Small to medium tree, branches more, leaves ovate to oblong, obtuse, length double than the bredth; flowers deep crimson	1. *E. fusca Lour.*
1b	Small tree, branches less spreading; leaves broad ovate to rhombic-ovate, nearly as long as the broad	2. *E. variegata* L.

16.1. ***Erythrina fusca*** Lour. Fl. Cochinch. 427.1790; Wlh. Ind. 3: 198.1952. *E. ovalifolia* Roxb. Fl. Ind. 3: 254.1832; Baker, Hook.f. Fl. Brit. India 2: 189.1876; Prain Beng. Plants 1: 284.1903 (Rep. ed. 1963), Rec. bot. Surv. India 3(2): 198.1905; Benthall trees Calc. 158.1946. *(Vern. Name: Harikakra).*

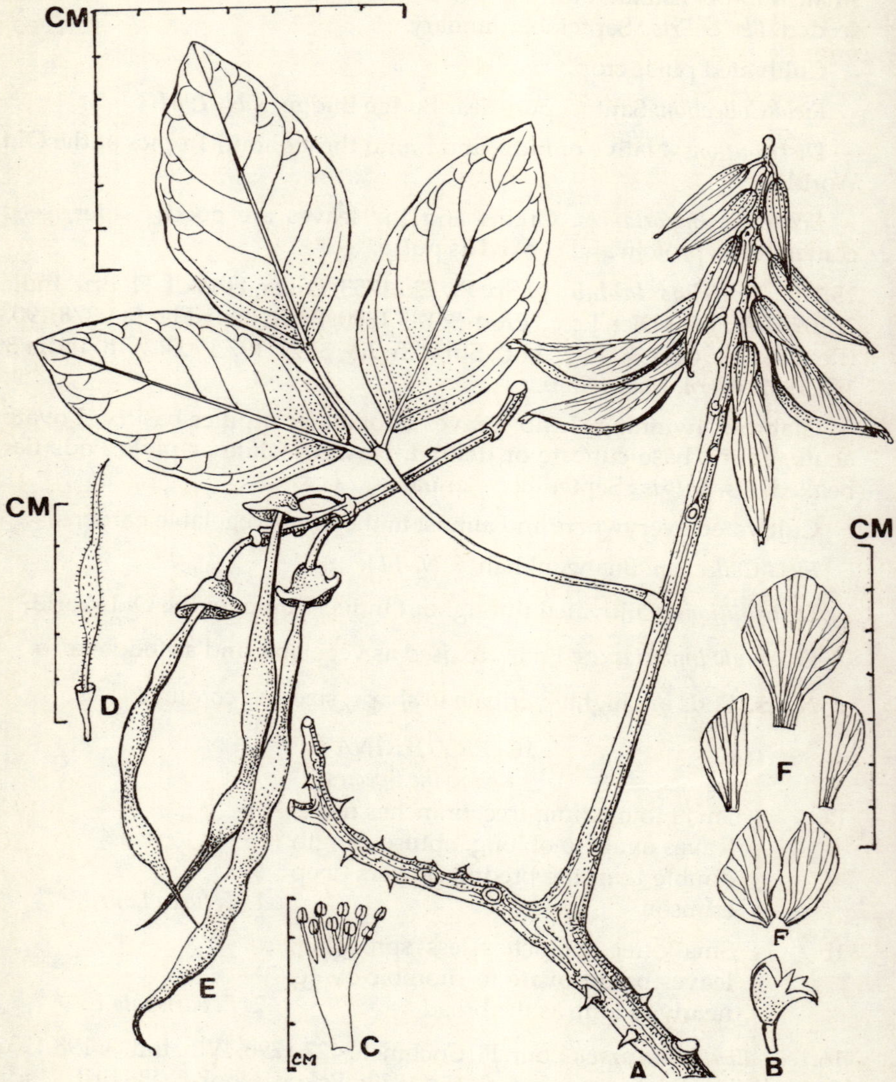

Fig. 13. ERYTHRINA FUSCA Lour.

A. Flowering branch, B. Calyx tube, C. Androecium with anthers, D. Young pod, E. Mature Pods, F. Petals.

Small to medium tree, deciduous, much spreading branches; trunk armed with numerous prickles. Deep crimson flowers on the cluster. Pods numerous, cluster 10 -15 cm long. *Fls. & Frts.:* February-May. (Fig. 13).

Occasionally on the village shrubberies and roadsides.

Field Collection: Subhashgram, *K.N.-1340.*

Distributions: Bengal and Assam. In Malaya and Bangla Desh.

Economic Importances: Quick growing plant so useful as fire wood; grow as shade plant for coffee and other crops.

16.2. ***Erythrina variegata*** L. Herb. Amb. 10. 1754; Bullock, Kew Bull. 20: 294.1966. *E. variegata* L. var. *orientalis* (L.) Merr. Inter. Herb. Amb. 276.1917; Wlh. Ind. 3: 197.1952; Santapau, Rec.bot. Surv. India 18: 59.1960. *E. indica* Lamk. Encycl. 2:188.1876; Baker, Hook. f. Fl. Brit. India 2:188.1876; Watt Dict. Econ. Prod. 3:269. 1890; Prain Beng. Plants 1:298.1903 & Rec. bot. Surv. India 3(2): 198.1905. *(Vern.Name: Palita Mandar).*

Moderate sized, deciduous tree; trunk armed with numerous prickels. Leaflets broad ovate to rhombic ovate, nearly as long as broad. Flowers bright scarlet, showy, numerous; calyx spathaceous. Pods in clusters. *Fls. & Frts.:* March-June.

Planted in gardens for ornamental purposes. Sometimes grow wild. The flowers appear on leafless branches.

Field Collections: Haroa, *K.N.-584;* Malancha, *K.N.-1350.*

Distributions: Almost throughout India; native of the coast of India; Java and Polynesia.

Economic Importances: Often planted for boundering the village gardens; the leaves have some medicinal values.

17. *FLEMINGIA* Roxb. ex Aiton *nom. cons. recomm.*

17.1. ***Flemingia macrophylla*** (Willd.) Prain ex Merrill, Philipp. J. Sci. Bot. 5: 130. 1910. *Crotalaria macrophylla* Willd. Sp. Pl. 3: 982.1802. *Flemingia congesta* Aiton, Hort. Kew (ed. 2) 4: 349.1812: Roxb. Fl. Ind.3: 340.1832; Baker, Hook. f. Fl. Brit. India 2: 228. 1876; Watt Dict. Econ. Prod. 3: 400. 1890; Prain Beng. Plants 1:268. 1903 (Rep. ed. 1963), Rec. bot. Surv. India 2: 300.1903 & Ibid 3(2): 194.1905. *Moughania macrophylla* (Willd.) Kuntze, Rev. Gen.: 1.1891: Backer & Bakh.f. Fl. Java 1: 638.1903 *(Moghania);* Naskar, *et al.* Bull. bot. Soc. Bengal. 33: 39-43.1979. *(Vern.Name: Bara salphan).*

Erect shrub 1-2 m high, young branches angular, sulcate, appressedly pubescent. Leaves trifoliate, leaflets subcoriaceous, ovate oblong, acute, dark-green. Flowers in dense, axillary velvety congested racemes shorter than the petioles, pedicels very short, hairy; bracts lanceolate, densely velvety; corolla broad; white streaked with pink wing purplish. Pods oblong, turgid, apiculate, finely pubescent, not glandular. Seed 2. *Fls. & Frts.:* December-March.

On the hedges and thickets of the road sides.

Field Collection: Jugdia-Benipur, *K.N.216.*

Distributions: Bihar and W. Peninsula; Sri Lanka and Burma.

18. *GLIRICIDA* H.B.& K.

18.1. ***Gliricida sepium*** (Jacq.) Kunth ex Steud. *Nom Bot* (ed.2), 1: 688.1840; Wlh. India 4: 137.1956. *Robinia sepium* Jacq. Enum. P. Carib. 28. 1760. *G. maculata* H.B. & K. Nov. Gen. & Sp. Pl. 6: 393.1823.

Deciduous tree. Leaves alternate, grey pubescent when young; leaflets opposite, 12-14 pairs, elliptic ovate, entire, sub-acute. Inflorescence a branched raceme; bracts small, caducous. Calyx cup-shaped glabrous; corolla much exerted, rosy white or pink. Pods thick, woody, narrow at both ends, pale yellow. *Fls. & Frts.:* January-March.

Occasionally cultivated as an ornamental plant.

Field Collection: Harinbari/Sagar Islands, *K.N.-472.*

Distributions: This tree is introduced in India and now common in Madras, Mysore, Bombay and Travancore-Cochin and other parts of India. Native of Tropical America.

Economic Importance: Ornamental, avenue tree; source of green manure for paddy in Madras and has recommended for cultivation on the bunds of the agriculture fields. Seed powder is pesticide.

Notes: D. Prain (1903a & 1905) has not mentioned this taxon from Bengal as it is a recently introduced plant in India.

19. *GLYCINE* L.

19.1. ***Glycine max*** (L.) Merrill., Interpr. Rump. Herb. Amboin. 274. 1917; Wlh. Ind. 4: 142.1956. *G. hispida* Maxim., Bull. Acad. Peterb. 18: 398. 1873; Prain Beng. Plants 1: 279.1903 (Rep. ed. 1963). *G. soja* Sieb. & Zucc. Abh. Akad. Nuench. 4 (2): 119. 1843; Baker, Hook.f.Fl. Brit. India 2:

184.1876.(not of Sieb. & Zucc.) ; Watt Dict. Econ. Prod. 3: 509.1890. *Dolichos soja* Roxb. Fl. Ind. 3: 314.1832. *(Vern. Name: Soya Bean).*

Trailing or sub-erect herb, 0.3 - 0.6 m high; stem hirsute. Leaflets ovate or oblong-lanceolate, hairy on both the side; petioles 12-13 cm long. Flowers blue in sessile very short and hirsute receme. Pods oblong, narrow at the base, straight or slightly curved. Seeds compressed, ellipsoid; deep brown or black, *Fls. & Frts.:* October- May.

Sometime cultivated for its nutritive fruits.

Field Collection: Narendrapur/public garden, *K.N.-1986.*

Distributions: Cultivated almost throughout India. Native of South Asia and its first cultivation was recorded in China in 2838 B.C.

Economic Importances: Pods and seed oil is edible. Fruits or pods contain high value protein (35% on dry weight basis); rich source of vitamin-C. Soyabean milk and meat are valuable and nutritious food for all age. Seed-oil is used in various industrial raw materials.

20. *INDIGOFERA* L.

Key to the Species

1a	Leaves pinnately compound	(2)
1b	Leaves simple, pod globose, not spiny	2. *I. linifolia* Retz.
2a (1)	Leaves 5-11 foliolate, flowers small in axillary, sessile, bright red; pods oblong, cylindrical	1. *I. enneaphylla* L.
2b (1)	Leaflets many foliolate; flowers pinkish numerous in short racemes	(3)
3a (2)	Pods many, stout, 8-10 seeded, blunt	3. *I. sumatrana* Gaertn.
3b (2)	Pods linear, straight, glabrous	4. *I. tinctoria* L.

20.1. ***Indigofera enneaphylla*** L. Mant. Ait. 272.1711 *nom. superfl.;* Baker, Hook.f. Fl. Brit. India 2: 94.1876; Watt Dict. Econ. Prod. 4: 385.1890; Prain Beng. Plants 1: 310. 1903 (Rep. ed. 1963); Wlh. India 5: 178.1959. *I. prostrata* (L.) Domin., Bibl. Bot. Stuttgrt 187.1926, *nom* Willd. 1803.

Prostrate, much branched herb, with trailing branches. Stems appressed hairy. Leaves 5-9 foliolate, stipules ovate, cuspidate; leaflets small, oblanceolate, retuse, mucronate. Flowers small, bright red, in dense heads. Pods oblong, cylindrical, hispid. Seeds 2. *Fls. & Frts.:* August-January.

Commonly found during rainy season, forming a dense mat in lawns. An effective soil binder, due to its prostrate creeping habitats.

Field Collection: Subhashgram, *K.N.-1331.*

Distributions: Throughout the plains of India; Sri Lanka, Angola, Malay Islands and North Australia.

Economic Importances: Good green manure and act as soil binder.

20.2. ***Indigofera linifolia*** (Linn.f.) Retz., Obs. Bot. fasc. 4: 29.1786 & Ibid. 6: 33.t.2.1791; Baker, Hook.f. Fl. Brit. India 2: 92. 1876; Watt Dict. Econ. Prod. 4: 386.1890; Prain Beng. Plants 1: 309. 1903 (Rep. ed. 1963) & Rec. bot. Surv. India 3 (2): 203. 1903; Majumdar, Ind. Agr. 6 (1 & 2): 140. 1962 (Rep. ed.); Wlh. India 5: 180. 1959; Dutta & Ganguly, Fl. Calc. & Neighb. 24.1966. *Hedysarum linifolium* Linn. f. Suppl. 331.1781. *(Vern. Name: Bhangra).*

Annual much branched herb; stem numerous. Leaves numerous, simple, linear, acute at both ends, mucronate, long; stipules minute setaceous. Flowers in dense sessile or very shortly peduncled axillary racemes. Petals caducous. Pod globose, apiculate, glistering white with appressed silky hairs. One seeded. *Fls. & Frts.:* May-August.

Common in waste grassy places, in association with *Alysicarpus* sp.

Field Collections: Dum Dum Cant., *K.N.-1394;* Sodpur, *K.N.-1469.*

Distributions: Throughout India; Sri Lanka, Afghanisthan, Abyssinia and North Australia.

Economic Importances: Used medicinally in febrile eruptions; also useful as soil fertilizer.

20.3. ***Indigofera sumatrana*** Gaertn; Baker, Hook.f. Fl. Brit. India 2: 99 (partly). 1876; Roxb. Fl. Ind. 3: 379.1832; Prain Beng. Plants 1: 310.1903 (Rep. ed. 1963) ; Wlh. India 5: 182.1959. *(Vern. Name: Nil).*

Shrub with twiggy woody branches. Leaves compound, odd-pinnate, leaflets numerous, opposite. Racemes many flowered, elongated; flowers small. Pods many, stout, 8-10 seeded, blunt. *Fls. & Frts.:* January-April.

Common on the dry waste lands, roadsides and railway tracks.

Field Collection: Santoshpur/Budge Budge, *K.N.-741.*

Distributions: Mostly in South India & Java.

Economic Importances: Principal source of indigo in Bihar and Bengal; act as green manure.

20.4. ***Indigofera tinctoria*** L. Sp. Pl. 751.1753; Baker, Hook.f. Fl. Brit. India 2: 99.1876; Watt Dict. Econ. Prod. 4: 387. 1890; Prain Beng. Plants 1: 310.1905 (Rep. ed. 1963); Wlh. India 5: 182.1959. *(Vern. Name: Nil).*

Cultivated shrub. Stipulated leaves; leaflets opposite, membranous, green but after drying greyish black. Flowers numerous, in nearly sessile lax spicate racemes. Pods linear, straight, apiculate, thickened at the sutures, glabrous. *Fls. & Frts.:* December-March.

Common on the railway tracks and moist waste lands and roadsides.

Field Collections: Salipur/Baruipur, *K.N.-85,* Kadampur, *K.N.-634.*

Distributions: Considered to be Asian Origin; cultivated almost throughout India. Occurring wild in Africa.

Economic Importance: Cultivated as the source of indigo; also as green manure crop in coffee plantation and rice fields in South India. Plant extract used in epilepsy and nervous disorders.

21. *LATHYRUS* L.

Key to the Species

1a	Stem not winged, leaflet-o; stipules leaf-like; flowers yellow	1. *L. aphaca* L.
1b	Stem winged, leaflets -2; flowers blue or reddish purple	2. *L. sativus* L.

21.1. ***Lathyrus aphaca*** L. Sp. Pl. 729.1753; Baker, Hook.f. Fl. Brit. India 2: 170. 1876; Watt Dict. Econ. Prod. 4: 590.1890; Prain Beng. Plants 1: 261.1903 (Rep. ed. 1963) & Rec. bot. Surv. India 3 (2): 192. 1905; Wlh. India 5: 41.1962; Majumdar, Ind. Agr. 6: 139. 1962 (Rep. ed.). *(Vern. Name: Jangli-Matar).*

Annual herb with slender wing less much branched stem. Leaves reduced to a tendril. Flowers yellow, 1-2 on a long peduncle. Pods linear oblong, wingless; seeds 4-6. *Fls. & Frts.:*January- March.

Cultivated as winter crop, but wild and self sown in the waste lands.

Field Collection: Narendrapur/R.K.M. Ashram, *K.N.-1974.*

Distributions: Almost throughout India. An Eurasian species.

Economic Importances: Used as fodder and low quality pulses.

21.2. ***Lathyrus sativus*** L. Sp. Pl. 730.1753; Baker, Hook.f.Fl. Brit. India 2: 179.1876; Watt Dict. Econ. Prod. 4: 590. 1890; Prain Beng. Plants 1: 261. 1903 (Rep. ed. 1963); Wlh. India 5: 37.1962 *(Vern. Name: Khensari).*

Annual herb with slender wingless, much-branched stems. Leaves reduces to a tendrils. Stipules modified into leaves, about 3 cm long, in pairs, appressed to the stem, hastate, truncate, entire. Blue, reddish purple, 1-2 cm on a long peduncle. Pods linear oblong, wingless, seeds 4-6. *Fls. & Frts.:* January-March.

Weed of winter season crop fields; sometime cultivated in low lands.

Field Collections: Maradhat/Baruipur, *K.N.-68;* Garia, *K.N.-324.*

Distributions: Almost throughout India. Iran, Middle East, South Europe and South America; considered as a native of South Europe.

Economic Importance: The plant has a bad reputation for causing paralysis and considered as low quality pulses. It contain lathyrine, a causative factor for lathyrism. Contain toxic amino acid.

22. *LENS* Mill. *nom cons.*

22.1. ***Lens culinaris*** Medik, Vorles. Chrupf. Phys. Ges. 2: 361 1787. *Lens. esculenta* Moench. Meth. 131.1794; Watt Dict. Econ Prod. 4: 621. 1890; Prain Beng. Plants 1: 260.1903 (Rep. ed. 1963) & Rec. bot. Surv. India 3(2): 192. 1905. *(Vern. Name: Masur).*

Small, erect, softly pubescent annual herb, partially twining, mostly branching from the base. Leaves pinnate, the rachis ending in a short bristle; leaflets 4-6 pairs, sessile lanceolate, often mucronate. Flowers pale purple, in 2-4 flowered racemes. Pods compressed, smooth; seeds 2, compressed, grey, with minute spots, *Fls. & Frts.:* December-March.

Cultivated in the winter season in low lands or rice fields.

Field Collection: Falta, *K.N.-1996.*

Distributions: Cultivated throughout India, introduced into India by the early Aryan settlers. Native of Central Europe, Mediterranean region and Western Asia, including Afghanisthan.

Economic Importances: Very important pulse crop. It is considered to be one of the most nutritious pulse food. The leaves and stems are given as fodder to cattles.

23. *MEDICAGO* L.

Key to the Species

1a	Pods reniform, unarmed, 1-seeded	1. *M. lupulina* L.
1b	Pods spirally twisted, muricate, 2-more seeded	2. *M. polymorpha* L.

23.1. ***Medicago lupulina*** L. Sp. Pl. 779. 1753; Baker, Hook .f. Fl. Brit India 2: 90.1876; Watt Dict. Econ. Prod. 5: 199. 1891; Prain Beng. Plants 1: 296. 1903 (Rep. ed. 1963); Wlh. India 6: 312.1962; Dutta & Ganguly Fl. Calc. & Neighb. 23. 1966; Naskar, Aq. & Semi. Plant 115.1990.

Procumbent, much-branched herb. Flowers yellow in dense ovoid head. *Fls. & Frts.:* January-May.

Economic Importances: Green manure and fodder plant. Contain lenitive properties. Naskar (l.c.) described this plant in respect of its distribution and ecological adaptations.

23.2. ***Medicago polymorpha*** L. Sp. Pl. 779. 1753; Naskar, Aq. & Semi. Plants 116.1990. *(Vern. Name: Maina).*

Sub-glabrous annual herb with prostrate branches. Flowers yellow. *Fls. & Frts.:* December-March.

Economic Importances: Green manure and fodder plants. Leaves are rich source of ascorbic acid. Naskar (l.c.) described this plant in respect of its distribution, country of origin etc.

24. *MELILOTUS* Miller

Key to the Species

1a	Flowers white; pods usually 2-seeded	1. *M. alba* Medik.
1b	Flowers yellow; pods usually 1-seeded	2. *M. indica* (L.) All.

24.1. ***Melilotus alba*** Medik. Vorles. Chrupf. Phy. Ges. 2: 382.1797; Naskar, Aq.& Semi. Plants 117.1990. *(Vern.Name: Safed Ban Methi).*

Annual, decumbent herb. Flowers white, half to full pendulous. *Fls. & Frts.:* December-April.

Economic Importances: Fodder and bee necter. Naskar (1.c.) described this plant in respect of its distributions and ecological notes.

24.2. ***Melilotus indica*** (L.) All., Fl. Podom. 1: 308.1785; Naskar, Aq. & Semi. Plants 117.1990. *(Vern. Name: Ban Methi).*

Erect annual herb. Flowers yellow; corolla twice the calyx. *Fls. & Frts.:* February-June.

Naskar (1.c.) described the plant in respect of its distributions, adaptations and economic importance.

25. *MILLETTIA* Wt. & Arn.

25.1. ***Millettia peguensis*** All. Kew Bull. 21: 489.1968 *nov.* sp. *M. ovalifolia* sensu Kurz. J. As. Soc. Bengal 42(2): 68.1873; Baker, Hook.f. Fl. Brit. India 2: 107. 1876. *Pongamia ovalifolia* Wt. & Arn.

Medium sized tree. Leaves compound, leaflets 7, ovate elliptic, glaucous. Flowers mauve or lilac, in pendulous racemes. Pods 2-3 seeded, narrowed to the base; compressed, greenish-yellow, beaked. *Fls. & Frts.:* August-September.

Planted in the garden for its beautiful flowers.

Field Collection: Brace Bridge, *K.N.-2231.*

Distributions: Cultivated almost throughout India; native of Burma.

Economic Importances: As an ornamental plant; avenue tree.

26. *MUCUNA* Adans. *nom. cons.*

26.1. ***Mucuna pruriens*** DC. Prodr. 2: 404.1825; Baker, Hook.f. Fl. Brit. India 2: 187.1876; Watt Dict. Econ. Prod. 5: 286. 1891; Prain Beng. Plants 1: 285. 1903 (Rep. ed. 1963) & Rec. bot. Surv. India 3(2): 198. 1905; Wlh. India 6: 442. 1962. *M. prutita* Hook., Bot. Misc. 2: 248. 1830-1832; Santapau, Rec. bot. Surv. India 6(1): 64.1967. *(Vern. Name: Alkushi).*

Extensively climbing, annual herb. Leaves compound; leaflets 3, ovate-rhomboid, green and glabrous above, grey silky beneath. Flowers dark purple, in large drooping racemes. Pods turgid, linear with the ends, curved in opposite directions. *Fls. & Frts.:* September-November.

Frequently found climbing upon trees, shrubs, hedges or bushes. The skin irritation caused by the fine bristles of the pods is very intense if touched to body and may last for several hours. Application of Ammonia considerably reduce the burning sensation produced by this plant.

Field Collection: Diamond Harbour, *K.N.-1948.*

Distributions: Throughout India. In the Tropics.

Economic Importances: Said to be useful as medicine; the fine bristles were introduced into European pharmacy as a vermifuse. Useful as green manure and cover plant of the crop. Seeds are used as nervine, tonic, strong infusion of roots, mixed with honey is given in cholera.

27. PACHYRHIZUS Rich.

27.1. ***Pachyrhizus angulatus*** Rich. ex DC. Prod. 2: 402.1825; Baker, Hook.f. Fl. Brit. India 2: 207.1876; Watt Dict. Econ. Prod. 6(1): 1893;

Prain Beng. Plants 1: 277.1903 (Rep. ed. 1963) & Rec. bot Surv. India 3(2): 196.1905. *P. erosus* (L.) Urban.; Wlh. India 7: 208.1966. *P. trilobus* DC. Prodr. 2: 402.1825. *(Vern. Name: Sankaalu).*

Large-climbing herb. Root large tuber like, a turnip. Leaves pinnately trifoliate with stipulate lobed leaflets. Racemes long, with tumid nodes and fascicled pedicles; bracts and bracteoles setaceous, caducous. Calyx as long as the pedicel. Corolla reddish 2.5 cm or more long. Pods straight, glabrous, 8-12 seeded; poisonous. *Fls. & Frts.:* December-March.

Commonly cultivated in the village gardens.

Field Collection: Subuddhipur/Baruipur, *K.N.-357.*

Distributions: Cultivated throughout India, but not as wild. Native of America, everywhere in the Tropics.

Economic Importance: Tuberous roots are largely edible either raw or cooked.

28. *PHASEOLUS* L.

1a	Pods clothed with long silky hairs; seeds usually green	1. *P. aureus* Roxb.
1b	Pods linear straight, turgid glabrous	2. *P. vulgaris* L.

28.1. ***Phaseolus aureus*** Roxb. (Hort. Beng.35.1814 *nom. nud.*) & Fl. Ind. 3: 297.1832; Watt Dict. Econ. Prod. 6(1): 188.1893. *P. radiatus* L. Sp. Pl. 725.1753 var. *aureus* Prain, J. As. Soc. Bengal 66: 422.1898. *(Vern. Name: Sona Mug).*

Much branched, sub-erect, hirsute annual herb. Leaves dark green; stipules conspicuous; leaflets deltoid or rounded at the base. Flowers yellow, in capitate racemes. Pods clothed with long silky hairs, seeds usually green. *Fls. & Frts.:* September- October.

Cultivated as a rainy season crop.

Field Collection: Brace Bridge, *K.N.-2232.*

Distributions: Cultivated throughout India. In the Tropics.

Economic Importances: Valuable pulse crop, specially given to the patients as it is very easily digestable.

28.2. ***Phaseolus vulgaris*** L. Sp. Pl. 723.1753; Baker, Hook.f. Fl. Brit. India 2: 200.1876; Watt Dict. Econ. Prod. 6(1): 276. 1893; Prain Beng. Plants 1: 261. 1903 (Rep. ed. 1963): Wlh. India 8: 8. 1969. *(Vern Name: Bakla).*

Sub-erect or twining, glabrous, annual herb. Leaves trifoliolate. Flowers white, pink or lilac. Pods linear, straight, turgid; glabrous. *Fls. & Frts.:* February-April.

Cultivated for its edible pods.

Field Collection: Shyamnagar, *K.N.-2086.*

Distributions: Largely grown in the gardens throughout India. A Native of South America.

Economic Importance: Pods are vegetable.

29. *PISUM* L.

Key to the Species

1a	Seeds compressed	1. *P. arvense* L.
1b	Seeds globose	2. *P. sativum* L.

29.1. ***Pisum arvense*** L. Sp. Pl. 227.1753; Baker, Hook.f. Fl. Brit. India 2: 181. 1876; Watt Dict. Econ. Prod. 6(1): 276. 1893; Prain Beng. Plants 1: 261. 1903 (Rep. ed. 1963) & Rec. bot. Surv. India 3(2): 192. 1905. *(Vern. Name: Matar).*

Smooth, annual herb. Leaflets 4-6; rachis ending in branched tendril, stipules large, ovate. Peduncles 1-few flowered; standards petal pink, wings deep purple. Pods reticulate; seeds angled, greenish yellow, spotted with red. *Fls. & Frts.:* January- March.

Cultivated as winter crop, both in high and low lands.

Field Collections: Shyamnagar, *K.N.-2100.*

Distributions: Almost throughout India; a native of India. In most Tropical countries of the world.

Economic Importances: Pulses and vegetable yielding crop.

29.2. ***Pisum sativum*** L. Sp. Pl. 727. 1753; Baker, Hook.f. Fl. Brit. India 2: 181.1876; Watt Dict. Econ. Prod. 6(1): 277. 1893; Prain Beng. Plants 1: 261.1903 (Rep. ed. 1963) & Rec. bot. Surv. India 3(2): 192. 1905; Wlh. India 8: 124. 1969. *(Vern. Name: Kabuli Matar).*

Extensively cultivated as winter crop, throughout.

Field Collections: Garia, *K.N.-314;* Shibrampur, *A. Gupta-21* (CAL).

Distributions: Cultivated almost throughout India. Native of Western Asia.

Economic Importances: The flower buds and leaves are eaten as vegetable and the pods are sold in the market for peas, which are eaten in the form of various preparation. Used as fodder. The seed oil has sterility effect.

30. *PSOPHOCARPUS* Neck.

30.1. ***Psophocarpus tetragonolobus*** DC. Prodr. 2: 403.1825; Baker, Hook.f. Fl. Brit. India 2: 211.1876; Prain Beng. Plants 1: 279.1903 (Rep.ed.1963). *Dolichos tetragonolobus* L. Syst. (ed.10) 1162.1758-1759; Roxb. Fl.Ind. 3: 305.1832. *(Vern. Name: Chara-koni-seem).*

Twining plant with squared pods, 15-25 cm long with much crisped and toothed wings on the angles. *Fls. & Frts.:* December- February.

Very rare on the village shrubberies.

Field Collection: Shibrampur/Thakurpukur, *A. Gupta-38* (CAL).

Distributions: Dampy areas of India, likely to have originated in the African regions.

Economic Importances: The young fruits are boiled as vegetable.

31. *PSORALEA* L.

31.1. ***Psoralea corylifolia*** L. Sp. Pl. 764.1753; Naskar, Aq. & Semi. Plants 118.1990. *(Vern. Name: Barachi).*

Erect annual herb. Naskar (1.c.) described the plant in respect of its taxonomy, key characters, distributions and economic importances.

Economic Importances: Contain oleo-rasinous substances, which is useful for leucoderma, leprosy and other skin diseases. Seeds are also useful for anthelmintic and for promoting urination.

32. *PTEROCARPUS* L.

32.1 ***Pterocarpus indicus*** Willd. Sp. Pl. 3: 729. 1800.

Tall erect tree; occasionally planted as avenue tree. Distributed in India, mostly planted; indigenous in Malaya and Burma. Seed carnel is emetic.

33. *RHYNCHOSIA* Lour. *nom. cons.*

Key to the Species

1a	Seeds with a waxy strophiole	1. *R. aurea* DC.
1b	Seeds without a strophiole	2. *R. minima* (L.) DC.

33.1. ***Rhynchosia aurea*** DC. Prodr. 2: 386. 1825; Baker, Hook.f. Fl. Brit. India 2: 221.1876. *Nomismia aurea* W. & A. 237. *N. capitata* W. & A.

Trailing herb; stems clothed with fine spreading hairs. Leaves trifoliolate; rachis prolonged; leaflets rhomboid-ovate or rhomboid-obovate. Flowers in close 6-20 flowered racemes, yellow. Pods orbicular, slightly compressed. Seeds 2 oblong, compressed rounded at the ends. *Fls. & Frts.:* December-March.

Occasionally on the waste lands, road-side hedges and village thickets.

Field Collection: Subuddhipur/Baruipur, *K.N.-329.*

Distributions: Throughout India. The Tropical species.

Notes: Prain (1903a, 1905) has not mentioned this taxon from Bengal.

33.2. ***Rhynchosia minima*** (L.) DC. Prodr. 2: 385.1825; Baker, Hook.f. Fl. Brit. India 2: 223. 1876; Prain Beng. Plants 1: 270.1903 (Rep. ed. 1963); Wlh. India 9: 23. 1972. *Dolichos minima* L. Sp. Pl. 726. 1753. *Rynchosia medicaginea* DC. Prodr. 2: 386.1825.

Spreading or twisting glabrous, slender annual herb. Leaves trifoliolate, leaflets deltoid at the base. Flowers yellow in lax racemes. Pods turgid, slightly recurved, usually 2-seeded. *Fls. & Frts.:* May-November.

Commonly found around the hedges and thickets of the villages.

Field Collections: Shyamnagar, K.N.-2089.

Distributions: Throughout India; Sri Lanka, Tropical regions, Cape of the United States.

Economic Importance: Root nodules enrich the soil; the plant is useful as fodder, said to be toxic to fishes. Seeds are bitter and poisonous; leaves are also said to be used as abortifacient.

34. *SESBANIA* Adans. *nom. cons.*

Key to the Species

1a	Flowers usually small or medium upto 1.8 cm, but straight	(2)
1b	Flowers usually large 7.5 cm or more; buds falcately recurved	4. *S. grandiflora* (L.) Poir.
2a (1)	Pods twisted, pendulous; flowers 1.25 cm or more; stems and branches unarmed	(3)
2b (1)	Pods not twisted, often erect or pendulous; flowers 1 cm or less	(5)
3a (2)	Perennial, stem woody; flowers 1.3 cm, pods 15 cm long, suture undulated and valves widely depressed between the seeds	6. *S. sesban* (L.) Merr.

3b (2)	Annual, stem pith like; flowers 1.5 cm or more, pods 23-30 cm, suture straight and valves slightly abruptly depressed between the seeds	(4)
4a (3)	Leaves 5-10 cm long, leaflets 20-30 cm, shortly petioled; flowers deep yellow; pods pendulous, 15-25 cm long	1. *S. aegyptica* Poir.
4b (3)	Leaves 7.3-15 cm long; leaflets 21-41; flowers pale yellow, pods pendulous, 25-30 cm long	5. *S. roxburghii* Merr.
5a (2)	Stem woody; racemes 3-4 flowered; pods straight sutures and valves not depressed	3. *S. cannabina* (Retz.) Pers.
5b (2)	Stem semi-woody, racemes in lax axillary 3-6 flowered, pods slightly falcate, beaked, not torulose	2. *S. bispinosa* (Jacq.) Wight

34.1. ***Sesbania aegyptica*** Poir. Encyc. 7: 128. 1806 *(Sesban aegyptiacus);* Baker, Hook.f. Fl. Brit. India 2: 114. 1876. Watt Dict. Econ. Prod. 6(2): 543.1893; Prain Beng. Plants 1: 288.1903 (Rep. ed. 1963). *S. aegyptiaca* Poir var. *picta* Prain, J. As. Soc. Bengal 66: 367.1898. *(Vern. Name: Jayanti).*

Large shrub, 2-3 m high, unarmed; branches terete, straight glabrous, pale. Leaves 5-10 cm long; petioles 0.6- 2.0 cm long; leaflets 20-30, shortly petiolated, pale green, linear oblong, obtuse faintly apiculate, glabrous. Flowers yellow in lax slender pendulous 6-10 flowered racemes; buds straight; pedicels filiform; calyx 0.75 cm long, glabrous, membranous; corolla 1.25- 2cm long; standard as broad as long. Pods pendulous 15-25 cm long, twisted, weak, slightly torulose, sharply beaked. Seeds 20- 30. *Fls. & Frts.:* November-March.

Sometime on the village thickets and roadside hedges.

Field Collection: Naihati, *K.N.-2021.*

Distributions: Throughout India, probably native of India. Throughout the Tropics.

34.2. ***Sesbania bispinosa*** (Jacq.) Wight, U.S. Dept. Agr. Bur. Pl. Ind. Bull. 137: 15. 1909; Naskar, Aq. & Semi. Plants 120.1990.

An annual, unarmed aquatic plant. Naskar (1.c.) described this taxon in respect of its taxonomy, ecology and economic importances.

34.3. ***Sesbania cannabina*** (Retz.) Pers. DC. Prodr. 2: 215. 1825; Naskar Aq. & Semi. Plants 121. 1990.

An annual, tall erect undershrub. Naskar (1.c.) described this taxon in respect of its taxonomy, ecology and economic importances.

34.4. ***Sesbania grandiflora*** (L.) Poir., Lamk. Encycl. 7: 127. 1806; Baker, Hook.f. Fl. Brit. India 2: 115. 1876; Prain Beng. Plants 1: 288. 1903 (Rep. ed. 1963). *Aeschynomene grandiflora* L. Sp. Pl. (ed.2) 1050. 1763. *Agati grandiflora* Desu, J. Botany 1: 120. 1813; Watt Dict. Econ. Prod. 6(2): 544.1893.

Vern. Name: Agati or *Bak.*

Soft-wooded, spreading branched, small tree. Leaflets 10-35 pairs. Flowers showy, white or sometimes pinkish, in lax 2-4 flowered racemes. Pods 30-50 cm long, slender, straight or falcate. *Fls. & Frts.:* October-April.

Frequently cultivated or planted in the public gardens as ornamental tree.

Field Collection: Harinbari/Sagar Islands, *K.N.-478.*

Distributions: Planted almost throughout India. A Tropical Species.

Economic Importances: Flowers edible as vegetable; also as ornamental plants; reported to have some medicinal values like root juice and honey is useful as an expectorant. Leaves disinfect throat and mouth.

34.5. ***Sesbania roxburghii*** Merr. Philip. J. Sci (Bot.) 4: 269.1909; Naskar Aq. & Semi. Plants 122.1990.

Annual shrub; flowers pale yellow. Naskar (l.c.) described this taxon in respect of taxonomy, ecology, key notes and economic uses.

Leaves of this plant used as fodder and is also reported as poultices.

34.6. ***Sesbania sesban*** (L.) Merr. Philip. J. Sci (Bot.) 7: 235.1912; *Aeschynomene sesban* L. Sp. Pl. 714. 1753. *S. aegyptica* L. Sp. P. 713. 1753. *S. aegyptica* Poir. Encl. 7: 128.1806; Baker, Hook.f. Fl. Brit. India 2: 114. 1876; Watt Dict. Econ. Prod. 6(2): 543. 1893; Prain Beng Plants 1: 288. 1903 (Rep. ed. 1963). *(Vern. Name: Jainti).*

Soft wooded, short lived, unarmed shrub, 2-5 m high; branches pale, terete or obscurely angled, straight, glabrous or nearly so. Leaves 5-10 cm long; petioled, stipules scarious, linear, acute, caducous; leaflets petioled, pale green, linear oblong, obtuse. Flowers in lax slender pendulous 6-10 flowered racemes; buds straight, pedicels filiform; calyx glabrous membranous; corolla standard as broad as long. Pods pendulous, twisted, weak, sharply beaked. Seeds 25-30. *Fls. & Frts.:* July-December.

Occasionally cultivated as ornamental plant.

Field Collection: Sarisha/Diamond Harbour, *K.N.-388.*

Distributions: Cultivated throughout India. Throughout the Tropics of the Old World.

Economic Importances: The seeds and leaves are employed in indigenous medicines for the valuable remedy in cutaneous diseases.

Notes: The name *'sesban'* by Adanson (1763) was simply the Egyptian name of *S. aegyptiaca* – but the name *Sesbania* now universally accepted.

35. *SHUTERIA* W. & A. *nom. cons.*

35.1. ***Shuteria vestita*** W. & A. Prodr. 207. 1834; Baker, Hook.f. Fl. Brit. India 2: 181. 1876; Wlh. India 9: 322. 1872. *Glycine vestita* Grah, Wall. Cat. 5512. 1828. 1849. *G. involucrata* Wall. Cat 5506.1828-1849.

Stem slender, woody, glabrescent. Stipules lanceolate, rigid, petiole 2.5-7.5 cm, hairy or glabrous; leaflets membranous or sub-coriaceous, obscurely hairy on both sides or glabrescent. Racemes copious, close 2.5 - 10.0 cm long; bracts-rigid, calyx densely hairy, teeth shorter than tube; corolla reddish, twice the calyx. Pods 2.3 cm long, hairy, recurved. *Fls. & Frts.:* December- March.

Common on the dry waste lands, roadside thickets and railway tracks.

Field Collection: Dum Dum Cantt., *K.N.-2062.*

Distributions: Eastern Himalayas and W. Peninsula; Sri Lanka, Sikkim and Bhutan.

Notes: Prain (1903 a, 1903b and 1905) has not mentioned this taxon from Bengal.

36. *SMITHIA* Ait. *nom cons.*

36.1. ***Smithia sensitiva*** Ait. Hort. Kew. (ed. 1) 3: 496.1789; Baker, Hook.f. Fl. Brit. India 2: 148. 1876; Watt Dict. Econ. Prod. 6(3): 255. 1893; Prain Beng. Plants 1: 299. 1903 (Rep. ed. 1963); Wlh. India 9: 368. 1972. *(Vern. Name: Nala kashina).*

Diffused, much-branched herb, 0.3 -1.0 m high, slender. Leaves abruptly pinnate; petioles short. Flowers 2-6 in simple racemes, yellow. Pods flattened. *Fls. & Frts.:* November-March.

Very rare on the roadside waste lands and village thickets.

Field Collection: Salipur/Baruipur, *K.N.-87.*

Distributions: Throughout India, Andaman and Nicobar Islands. Triopical Africa, Madagascar, Java and China.

Economic Importance: Leaves used as pot herb and as fodder.

37. *TEPHROSIA* Pers. *nom. cons.*

Key to the Species

1a	Pods clothed with fulvous hairs	(2)
1b	Pods not covered with hairs	2. *T. purpurea* (L.) Pers.
2a (1)	Leaflets 20-30, slightly elliptic; flowers white with tinged red; pods clothed with brown silky hairs	1. *T. candida* DC.
2b (1)	Leaflets 11-19, oblanceolate; flowers red, pods velvety with dense spreading fulvous hairs	3. *T. villosa* (L.) Pers.

37.1. ***Tephrosia candida*** DC., Prodr. 2: 249. 1825; Baker, Hook.f. Fl. Brit. India 2: 11. 1876; Prain Beng. Plants 1: 289. 1903 (Rep. ed. 1963); Wlh. India 10: 151. 1976. *Robinia candida* Roxb. Fl. India 3: 327. 1832. *(Vern. Name: Bangara).*

Small shrub, branches grooved. Leaves pinnate, leaflets 20-30, narrowly elliptic, cuspidate at the apex. Flowers white or tinged with red. Pods slightly recurved, clothed with brown silky hairs. *Fls. & Frts.:* August-December.

Frequently found in moist places, near gardens and railway tracks.

Field Collection: Dhakuria, *K.N.-1965.*

Distributions: Almost throughout India; W. Indo-China and Malay Archipelago.

Economic Importances: Rarely as green manure and sometimes planted as ornamental plant. Reported as fish poison; the bark and leaves are reported to be chiefly used as fish poison in Bangladesh and Burma.

37.2. ***Tephrosia purpurea*** (L.) Pers. Syn. 2: 329. 1807; Baker, Hook.f. Fl. Brit. India 2: 112. 1876; Watt Dict. Econ. Prod. 6(4): 14. 1893; Prain Beng Plants 1: 289. 1903 (Rep. ed. 1963) & Rec. Bot. Surv. India 3(2): 200. 1905. *Cracca purpurea* L. Sp. Pl. 752. 1753. *(Vern. Name: Ban-Nil).*

Copiously branched, sub-erect, herbaceous. Leaves short petioled, stipule, linear. Racemes copious. Flowers red or purplish. Pod broad and recurved, 5-10 seeded. *Fls. & Frts.:* April-November.

Abundantly found on dry, open places, on the waste grounds, railway tracks and along road sides.

Field Collection: Fultala/Baruipur, *K.N.-44.*

Distributions: Throughout India. Everywhere in the Tropics.

Economic Importances: This taxon is considered as tonic, laxative, used as anthelmintic for children and used internally as blood purifier and considered a cardiac tonic. Root is bitter and used in tympanitis, dispepsia, cronic diarrhoea, in urinary troubles, fever and cold. An extract from pods is used in relieving body pain and inflation troubles.

37.3. ***Tephrosia villosa*** (l.) Pers. Syn. Pl 2: 329. 1807; Baker, Hook.f. Fl. Brit. India 2: 113. 1876; Watt Dict. Econ. Prod. 6(4): 15. 1893; Prain Beng. Plants 1: 289.1903 (Rep. ed. 1963); Wlh. India 10. 155. 1976. *Cracca villosa* L. Sp. Pl. 752.1753.

Annual gregarious, erect bushy herb; stem angular densely hairy. Leaves imparipinnate, 5-8 cm long; petioles 10-12 cm long; leaflets pale-green, more or less glabrous above, densely hairy beneath. Flowers large, red, in lax racemes. Pods 25-30 mm X 5-6 mm. Seeds 6-8. *Fls. & Frts.:* November-March.

Common on the village shrubberies and railway tracks.

Field Collection: Shyamnagar, *K.N.-2097.*

Distributions: Punjab, Rajasthan, Gujarat, Madhya Pradesh, U.P., Bihar and West Bengal; Sri Lanka, Mauritius and Tropical Africa.

Economic Importances: Reported as a good green manure; leaves juice is useful in dropsy.

38. *TERAMNUS* R. Br.

38.1. ***Teramnus labialis*** (Linn.f.) Spreng., Syst. 3: 235. 1826. Baker, Hook.f. Fl. Brit. India 2: 184. 1876; Prain Beng. Plants 1: 280.1903 (Rep. ed. 1963): Wlh. India 10: 157.1976. *Glycine labialis* L. f. Suppl. 325. 1781. *G. warreensis* Dalz., Kew J. Botany 3: 210.1851.

Widely spreading twining herb; stem slender, more or less appressedly hairy. Leaves 3- foliolate, ovate-lanceolate, acute, deciduous. Leaflets membranous and sub-coriaceous. Flowers in axillary few flowered lax racemes 5-15 cm long, solitary or fascicled along a slender more or less hairy rachis; corolla reddish. Pods narrowly linear, straight; or slightly incurved, hairy when young. Seeds 8-12, oblong,truncate or slightly rounded at the ends, smooth, dark brown. *Fls. & Frts.:* August- November.

Rarely grow on the village shrubberies.

Field Collection: Kachuberia/Sagar Islands, *A.K.M.-8759.*

Distributions: Throughout India, Sri Lanka and the Tropical countries in general.

Economic Importance: Used as cover pasture crop.

39. *TRIGONELLA* L.

Key to the Species

1a	Diffuse, slender herb; flowers many, closely racemose; pods flat, linear recurved	1. *T. corniculata* L.
1b	Erect, robust herb; flowers few, axillary; pods long, turgid, beaked	2. *T. foenum-graecum* L.

39.1. ***Trigonella corniculata*** L. Syst. (ed. 10) 1180; Baker, Hook.f. Fl. Brit. India 2: 88. 1876; Prain Beng. Plants 1: 296. 1903 (Rep. ed. 1963); Majumder, Ind. Agr. 6: 137. 1962; Wlh. Ind. 10. 299. 1976. *(Vern. Name: Piring).*

Diffuse, slender, terrestrial herb. Leaves pinnately trifoliate, toothed, stipules adnate. Flowers yellow, in raceme. Anthers uniform. Pods linear, flat, fulcately recurved. *Fls. & Frts.:* December-March.

Sometimes cultivated and very occasionally found as a garden escape.

Field Collections: Champahati, *K.N.-329;* Chemagari, *K.N.-456.*

Distributions: Western Himalayas, Bihar and West Bengal; Afghanisthan, Orient and South Europe.

Economic Importances: Fruits is applied to swelling and bruises; leaves are cooked as vegetable.

39.2. ***Trigonella foenum-graecum*** L. Sp. Pl. 777.1753; Baker, Hook.f. Fl. Brit. India 2: 87.1876; Watt Dict. Econ. Prod. 6(4): 86.1893; Prain Beng. Plants 1: 296. 1903 (Rep. ed. 1963) & Rec. bog. Surv. India 3(2): 201.1905; Wlh. India 10: 299.1976. *(Vern. Name: Methi).*

Nearly smooth, erect annual cultivated herb. Leaves tri-foliate, oblanceolate oblong, toothed. Flowers 1-2 axillary, sessile. Calyx teeth linear. Corolla exerted. Pods long, turgid, beaked. *Fls. & Frts.:* December-February.

Sometime cultivated and rarely as self-grown on the waste lands.

Field Collection: Shibrampur/Thakurpukur, *A. Gupta-8* (CAL).

Distributions: Kashmir, Punjab and Upper Gangetic Plains; South Europe and Orient, widely cultivated.

Economic Importances: Seeds contain 30% protein, 6-8% fatty-oil with a foetid odour and bitter taste; it also contain a small quantity of brown coloured essential oil of intense odour; leaves also contain high value of vitamins. In villages the seeds used with coconut oil for scent. Prepared fish food for angling. Leaves used as vegetable; seed spice and condiment, carminative and tonic and externally used for boils, ulcers and abscesses. Seeds contain diosgenin, precursor of steroids, like sex hormones and oral contraceptives.

40. *URARIA* Desv.

40.1. ***Uraria picta*** (Jacq.) Desv., J. Bot. 1: 123. t.5, f.19.1813; Baker, Hook.f. Fl. Brit. India 2: 155. 1876; Watt Dict. Econ. Prod. 6(4): 212. 1893; Prain Beng. Plants 1: 301. 1903 (Rep. ed. 1963) & Rec. bot. Surv. India 3(2): 202. 1905; Wlh. India 10: 413. 1976. *Hedysarum picta* Jacq., Coll. 2: 262. 1788. *(Vern. Name: Sankar-Jata).*

Erect, pubescent, perennial undershrub. Leaves variable, lower simple or 3-foliolate, rounded or oblong, upper 5-9 foliolate, rachis downy, linear lanceolate, acuminate, leaflets 6-20 cm long, linear-lanceolate. Flowers purple in terminal racemes; pedicels incurved and clothed with hooked hairs; calyx with long spreading hairs. Pods 3-6 jointed. *Fls. & Frts.:* June-January.

Commonly found in waste shady places and along the roads.

Field Collections: Nurpore, *K.N.-1931.*

Distributions: Throughout India; Sri Lanka, Tropical Africa, Malaya Islands and Philippines.

Economic Importances: Root is having aphrodisiac properties; decoction used for cough, chills and fevers; leaves antiseptic and used in gonorrhoea.

41. *VICIA* L.

Key to the Species

1a Leaflets 1-3 pairs 1. *V. faba* L.

1b	Leaflets 5-12 pairs	(2)
2a (1)	Flowers whitish, small, pods 2-seeded	2. *V. hirsuta* (L.) Gray
2b (1)	Flowers reddish blue, larger, pods more than 2-seeds	3. *V. sativa* L.

4.1. ***Vicia fava*** L. Sp. Pl. 737.1753; Baker, Hook.f. Fl. Brit. India 2: 179. 1876; Watt Dict. Econ. Prod. 6(4): 234.1893; Prain Beng. Plants 1: 260. 1903 (Rep. ed. 1963); Wlh. India 10: 460.1976.

Erect, glabrous, annual herb. Stem quadrangular. Leaflets 4-6 oval to oblong; stipules nectaries conspicuous. Flowers about 2 cm long, dull white with a purple spot. Pods turgid, sub-cylindric, glabrous, beaked. *Fls. & Frts.:* December-March.

Cultivated during cold season for its young pods.

Field Collection: Sarisha/Diamond Harbour, *K.N.-1943.*

Distributions: Cultivated almost throughout India; introduced in India. A native of North America and North West Asia.

Economic Importances: It is planted in ornamental pots for showy flowers; pods and flowers are edible as vegetable.

41.2. ***Vicia hirsuta*** (L.) Gray, Nat. Arr. Brit. Pl. 2: 614. 1821; Baker Hook.f. Fl. Brit. India 2: 177. 1876; Watt Dict. Econ. Prod. 6(4): 234. 1893; Prain Beng. Plants 1: 260. 1903 (Rep. ed. 1963): Wlh. India 10: 465. 1976. *Eryum hirsurum* L. Sp. Pl. 738.1753. *(Vern. Name: Masur Chana).*

Fairly climbing, annual herb. Leaves paripinnate ending in twisted tendrils; stipules semi-sagittate; leaflets 5-8 paired. Flowers very small, whitish; corolla scarcely exserted. Pods oblong-rhomboid, black, 2-seeded. *Fls. & Frts.:* December-March.

Commonly cultivated during winter.

Field Collection: Subuddhipur/Baruipur, *K.N.-341.*

Distributions: Almost throughout India. Native of Europe, West Africa and Continental Asia.

Economic Importances: Low quality pulses; mainly grown for fodder.

41.3. ***Vicia sativa*** L. Sp. Pl. 736. 1753; Baker, Hook.f. Fl. Brit. India 2: 175. 1876; Watt Dict. Econ. Prod. 6(4): 234. 1893; Prain Beng. Plants 1: 239.1903 (Rep. ed. 1963) & Rec. bot. Surv. India 3(2): 191.1905. *(Vern. Name: Ankari).*

Annual herb with slender, sub-erect stems. Upper leaves 2-2.5 cm long, ligulate; lower shorter and broader; leaflets 6-12 pairs. Flowers reddish, solitary or paired; corolla twice the calyx. Pods 8-10 seeded. *Fls. & Frts.:* December-March.

Common in the cultivated fields and moist waste grounds.

Field Collection: Madhyamgram, *R.N. Mukherjee s.n.* (CAL).

Distributions: Introduced and cultivated everywhere in India. Native of Europe, North Africa and West Asia.

Economic Importance: Introduced as a forage and cover crop in many parts of India.

42. *VIGNA* Savi.

Key to the Species

1a	Annual or perennial, slender sub-erect, diffuse herbs; flowers bright yellow pods 2.5-5.0 cm long straight with 5-12 seeds	(2)
1b	Annual twining herb; flowers yellow with purple wing; pods linear 25-30 cm long with 15-25 seeds, seeds generally white	3. *V. sinensis* Savi. ex Hassk.
2a (1)	Branches from the woody root stocks; leaflets vary much in shape and size; leaflets 3-foliolate	4. *V. trilobatum* Verd.
2b (1)	Branches not from the woody root stock; slender sub-erect or diffusely trailling herbs; leaflets not vary much	(3)
3a (2)	Flowers yellow, minute	1. *V. aconitifolia* Mare.
3b (2)	Flowers deep yellow but larger	2. *V. mungo* Hepper

42.1. ***Vigna aconitifolia*** (Jacq.) Marechal, Bill. Jard. Bot. Nat. Belge 39: 160 1969; Wlh. India 10: 471. 1976. *Phaseolus aconitifolius* Jacq., Obs. 3: 2.t.52.1768; Baker, Hook.f. Fl. Brit. India 2: 202.1876; Watt Dict. Econ. Prod. 6(1): 182. 1893; Prain Beng. Plants 1: 273. 1903 (Rep. ed. 1963). *(Vern. Name: Kheri).*

Slender, sub-erect, diffuse or trailing, slightly hairy herb. Leaflets deeply 3-lobed linear or lanceolate, acute. Flowers yellow, very small, usually in capitate racemes; bracteoles linear. Pods 2.3-5.0 cm long, 7-8 seeded. *Fls. & Frts.:* July-October.

Grown as rainy season crop, often mixed with paddy crop; also found self-grown in grass lands, climbing on large bushy grasses.

Field Collection: Dhopagachi/Baruipur, K.N.-2136.

Distributions: A Native of India. Tropical species.

Economic Importance: The grains and leaves are used as fodder.

42.2. ***Vigna mungo*** (L.) Hepper, Kew Bull. 11: 128.1956; Wlh. India 10: 476.1976. *Phaseolus mungo* L. Mantiss 101.1767. var. *roxburghii* Prain J. As. Soc. Bengal 45: 423. 1898 & Beng. Plants 1: 275.1903 (Rep. ed. 1963). *P. radiatus* Roxb. Fl. Indica 3: 296.1832 (non L.) *P. roxburghii* W. & A. Prodr. 245.1834. *P. mungo* var. *radiatus* Baker, Hook.f. Fl. Brit. India 2: 203.1876; Watt Dict. Econ. Prod. 6(1): 191. 1893. *(Vern. Name: Mash-kalai).*

Stem hirsute, elongate, diffuse spreading, but not scandent. Flower bright yellow. Seeds grey. *Fls. & Frts.:* December- February.

Cultivated during winter season.

Field Collection: Taki, *K.N.-1034.*

Distributions: Cultivated throughout India. In the Tropics.

Economic Importance: One of the most important pulse, cultivated in India.

42.3. ***Vigna sinensis*** (L.) Savi. ex Hassk. Cat Hort. Bogor. 279.1844. *V. catijang* Endl. Walp., Linnaea 13: 533.1839; Baker, Hook.f. Fl. Brit. India 2: 205.1876; Watt Dict. Econ. Prod. 6(4): 236.1893. *Dolichos sinensis* L. Cent. Pl. 2: 28.1765. *(Vern. Name: Barbati).*

Twining herb; stem striate hispidulous or glabrous. Flowers from the nodes, pale yellow with purple wings. Pods linear, generally 25-40 cm long with 15-25 seeds. Seeds white. *Fls. & Frts.:* July-December.

Commonly cultivated in the village gardens along with *Dolichos* sp.

Field Collection: Champahati, *K.N. 1251.*

Distributions: Cultivated almost throughout India. Tropics of the Old World.

Economic Importances: Pods are common vegetable and very nutritous with high protein contents.

42.4. ***Vigna trilobata*** (L.) Verdcourt, Taxon 17. 172. 1968. *Phaseolus trilobatus* Ait., Hort. Kew (ed. 1) 3: 30.1789; Baker, Hook.f. Fl. Brit. India 2: 201.1876; Watt Dict. Econ. Prod. 6(1): 194.1893: Prain Beng. Plants

1: 275.1903 (Rep. ed. 1963); Wlh. India 10: 495.1976. *(Vern. Name: Jangi Kalai* or *Rakhal Kalai).*

Annual herb. Stem numerous from a woody rootstock. Leaflets vary much in form and size, sometime deeply lobed leaflets, generally 3-foliolate. Flowers in sub-capitate few flowered racemes; calyx glabrous; corolla yellow. Pods straight, 3-5 cm long, sub cylindric. Seeds 6-12. *Fls. & Frts.:* August-January.

Rarely cultivated but mostly wild in the waste lands or paddy fields.

Field Collections: Dhap Dhapi, *K.N.-34;* Taki, *K.N.-950.*

Distributions: Very common throughout India; whole World.

Economic Importance: Low quality pulses and fodder crop.

Remarks: There are several other cultivated/ornamental species distributed throughout the lower Ganga Delta regions of West Bengal.

XLVIII. CAESALPINIACEAE

Key to the Genera

1a	Leaves simple, shallowly or deeply cleft		(2)
1b	Leaves compound		(3)
2a (1)	Flowers bi-sexual, pods dehiscent	2. *BAUHINIA* L.	
2b (1)	Flowers unisexual, pods indehiscent	9. *PILIOSTIGMA* Hochet.	
3a (2)	Leaves unipinnate		(4)
3b (2)	Leaves bi-pinnate		(7)
4a (3)	Petals absent, calyx petaloid	11. *SARACA* L.	
4b (3)	Petals present, calyx not petaloid; stamens 10 or less		(5)
5a (4)	Anthers basifixed; fertile stamens more than 3	4. *CASSIA* L.	
5b (4)	Anthers versatile; fertile stamens 3		(6)
6a (5)	Stamens monadelphous	12. *TAMARINDUS* L.	
6b (5)	Stamens diadelphous	1. *AMHERSTIA* Wall.	

7a (3)	Leaves rachis ending in spine; the rachilla long, flat	7. *PARKINSONIA* L.
7b (3)	Leaves not as above	(8)
8a (7)	Shrubs	(9)
8b (7)	Trees	(10)
9a (8)	Pods wingless	3. *CAESALPINIA* L.
9b (8)	Pods winged	6. *MEZONEURON* Desf.
10a (8)	Flowers yellow	8. *PELTOPHORUM* Walp.
10b (8)	Flowers orange red or bright scarlet	(11)
11a (10)	Pods flat, straight and woody	5. *DELONIX Raf.*
11b (10)	Pods thin, flat, broadly linear, strap and with black seed	10. *POINCIANA* L.

1. *AMHERSTIA* Wall.

1.1. ***Amherstia nobilis*** Wall., Pl. As. Rar. 1: 1.t, 1-2. 1829; Baker, Hook.f. Fl. Brit. India 2: 272. 1876; Benthall Trees Calc. 181.1946.

Small spreading tree; leaves very large, abruptly pinnate, dark green, pinnately compound, leaflets entire, acuminate. Flowers large, showy in ample long pendulous racemes with a pair of bright scarlet bracteoles; sepals 4; petals 3-developed. Pods broad, glabrous, truncate 4-6 seeded. *Fls. & Frts.:* December-April.

Commonly cultivated as ornamental plants in the gardens, in lawns.

Field Collection: Ram Nagar/Metiabruz, *K.N.-1382.*

Distributions: Cultivated throughout India; most humid Tropics of the World. Native of Burma.

Economic Importances: A beautiful ornamental tree, commonly planted for its pendulous bright scarlet flowers and dark foliages.

Notes: D. Prain (1903a and 1905) has not mentioned this taxon.

2. *BAUHINIA* L.

Key to the Species

1a	Fertile stamens 10; calyx with short tube and spathaceous 5-toothed limb; pods narrow	(2)

1b	Fertile stamens 3-5; calyx tube elongated, pods flat	(3)
2a (1)	Flowers white in short axillary racemes; calyx-limb subulately 5-toothed at the tip; pod ribbed near upper suture	1. *B. acuminata* L.
2b (1)	Flowers yellow, 1-3 axillary; calyx limb entire; pod not ribbed near upper suture	3. *B. tomentosa* L.
3a (1)	Calyx limb entire, spathaceous, equalling the cylindric tube, flowers appear when the tree is nearly leaflets	4. *B. varigata* L.
3b (1)	Calyx limb irregularly 3-5 cleft, longer than the turbinate tube; flowers appearing with leaves	2. *B. purpurea* L.

2.1. ***Bauhinia acuminata*** L. Sp. Pl. 375. 1753; Baker, Hook.f. Fl. Brit. India 2: 275. 1876; Watt Dict. Econ. Prod. 1: 419. 1889; Prain Beng. Plants 1: 316.1903 (Rep. ed. 1963). *(Vern. Name: Kanchan).*

Cultivated large shrub or small tree. Leaves unifoliolate, bilobed, acute. Flowers 10-15 cm long, white, fragrant, in close axillary racemes. Pods dehiscent. *Fls. & Frts.:* November-April.

Very common throughout in the road sides or ornamental gardens.

Field Collection: Padmapukur/Baruipur, *K.N.-88*

Distributions: All over India; Malaya, Japan and China.

Economic Importance: Ornamental tree.

2.2. ***Bauhinia purpurea*** L. Sp. Pl. 375. 1753; Baker, Hook.f. Fl. Brit. India 2: 284. 1876; Watt Dict. Econ. Prod. 1: 421.1889; Prain Beng. Plants 1: 317.1903 (Rep. ed. 1963); Wlh India 1: 160.1948. *B. triandra* Roxb. (Hort. Beng. 31. 1814) & Fl. Ind. 2: 320.1832. *(Vern. Name: Rakta Kanchan).*

Medium sized, deciduous tree. Leaves 10-15 cm long, bilobed, sometime overlapping, tapering and broadly acute, sub-coriaceous, cordate or truncate at base. Flowers purple to deep-rose, few in terminal and axillary, short-peduncled corymbose racemes; calyx splitting into two parts. Pods 15-20 cm long, linear, flat, 10-20 seeded. *Fls. & Frts.:* October-April.

Often planted as ornamental plant and as avenue tree.

Field Collection: Chaital/Basirhat, K.N.-722.

Distributions: Native of the foot-hills of Himalaya; planted throughout India, China and Malaya.

Economic Importances: The bark is used for dyeing and tanning and from the inner portion, a fibre is prepared.

2.3. ***Bauhinia tomentosa*** L. Sp. Pl. 375.1753; Baker, Hook.f. Fl. Brit. India 2: 275.1876; Watt Dict. Econ. Prod. 1: 423.1889; Prain Beng. Plants 2: 316.1903 (Rep. ed. 1963). *(Vern. Name: Kanchan).*

Handsome shrub with downy branches; leaves shinning and glabrous above, pubescent and whitish beneath; petiole swollen at both ends. Flowers yellow, axillary; calyx spathaceous. Pods stalked 10-13 cm long, 10-15 seeded. *Fls. & Frts.:* August-March.

Very good ornamental shrubs distributed throughout as avenue tree.

Field Collection: Rabindra Sarabar/Calcutta, *K.N.-1504.*

Distributions: Throughout India. Native of Malabar and Sri Lanka, Guina and Tropical Africa.

Economic Importances: Ornamental tree. Bark is used as cordage.

2.4. ***Bauhinia varigata*** L. Sp. Pl. 375. 1753; Baker, Hook.f. Fl. Brit. India 2: 284.1876; Watt Dict. Econ. Prod. 1: 425.1889; Prain Beng. Plants 1: 317.1903 (Rep. ed. 1963); Wlh. India 1: 161. 1948. *B. candida* Roxb. Fl. Ind. 2: 318. 1932. *(Vern. Name: Swet Kanchan).*

Small deciduous tree. Leaves usually deeply cordate, rigidly sub-coriaceous, 9-10 nerved. Flowers few, large fragrant, arranged in lateral corymbose, calyx spathaceous, purple or all of them reddish with upper one more deeply coloured and varigated with cream and purple. Pods 15-25 cm long, hard, flat, slightly curved with long stige about 2.5. cm long; seed 10-20. *Fls. & Frts.:* January-May.

Often planted as ornamental avenue tree in roadsides and railway stations.

Field Collection: Titagarh, *K.N.-2210*

Distributions: Throughout India; Malaya and China.

3. *CAESALPINIA* L.

Key to the Species

1a	Medium to large tree; flowers yellow or, greenish; pods thick, spirally twisted	1. *C. coriaria* Willd.
1b	Scandent or straggling shrub	(2)
2a (1)	Pods armed on the face with wiry prickles petals narrow	2. *C. bonduc* (L.) Roxb.
2b (1)	Pods not armed with prickles; petals broad	(3)
3a (2)	Leaflets 2-3 pairs; coriaceous, large, pods short; seeds solitary	3. *C. crista* L.
3b (2)	Leaflets 8-20 pairs; pods 3-8 seeded; a climbing shrub	4. *C. decapetala* Alston

3.1. ***Caesalpinia coriaria*** (Jacq.) Willd. Sp. Pl. 2: 533.1799; Watt Dict. Econ. Prod. 2: 6.1889; Prain, Rec. bot. Surv. India 3(2): 207.1905; Benthall Trees Calc. 165.1946; Bor & Raizada, Beaut. Ind. Climb. & Shrub. 64, t. 42.1954; Wlh. India 12: 2.1950. *(Vern. Name: Dibi-dibi).*

Low tree. Leaves bipinnate. Flowers yellow or greenish, scented in terminal panicles. Pods thick, spirally twisted. *Fls. & Frts.:* April -June.

Commonly planted in the ornamental gardens or road sides.

Field Collection: Dum Dum Cantt., *K.N.-1388.*

Distributions: Planted throughout India. Native of West Indies and South America.

Economic Importance: Pods contain a large quantity of tannin.

Notes: D. Prain (1903a and 1903b) has not mentioned this taxon.

3.2. ***Caesalpinia bonduc*** (L.) Roxb. (Hort. Beng. 32.1818), *nom nud* & Fl. Ind. 2: 262.1832 emed. (except Pl. descs.). *Guilandina bonduc* L. Sp. Pl. 381.1753. *G. bonducella* L. Sp. Pl. (ed. 2) 545.1762. *Caesalpinia bonducella* (L.) Fleming, As. Res. 11:152. 1810; Baker, Hook.f. Fl. Brit. India 2: 254.1876; Watt Dict. Econ. Prod. 2:3.1889; Prain Beng. Plants 1: 323. 1903 (Rep. ed. 1963)& Rec. bot. Surv.India 2: 301.1903. *(Vern. Name: Nata Karanja).* (Fig.14).

Scrambling vine; leaves 2-pinnate, 6-11 pairs of pinnae; leaflets 16-24 in each pinnae, 2-5 cm long. Rachis and stem with hooked prickles. Stipules conspicous, pinnate. Inflorescence lateral, supra axillary, flowers bisexual,

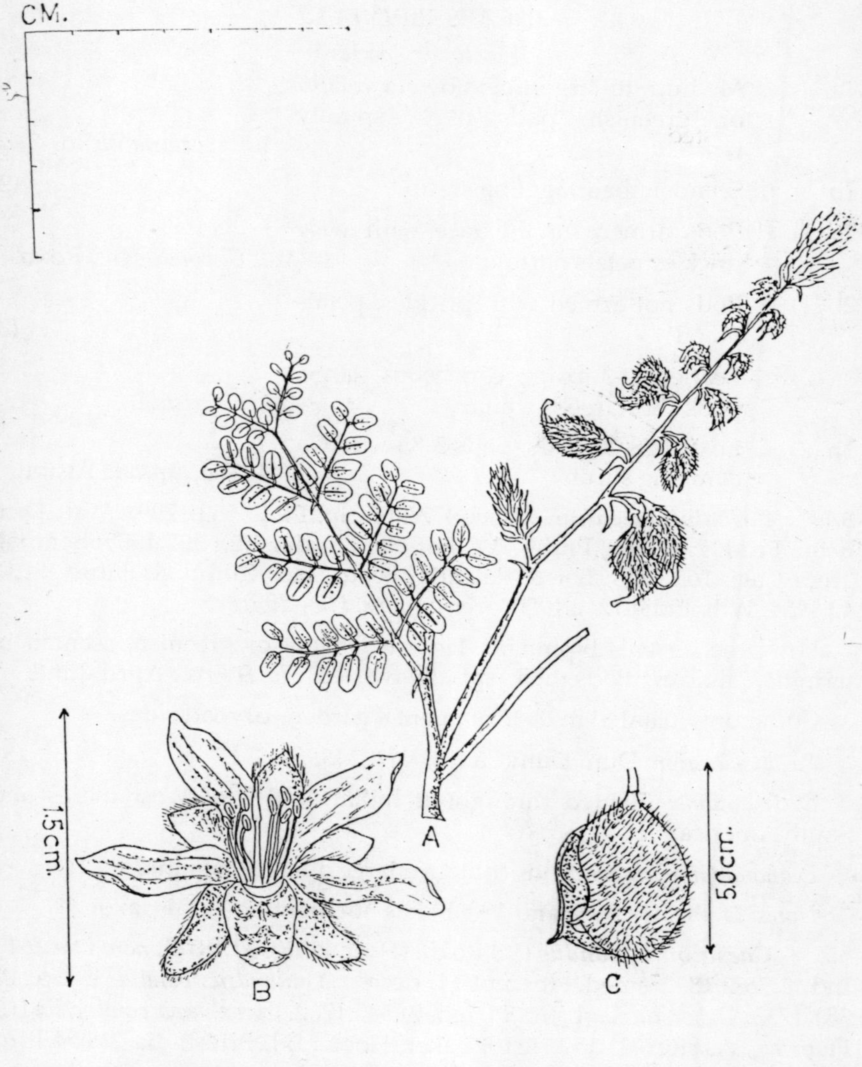

Fig. 14. CAESALPINIA BONDUC (L.) Roxb.

A. Flowering twig, B. Single flower, C. Single spiny pod.

petals yellow with reddish streaks, stamens 10, anther hairy. Fruit 1-2 seeded, ellipsoid. Seeds ovoid, smooth grey. *Fls. & Frts.:* July-March.

Common on the hedges and thickets and in the open waste lands.

Field Collection: Surjyapur/Raidhighi, *K.N.-1403.*

Distributions: Almost throughout India.Pantropic species.

Economic Importance: Seeds are powerful tonic.

3.3. ***Caesalpinia crista*** L. Sp. Pl. 380.1753. *C. nuga* (L.) Ait. f. Hort. Kew (ed. 2) 3: 32.1811; Baker, Hook.f. Fl. Brit. India 2: 255.1753; Watt Dict. Econ. Prod. 2: 10.1889; Prain Beng. Plants 1: 323.1903 (Rep. ed. 1963); Wlh. India 2: 5.1950. *C. paniculata* Roxb. Hort. Beng. 32. *(Vern. Name: Shingri Lata).*

Climbing upto 15m; leaves 2-pinnate, pinnae 2-4 pairs. Racemes axillary or aggregated into terminal axis. Flowers bisexual, petals yellow; stamens-10, with filaments. Pods ellipsoid, 4-8 cm long, beaked with 1 seed, occasionally 2-seed. Seed ovoid and black. *Fls. & Frts.:* January-March. (Fig. 15).

Common on the Sundarbans littoral forests and hedges or thickets.

Field Collection: Kapilanda Ashram/Sagar Islands, *K.N.-437.*

Distributions: Common on the coastal regions of India. Sri Lanka, Malaya, Philippines, China, N. Australia and Polynesia.

Economic Importance: Root powder is said to be diuretic; tonic and useful in the treatment of stones in the bladder; it is also said to be a fish poison.

3.4. ***Caesalpinia decapetala*** (Roth) Alston, Trimen, Handb. Fl. Ceyl. 6 (Suppl.): 89.1931. *Reichardia decapetala* Roth, Nov. Pl. Sp. 212.1821. *Caesalpinia sepiaria* Roxb. (Hort. Beng. 32.1814) *nom nud* & Fl. Ind. 2: 360. 1832; Baker, Hook.f. Fl. Brit India 2: 256.1876; Watt Dict. Econ. Prod. 2: 12.1889; Prain Beng. Plants 1: 323. 1903 (Rep.ed. 1963).

Large straggling shrub; leaves abruptly bipinnate, stipules semisagittate, deciduous, pinnae 6-10 pairs; leaflets opposite, 8-12 pairs. Flowers bright yellow, in terminal or supra axillary racemes. Pods 5-9 cm long on stout pedicels, smooth, beaked. Seeds 6-8. *Fls. & Frts.:* December-September.

Common in the hedge of the villages.

Field Collection: Lakshmikantapur, *K.N.-1774.*

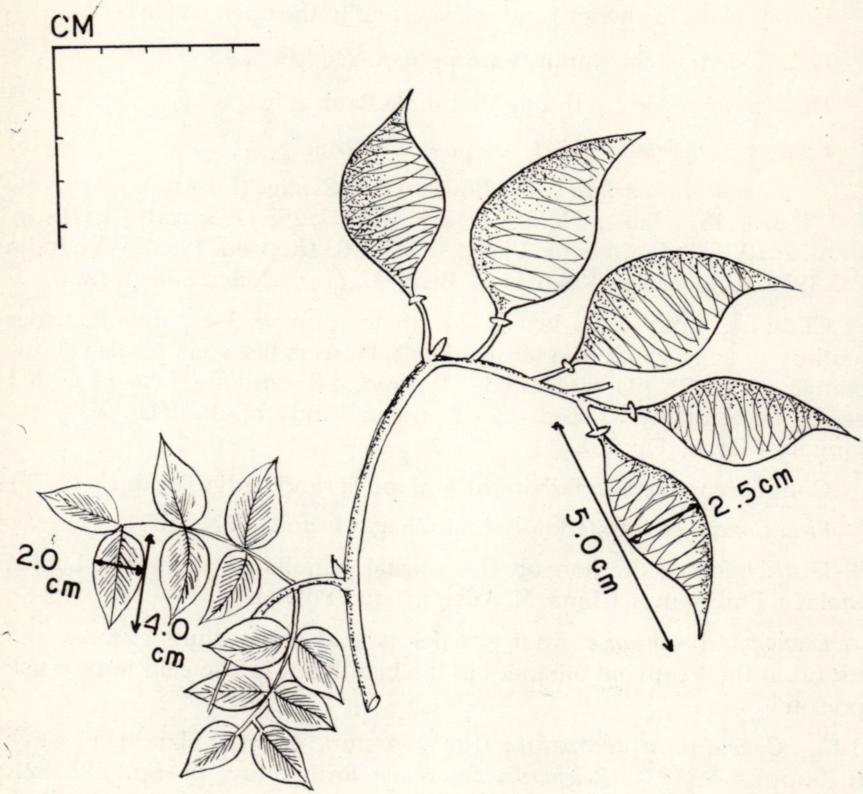

Fig. 15. CAESALPINIA CRISTA L.

Distributions: Throughout India; Sri Lanka, Malaya, China, Japan and most part of the South East Asian countries.

4. *CASSIA* L.

Key to the Species

1a	Small or large trees	(2)
1b	Herbs or shrubs	(6)
2a (1)	Flowers red or pink; anther filaments with globose swelling	5. *C. nodosa* Buch. Ham.
2b (1)	Flowers yellow; anther filaments without globose swelling	(3)
3a (2)	Stamens -10,6-7 fertile	9. *C. siamea* Lamk.
3b (2)	Stamens 10, all fertile	(4)
4a (3)	Stamens equal, pods indehiscent	4. *C. fistula* L.
4b (3)	Stamens nearly equal in length or unequal; pods dehiscent	(5)
5a (4)	Corolla bright yellow	11. *C. surattensis* Burm.f.
5b (4)	Corolla yellow but distinctly veined	2. *C. alata* L.
6a (1)	Fertile stames 5	(7)
6b (1)	Fertile stamens 6-7	(8)
7a (6)	Leaflets only two pairs	1. *C. absus* L.
7b (6)	Leaflets many	8. *C. pumila* Lamk.
8a (7)	Leaves with 1 gland just above the base of the petiole, none between the leaflets	(9)
8b (7)	Rachis usually with 1—several glands between the leaflets	(10)
9a (8)	Leaflets 3-5 pairs, ovate-acuminate, pods flat	7. *C. occidentalis* L.
9b (8)	Leaflets 4-10 pairs, oblong-lanceolate; pods more or less turgid	10. *C. sophera* L.
10a (8)	Large shrubs, leaflets 8-12 pairs	3. *C. auriculata* L.
10b (8)	Undershrubs, leaflets 3 pairs	(11)

11a (10)	Glands between each of the first 2 pairs of the leaflets, pedicels 1-1.5 cm long for pods	12. *C. tora* L.
11b (10)	Glands only between the first pairs of leaflets, pedicels 2-4 cm long for pods	6. *C. obtusifolia* L.

4.1. ***Cassia absus*** L. Sp . Pl. 376.1753; Baker, Hook.f. Fl Brit. India 2: 265.1878; Watt Dict. Econ. Prod. 2: 210.1889; Prain Beng. Plants 1: 315.1903 (Rep.ed.1963); Wlh. India 2: 93.1950. *Senna absus* Roxb. Fl . Ind. 2: 340.1832. *(Vern. Name: Chakan).*

Erect annual herb, clothed with grey bristly viscous hairs. Leaves linear, acute, persistent; leaflets 4, membranous. Flowers reddish-yellow, in few flowered racemes; stamen 5, equal, all fertile. Pods 2-5 cm long, oblique, seeds 5, compressed, dark brown, shinning. *Fls. & Frts.:* August-October.

Abundantly found during rainy season in the waste lands.

Field Collection: Ghoshpur/near Kulti, *K.N.-2127.*

Distributions: Throughout India; Sri Lanka, Tropical Africa.

Economic Importance: Seeds are used as a remedy for opthalmia.

4.2. ***Cassia alata*** L. Sp. Pl. 378.1753; Hook.f. Fl. Brit India 2: 264.1878; Watt Dict. Econ. Prod. 2: 211.1889; Prain Beng. Plants 1: 315. 1903 (Rep. ed. 1963). *Senna alata* Roxb. Fl. Ind. 2: 349.1832. *(Vern. Name: Dadmardan).*

Small to medium shrub; leaflets 8-14 pairs. Racemes narrow, peduncled. Flowers large, yellow. Pods 10-20 cm long, membranous, dehiscent, straight glabrous. Seeds 30 or more in each pod. *Fls. & Frts.:* November -March.

On the village scrub-jungles, road side hedges and not uncommon.

Field Collection: Shyamnagar, *K.N.-2098.*

Distributions: W. Peninsula. It is reported as introduced from West Indies. Malaya Islands, Sri Lanka and Siam.

Economic Importances: Leaves have a high repute as a remedy in skin diseases, notably ring-worms.

4.3. ***Cassia auriculata*** L. Sp. Pl. 379.1753; Baker, Hook.f. Fl. Brit. India 2: 263.1878; Watt Dict. Econ. Prod. 2: 213.1889; Prain Beng. Plants 1: 314.1903 (Rep. ed. 1963); Wlh. Ind. 2: 96.1950. *Senna auriculata* Roxb. Fl. Ind. 2: 349.1832. *(Vern. Name: Tarwar* (Hindi)*).*

Tall shrub with finely downy spreading branches. Leaves nearly sessile; rachis grooved, pubescent, leaflets 6-12 pairs. Flowers yellow, in axillary peduncled corymbose or corymbose racemes; calyx yellow. Pods straight, thin, flexible, seeds 6-12, compressed. *Fls. & Frts.:* December -March.

Planted in the gardens for its beautiful flowers; often found as an escape near gardens.

Field Collection: Dhopagachi, *K.N.-2138.*

Distributions: Dry regions in the Central Provinces and W. Peninsula; Sri Lanka.

Economic Importances: The bark is mostly used for dyeing and tanning leather and the seeds are used medicinally.

4.4. ***Cassia fistula*** L. Sp. Pl. 377. 1753; Baker, Hook.f . Fl. Brit . India 2: 261. 1878; Watt Dict. Econ. Prod. 2: 217.1889; Prain Beng. Plants 1: 314. 1903 (Rep.ed. 1963); Wlh. India 2: 97.1950. *Cassia rhombifolia* Roxb. Hort. Beng. 31.1814. *(Vern. Name: Bandar Lati* or *Sandal)*

Medium tree. Leaves pinnately compound, simple; leaflets 4-8 pairs 10-20 cm long; flowers in lax peduncles racemes, showy, yellow. Pods cylindrical, very long upto 0.6m, blackish green when young, hard and strong, indehiscent. *Fls. & Frts.:* December-May.

Common on the village scrub-jungles, garden edges, roadside thickets and on the railway tracks; also largely planted as ornamental tree.

Field Collections: Gosaba, *K.N.-177;* Naihati, *K.N.-2019.*

Distributions: Native of India, distributed throughout India. Burma, at present Pantropic.

Economic Importances: Flowers are showy. In Kerala, people use to decorate with the flowers in the celebration of their New Years at Pongal. Seed coat occasionally edible. Fruit pulf is purgative and used in laxative preparation.

4.5. ***Cassia nodosa*** Buch. Ham ex Roxb. (Hort. Beng. 31.1814), *nom. nud.* Fl. Ind. 2: 336.1832; Baker, Hook.f. Fl. Brit. India 2: 261.1878; Prain Beng. Plants 1: 314.1903 (Rep. ed. 1963).

Evergreen avenue tree. Leaves pinnately compound, paripinnate; leaflets 10-12 pairs, medium. Flowers pink in short corymbose. Bracts large persistent. Pods cylindric, 30-40 cm long. *Fls. & Frts.:* March-August.

Planted in the garden for its ornamental flowers.

Field Collection: Champahati, *K.N.-927.*

Distributions: Planted almost throughout India. Native of Malaya, Burma and Bangladesh.

Economic Importance: Ornamental plants.

4.6. ***Cassia obtusifolia*** L. Sp. Pl. 377.1753; Baker, Hook.f. Fl. Brit. India 2: 263 (in part) 1878; Prain Beng. Plants 1: 314.1903 (Rep. ed. 1963). *C. tora* auct non L. *Senna toroides* Roxb. Fl. Ind. 2: 341.1832. *(Vern. Name: Chakunda).*

Annual herb; 0.5 -2.0 m high, not foetid. Leaves 7-9 cm long, rachis grooved, more or less pubescent, with a conical gland between the lowest pair of leaflets only; stipules linear-subulate caducous. Leaflets 3 pairs, opposite. Flowers usually in sub-sessile pairs. Calyx glabrous, divided to the base; segments ovate, acute, spreading. Petals 5, bright yellow, sub-equal. Stamens 10, pods sub-terete, obliquely septate, glabrous. Seeds 30-35. *Fls. & Frts.:* July-October.

A dominant weed during rainy season, very often been found mixed with *C.tora* L.

Field Collection: Mollakhali, *K.N.-1844.*

Distributions: More or less naturalised throughout India. Native of Tropical America.

4.7. ***Cassia occidentalis*** L. Sp. Pl. 377.1753; Baker, Hook.f. Fl. Brit. India 2: 262.1878; Watt Dict. Econ. Prod.2: 221.1889; Prain Beng. Plants 1: 314.1903 (Rep.ed.1963); Wlh.India 2: 97.1950. *(Vern. Name: Kalkasunda).*

Undershrub; leaves paripinnate, leaflets ovate-oblong, petiolar gland just above the thickened base of the petiole. Flowers yellow in short axillary, short peduncled corymbose. Seeds arranged in a single row. *Fls. & Frts.:* May-March.

Abundantly found along the road sides, railway tracts and moist waste lands.

Field Collection: Dhosha Chandaneswar, *K.N.-891.*

Distributions: Almost throughout India; introduced long before the publication of the Roxburgh's Flora Indica (1814). Native of the Tropical America, at present Pantropic.

Economic Importances: The leaves, roots and seeds are used medicinally.

4.8. ***Cassia pumila*** Lamk. Encycl. Method. 1: 651.1784; Baker, Hook.f. Fl. Brit. India 2: 266.1878; Prain Beng. Plants 1: 315.1903 (Rep.ed.1963). *Senna prostrata* Roxb. Fl. Ind. 2: 352.1832. *(Vern. Name: Chota aura* (Hindi)*)*.

Prostrate, ascending or sub-erect, deep rooted, stout herb. Leaves 2.5-5.0 cm long, gland at the base of petiole; stipule semi-lanceolate, rigid, persistent, leaflets 10-20 pairs, linear oblong, rigidly coriaceous, mucronate. Flowers yellow, usually solitary. Pods 2.5-4.0 cm long, straight, flat, linear, 6-12 seeded. *Fls. & Frts.:* September-October.

Common during rainy season in the open fields and also in shades of trees.

Field Collection: Bireshwarpur/Lakshmikantapur, *K.N.-1759.*

Distributions: Throughout India. Tropical Asia and Australia.

4.9. ***Cassia siamea*** Lamk. Encycl. 1: 648.1785; Baker, Hook.f. Fl. Brit. India 2: 264.1878; Watt Dict. Econ. Prod. 2: 223.1889; Prain Beng. Plants 1: 315.1903 (Rep. ed. 1963); Datta & Majumder, Bull. bot. Soc. Bengal 20(2): 63.1966. *Senna sumatrana* Roxb. (Hort. Beng . 31.1814) & Fl. Ind. 2: 347.1832.

Medium, evergreen avenue tree. Leaves paripinnate, leaf rachis without gland. Flowers yellow in panicled racemes. Fertile stamens 7. Pods flat, thickened at sutures. *Fls. & Frts.:* June-January.

Planted as avenue tree and in the lakes and gardens.

Field Collection: Naihati, *K.N.-2022.*

Distributions: Mostly on the southern parts of India. Malaya, Burma; Native of South East Tropical Asia.

Economic Importance: Wood is reported to have very good paper pulp.

4.10. ***Cassia sophora*** L. Sp. Pl. 379.1753; Baker, Hook.f. Fl. Brit. India 2: 262.1878 (incl. var. *purpurea* (Roxb.) Baker); Watt Dict. Econ Prod. 2: 223.1889; Prain Beng. Plants 1: 314.1903 (Rep.ed. 1963). *(Vern. Name: Kalkasunda).*

Small shrub. Leaves 15-20 cm long, rachis grooved with a solitary dark gland near the base; leaflets opposite 6-10 pairs. Flowers yellow in axillary distinctly peduncled. Pods solitary turgid, torulose. Seed dark brown. *Fls. & Frts.:* July-February.

Common in the waste lands, along the roadsides and railway tracks.

Field Collection: Piyalinagar/Baruipur, *K.N.- 49.*

Distributions: Throughout India; introduced from America. Now naturalised in the waste lands throughout. A pantropic species.

Economic Importances: The bark, leaves and seeds are used in different medicines.

4.10./1. ***Cassia sophora*** var. *purpurea* (Edw.) Voigt Hort. Sub. Calc. 248. 1845; Baker, Hook.f. Fl. Brit. India 2: 263.1878; Prain Beng. Plants 1: 314. 1903 (Rep. ed. 1963). *Cassia purpurea* Edw. Bot. Reg. t. 856.1825. *(Vern. Name: Lal Kalkashunda).*

Shrub, usually greennish, sometime purplish. Leaves pinnate, leaflets 6-10, oblong lanceolate. Stamens 10, only 7 are fertile, dehiscence-porus, legume terete or compressed, solid. *Fls. & Frts.:* July-September.

Grow profusely on the roadside waste lands, high cultivated lands.

Field Collection: Malatipur/near Dey Ganga, *K.N.-1580.*

Distributions: Almost throughout India. In the Tropics.

4.11. ***Cassia surattensis*** Burm.f. Fl. Ind. 97.1768; De Witt, Webbia 11: 169.1956. *C. glauca* Lamk. Encycl. 1: 647.1785; Baker, Hook.f. Fl. Brit. India 2: 263.1878; Watt Dict. Econ. Prod. 2: 219.1889; Prain Beng. Plants 1: 314.1903 (Rep.ed.1963).

Large shrub or small tree. Leaves 15-23 cm long; rachis slightly channelled above and furnished with long conical glands between each of the 2-3 lowest pairs. Flowers bright yellow, in axillary peduncled corymbose; stamens 10, unequal. Pods 10-20 cm long, linear straight, thin flexible; seeds many, flattened. *Fls. & Frts.:* December-March.

Planted as hedge plants for ornamental purposes.

Field Collection: Mazilpur/Joynagar, *K.N.-1617.*

Distributions: Mainly in the South India. Native of Malaya, Burma, Tropical Asia and Australia generally, but usually cultivated.

4.12. ***Cassia tora*** L. Sp. Pl. 376.1753; Baker, Hook.f. Fl. Brit. India 2: 263.1878; Watt Dict. Econ Prod. 2: 224.1889; Prain Beng. Plants 1: 314. 1903 (Rep. ed. 1963). *Senna tora* Roxb. Fl. Ind. 2: 340.1832. *(Vern. Name: Chakunda).*

Annual weed; leaves paripinnate, obovate-oblong. Flowers bright yellow, usually in pairs. On very short axillary peduncles, fertile stamens 7, staminodes 3. Pods 15-25 cm long, stout obliquely septate. *Fls. & Frts.:* July-March.

Very common on the waste lands, railway tracks, roadsides.

Field Collection: Piyalinagar/Baruipur, *K.N.-1002.*

Distributions: Widespread in India; introduced in the early part of 17th Century . Native of America, Sri Lanka.

Notes: This taxon and C. obtusifolia L. were treated as conspecific by Maheshwari (1963); but De Wit and Brenan (1956) have shown clearly that they are distinct species.

5. *DELONIX* Raf.

5.1. ***Delonix regia*** (Boj.ex Hook.) Raf. F. Tellur. 292.1836; Wlh. India 3:30.1952. *Poinciana regia* Boj. ex Hook. Bot. Mag. t. 2884. 1829; Baker, Hook.f. Fl.Brit. India 2: 260 1878; Watt Dict. Econ. Prod. 6(1): 309. 1893; Prain Beng. Plants 1: 321.1903 (Rep. ed. 1963). *(Vern. Name: Krishnachura).*

Exotic horticultural shrub or tree. Leaves bipinnate. Flowers deep crimson or bright scarlet in terminal corymbs. Pods flat, straight, woody. *Fls. & Frts.:* July-February.

Commonly planted as ornamental species.

Field Collections: Palta, *K.N.-1835;* Naihati, *K.N.-2031.*

Distributions: Cultivated or planted throughout India; introduced into India from Madagascar. A native of Madagascar.

Economic Importance: Beautiful ornamental plants.

6. *MEZONEURON* Desf.

6.1. ***Mezoneuron cucullatum*** Wight. & Arn. Prodr. 283.1834; Naskar, Aq. & Semi. Plants 123.1990. *Caesalpinia cucullatum* Roxb. Fl Ind. 2: 358.1832. *(Vern. Name: Ragi).*

Large climber in the moist evergreen forest, not very common. Naskar (l.c.) described the plant in respect of its taxonomy, key characters, and ecological notes.

7. *PARKINSONIA* L.

7.1 ***Parkinsonia aculeata*** L. Sp. Pl. 375.1753; Naskar & Guha Bakshi Mang. Swamps of Sundarbans 118. 1987. *(Vern. Name: Bilati Kikar).*

An exotic small tree or shrubs; native of Tropical America. Naskar & Guha Bakshi (1987) described this taxon in respect of its taxonomy, ecology and economic importances.

8. *PELTOPHORUM* Walp.

8.1. ***Peltophorum pterocarpum*** (DC.) Baker ex Heyne, Nutt. Pl. Ned. Ind. (ed-2) 2: 755.1927. *Inga pterocarpa* DC. Prodr. 2: 441.1825. *Caesalpinia inermis* Roxb. Fl. Ind.2: 357.1832. *C. ferrugineus* Decne, Nouv. Ann. Muss. Paris. 3: 462.1834. *(Vern. Name: Gul Mohar).*

Tall evergreen tree. Leaves pinnately compound, bipinnate. Flowers yellowish in axillary and panicled terminal racemed panicle; stamens 10. Pod flat, winged, dehiscent, 1-4 seeded. *Fls. & Frts.:* March-September.

Planted as avenue tree and in the lawns or gardens.

Field Collections: Titagarh/Barrackpore, *K.N.-2200.*

Distributions: Planted almost throughout India; Native of Tropical Australia, Malaya and Sri Lanka.

Economic Importance: Ornamental avenue tree.

9. *PILIOSTIGMA* Hochst

9.1. ***Piliostigma malabaricus*** (Roxb.) Benth., Pl. Jungh. 261.1852; *Bauhinia malabarica* Roxb. Fl. Ind. 2: 321.1831; Baker, Hook.f. Fl.Brit. India 2: 277. 1878; Watt Dict. Econ. Prod. 1: 420.1889. *(Vern. Name: Amli* or *Karmai).*

Moderate evergreen tree with busy crown. Leaves 3.5-10.0 cm; long. 7-9, sometime 11 nerved; petiole swollen at both the ends, rusty pubescent. Flowers dull white, in axillary and terminal racemes, calyx brown, tomentose; stamens 10. Pods 16-35 cm long, flat, flexible, beaked, seeds 20-30, ovoid, globose, dark brown, polished. *Fls. & Frts.:* September-March.

As avenue tree or planted as ornamental garden plants.

Field Collection: Narendrapur, *K.N.-1987.*

Distributions: Throughout India; Burma, Thai and Malaysia.

10. *POINCIANA* L.

10.1. ***Poinciana pulcherrima*** L. Sp. Pl. 380.1753. *Caesalpinia pulcherrima* Swartz. Obs. 166.1791; Baker, Hook.f. Fl. Brit. India 2: 255.1878; Watt Dict. Econ. Prod. 2: 10.1889; Prain Beng. Plants 1: 323.1903 (Rep. ed. 1963). *(Vern. Name: Radha-Chura).*

Medium to large tree, deliquescent avenue tree. Leaves abruptly bipinnate. Flowers bright red, petals long-clawed, stamens 10, exserted.

Pods flat, thin broadly linear, strap end with black seeds. *Fls. & Frts.:* June-September.

Planted in the roadsides, lawns and gardens.

Field Collection: Joygram/Minakhan, *K.N.-301.*

Distributions: Planted almost throughout India; introduced from America. A pantropic species.

Economic Importance: Ornamental and avenue tree.

11. *SARACA* L.

11.1. ***Saraca indica*** L. Mant. 98.1767; Baker, Hook.f. Fl. Brit. India 2: 271.1878; Watt Dict. Econ. Prod. 6 (2): 475. 1893; Prain Beng Plants 1: 319.1903 (Rep. ed. 1963). *Jonesia asoca* Roxb. As. Res. 4: 335. 1795. *(Vern. Name: Asoke).*

Low deliquescent tree. Leaves pinnately compound; leaflets lanceolate, glabrous. Flowers vermillion, coloured in the dense corymbose; calyx 4-lobed. Stamens 4-8. Fruit flat, oblong, woody, 4-8 seeded. Seeds oblong, compressed. *Fls. & Frts.:* February-April.

Planted in the gardens or lawns for the sake of its very good looks.

Field Collection: Gobindapur, *K.N.-1377.*

Distributions: Native of India. Burma, Malaya and Sri Lanka.

Economic Importance: The bark is used in Ayurvedic medicine–as an astringent and uterine sedative. It has stimulating effect as endometrium and on the ovarian tissues and is commonly used in menorrhgia due to uterine fibroides; also used in haemorrhoides and dysentery; best remedy for all female diseases, especially for irregular menstruation with severe pain and leucorrhoea.

12. *TAMARINDUS* L.

12.1. ***Tamarindus indica*** L. Sp. Pl. 34.1753; Baker, Hook.f. Fl. Brit. India 2: 273.1878; Watt Dict. Econ. Prod. 6(3): 404.1893; Prain Beng. Plants 1: 320.1903 (Rep. ed. 1963). *(Vern. Name: Tetul).*

Large, handsome, sub-evergreen tree. Leaves 5-15 cm long, even pinnate; stipules persistent in young parts, linear lanceolate, leaflets 10-20 pairs, 1-2 cm long, linear oblong, slightly notched at the apex, coriaceous. Flowers pale-yellow or yellowish white, in few flowered terminal lax racemes; petals streaked with red, 3 only developed, lower most reduced to scales; fertile stamens only 3, rest reduce to setaceous staminodes. Pods

8-15 cm long, usually curved, pulpy inside, seeds dark brown, shining. *Fls. & Frts.:* May-December.

Planted in the gardens and in the roadsides.

Field Collection: Naihati, *K.N.-2034.*

Distributions: Throughout India.Native of tropical Asia and Africa; sometime said to be Sudanic Origin.

Economic Importances: Acidic fruits are eaten both ripe and green; an adhesive is made from the seeds. Seeds husk is used as fish toxicant. The wood is used both as fire wood and for furniture manufacturing. Unripe fruits contain high percentage of tartaric acid and the by-products like ethanol, lactic acid and citric acid. Seed pulp is refrigerent, carminative and laxative. Used for febrile diseases and bilious disorders. Fresh leaves poultice useful in swelling and boils and for relieving pains.

Remarks: The species *Afzelia retusa* Kurz. J. Asiat. Soc. Bengal 42: 2, 73.1873 (syn. *Intsia retusa* O. Kuntze), (Vern. Name: Hinga or Somdal) has been collected by me from Gosaba (Sundarbans) on 20.3.1977 *K.N.-674;*

Several other ornamental and garden plants are also available from the lower Ganga Delta regions, but all these plants were not studied and mentioned here.

XLIX. MIMOSACEAE

Key to the Genera

1a	Anther gland-tipped, stamens 5-10	(2)
1b	Anthers not gland-tipped	(3)
2a (1)	Small, aquatic herb; inflorescences capitate; heads small; pods thin smaller	6. *NEPTUNIA* Lour.
2b (1)	Large, terrestrial, spreading shrub or small tree; inflorescence elongate; pods thick, turgid, large	8. *PROSOPIS* L.
3a (1)	Stamens indefinite	(4)
3b (1)	Stamens definite; usually 10	(8)
4a (3)	Mostly spiny or prickly, shrubs or small tree; flowers spicate or capitate; stamens free	1. *ACACIA* Mill.
4b (3)	Usually unarmed tree (exception- *Pithocelobium dulci);* stamens monadelphous	(5)

5a (4)	Pods septated, indehiscent	9. *SAMANEA* Merr.
5b (4)	Pods not septated	(6)
6a (5)	Armed trees or shrubs; pods twisted with coriaceous valves	7. *PITHOCELOBIUM* Mart. *nom. cons*
6b (5)	Unarmed trees or shrubs; pods straight with thin valves	(7)
7a (6)	Large tree; stutures thin	2. *ALBIZIA* Durazz
7b (6)	Large shrub; stutures thick	3. *CALLIANDRA* Benth. *nom. cons.*
8a (3)	Large unarmed shrubs or small tree; leaves not closed with touch; pods straight	4. *LEUCAENA* Benth.
8b (3)	Small armed herb; leaves generally touched with touch; pods slightly curved	5. *MIMOSA* L.

1. *ACACIA* Mill.

Key to the Species

1a	Leaves reduced to phyllodea	1. *A. auriculiformis* A. Cunn.
1b	Leaves not reduced to phyllodea	(2)
2a (1)	Flowers in long pendulous spikes	(3)
2b (1)	Flowers in globose heads	(4)
3a (2)	Barks very dark; pinnae 2-24 pairs	2. *A. catechu* Willd.
3b (2)	Barks very light; pinnae 20-30 pairs	6.*A. polycantha* Willd.
4a (2)	Armature not stipular	3. *A. concinna* DC.
4b (2)	Armature stipular in origin	(5)
5a (4)	Shrubby, pods thick, cylindric black	4. *A. farnesiana* Willd.
5b (4)	Arboreous; pods flat, monifiform; with deep suture between the seeds	5. *A. nilotica* Willd.

1.1. ***Acacia auriculiformis*** A. Cunn. Hook. Lond. J. Bot. 1: 377. 1842. *(Vern. Name: Akashmoni).*

Medium, unarmed tree. Phyllodea large, falcate, coriaceous, parallel nerved, acute at both ends. Flowers small, yellow in short cylindrical, spikes. Pods much twisted and coiled. Seeds covered with yellow furniculus. *Fls. & Frts.:* March-December.

Planted in the gardens and roadsides as avenue tree.

Field Collection: Rabindra Sarabar, *K.N.-2243.*

Distributions: Planted almost throughout India. Native of Australia.

Economic Importances: Ornamental plants; avenue tree; wood is suitable for good paper-pulp.

Notes: D. Prain (1903a, 1903b and 1905) did not mention this taxon.

1.2. ***Acacia catechu*** Willd. Sp. Pl. 4: 1079.1806; Baker, Hook.f. Fl. Brit. India 2: 295. 1878; Watt Dict. Econ. Prod. 1: 27.1889; Prain Beng. Plants 1: 330.1903 (Rep. ed. 1963); Wlh. India 1: 9.1948. *(Vern. Name: Khayer).*

Medium tree. Leaves bipinnate; rachis prickly and downy, with one gland between the lowest pair or pinnae and at the insertions of the 3 uppermost pairs; pinnae 20 pairs or more, leaflets numerous, linear; stipular spines in pairs. Flowers creamy white, fragrant. Pods stalked, flat, dehiscent, dark-brown; 5-6 seeded. *Fls. & Frts.:* April-August.

Sometime on the village thickets and roadside hedges.

Field Collection: Garia, *K.N.-315.*

Distributions: Dried zones of India and on the coastlines. Plains and lower hill forests of the upper and lower Burma.

Economic Importances: An important commercial product *'catechu* or *Kattia'* is obtained from the wood, which is also valued for its hardness and durability. Produce vegetable gum for commercial purposes.

1.3. ***Acacia concinna*** DC. Prodr. 2: 464. 1825; Baker, Hook.f. Fl. Brit. India 2: 296.1878; Watt Dict. Econ. Prod. 1: 44.1889; Prain Beng. Plants 1: 331.1903 (Rep. ed. 1963) *(Vern. Name: Ban-Ritha).*

Extensive twining shrub, armed with numerous small hooked prickles; branches brown, dotted with white. Leaves 2-pinnate, 5- 10 cm long; main rachis armed with sharp hooked prickles; leaflets subsessile, sensitive, 12-25 pairs. Flowers in globose heads; bracts oblique, membranous, conspicuous. Pods shortly stalked, linear oblong, thick. Seeds 6-10. *Fls. & Frts.:* June-October.

Occasionally on the village thickets and roadside hedges.

Field Collections: Baruipur, K.N.-98.

Distributions: Throughout India; Malaya Islands and China.

Economic Importance: Pods are used for washing the hairs; pods contain 5% saponin, which is said to be important as fish toxicant.

Notes: Prain (1903a) did not mention this taxon from lower Bengal.

1.4. ***Acacia farnesiana*** (L.) Willd. Sp. Pl. 4: 1983.1806; Baker, Hook.f. Fl. Brit. India 2: 292.1878; Watt Dict. Econ. Prod. 1: 48.1889; Prain Beng. Plants 1: 330.1903 (Rep. ed. 1963); Wlh. India 1: 24.1948. *Mimosa farnesiana* L. Sp. Pl. 521.1753. *Vachellia farnesiana* (L.) W. & A. Prodr. 272.1834. *(Vern. Name: Guhiya Babla).*

Small tree. Leaves 2-8 pairs; leaflets 10-20 pairs, small linear oblong, sub-glabrous, green. Flowers golden yellow, frequent, in axillary, fascicled heads. Pods thick, cylindrical, dull brown. *Fls. & Frts.:* January-May.

Frequently found wild in the waste lands, road sides, railway tracks.

Field Collection: Shyamnagar, *K.N.-2092.*

Distributions: Introduced into India during the 18th Century; now naturalized throughout India, mainly in the hotter areas. Native of Tropical America.

Economic Importances: Planted as hedge plants and also for the ornamental purposes. Stem gum used in confectionery. Plants serves as host for lac insects.

1.5. ***Acacia nilotica*** (L.) Willd. ex Del., Fl. Aegypy. 79.1813. *Mimosa nilotica* L. Sp. Pl. 521.1753. *(Vern. Name: Babul* or *Babla).*

Small to medium tree, with long, straight 2-5 cm long white spines. Leaves bi-pinnate, 5-10 cm long. Flowers yellow, in globose heads; peduncles axillary, compressed, constricted at the sutures between the seeds, densely and persistently gray downy. *Fls. & Frts.:* Almost throughout the year.

Common on the village thickets, roadsides and railway tracks.

Field Collections: Brace Bridge, *K.N.-120.*

Distributions: Almost throughout India; Sri Lanka, Arbia, Egypt and Tropical Africa.

Economic Importance: Pods fodder; wood used as fuel etc.

1.5/1. ***Acacia nilotica*** (L.) Willd. ex Del. subsp. ***indica*** (Benth.) Brenan, Kew. Bull. 1957: 84.1957; Baker, Hook.f. Fl. Brit. India 2: 298. 1878; Prain Beng. Plants 1: 330.1903 (Rep. ed. 1963). *Mimosa nilotica* L. Sp. Pl. 521.1753.

Acacia arabica (Lamk.) Willd. var *indica* Benth., Hook. Lond. J. Botany 1: 500.1842; Wlh. India 1: 5.1948. *(Vern. Name: Babul* or *Babla).*

Small tree, stipules with fine spine, 1-5 cm long. Leaves 5-10 cm long; rachis brown showy; leaflets 15-30 pairs, 2-5 cm long, glabrous or slightly downy; linear oblong, sub-obtuse. Flowers in axillary, yellow; peduncles gray downy; bracteole above middle. Pods upto 20 cm long, gray downy, deeply constricted in between the seeds. *Fls. & Frts.:* All round the year.

Xerophytic habit, very common on the roadsides, railway tracks, hedges and thickets.

Field Collection: Falta, *K.N.-1834.*

Distributions: Throughout the hotter parts of India. N. Arabia; a native of Asia.

Economic Importances: Pods and seeds prefered by goat and sheeps; the bark contain tannin properties. Wood is commonly used as fire-wood, agricultural equipments and as handle of the cutting tools. Gums useful in diarrhoea, dysentery and in diabetes mollitus; bark is astringent and demulcent.

1.6. ***Acacia polyacantha*** Willd. Sp. Pl. 4: 1079. 1806; Maheshwari, Fl. Delhi 150.1963. *A. suma* (Roxb.) Buch Ham. Wall. Cat. 5227 C.1828; Baker, Hook.f. Fl. Brit. India 2: 294.1878; Watt Dict. Econ. Prod. 1: 60. 1889; Prain Beng. Plants 1: 330.1903 (Rep. ed. 1963); Wlh. India 1: 10.1948. *Mimosa suma* Roxb. Fl. Ind. 2: 563.1832. *(Vern. Name: Sai-Kanta).*

Small to medium tree; bark white, young shoots hanging. Leaves bi-pinnate 10-15 cm long, rachis pubescent with glands, the pairs of pinnae. Leaflets 30-50 pairs. Flowers sessile, whitish, peduncled spikes, rachis densely pubescent; calyx campanulate, downy. Pods 7.5-12.5. X 1.5 - 2.0cm, flat, glabrous. Seeds 6-8. *Fls. & Frts.:* June-November.

Occasionally on the railway tracks, roadside hedges and thickets.

Field Collection: Basanti, *K.N.-2270.*

Distributions: Bengal, Bihar and the Western Peninsula; S.E. Asian countries.

Economic Importances: Contain high value tannin and hard wood contain catechu.

2. ALBIZIA Durazz.

Key to the Species

1a Pods narrow almost cylindrical and longer; seeds comparatively larger ... 1. *A. lebbeck* (L.) Willd.

1b	Pods flat and broad; seeds comparatively smaller	2. *A. procera* (Roxb.) Benth.

2.1. ***Albizia lebbeck*** (L.) Willd. Sp. Pl. 4: 1066.1806; Baker, Hook.f. Fl. Brit. India 2: 293. 1878; Watt Dict. Econ. Prod. 1: 156.1889; Prain Beng. Plants 1: 332. 1903 (Rep. ed. 1963); Wlh. India 1: 43. 1948. *Mimosa lebbeck* L. Sp. Pl. 516.1753. *Acacia speciosa* Willd. Sp. Pl. 4: 1066. 1806. *(Vern. Name: Siris).*

Large deciduous tree; young shoots glabrous. Pinnae 2-4 pairs; rounded at apex, truncate at base, obliquely rounded in terminal pair, lateral leaflets elliptic-oblong, terminal pair obovate, oblong. Petioles very short. Flowers white, fragrant, in globose heads which are arranged in solitary racemes. Seeds elliptic-oblong, pale brown. *Fls. & Frts.:* July-March.

Ecological Notes: Mostly naturalized, commonly planted in the village gardens, roadsides and the embankments of ponds.

Field Collection: Magrahat, *K.N.-2521.*

Distributions: Throughout India. Burma, Sri Lanka, Africa, Indo-China, S. China, Formosa, Malaya and Australia.

Economic Importances: Bark and seeds astringent; useful in piles, diarrhoea and act as a tonic and restroactive. Root gum used as dental powder, strengthen gum. Leaves useful in night-blindness.

2.2. ***Albizia procera*** (Roxb.) Benth., Hook. London J. Bot. 3: 89. 1844; Baker, Hook.f. Fl. Brit. India 2: 299. 1878; Watt Dict. Econ. Prod. 1: 159. 1889 Prain Beng. Plants 1: 332. 1903. *Mimosa procera* Roxb. Pl. Cor. 2: 12, 6.121. 1799. *(Vern. Name: Kori* or *Koroi).* (Fig. 16).

Unarmed medium tree. Leaves bi-pinnate, glabrous 25-30 cm long; pinnae 4-6 pairs, 10-15 cm long; leaflets 5-12 pairs, dark green and glabrous above. Flowers sessile, in numerous small 12-20 flowered heads, 2 cm across, arranged in terminal panicles. Calyx glabrous, teeth distinct, triangular, corolla 5 mm long, pubescent outside, teeth lanceolate, half as long as the corolla tube. Filaments greenish yellow. Pods stalked. Seeds 5-10, broadly ovate, pale brown. *Fls. & Frts.:* December-April.

Very often planted in the newly settlement areas in the villages.

Field Collection: Subhashgram, *K.N.-1339.*

Distributions: Central and Eastern Himalayas, Bihar and Western Peninsula. Bangladesh, Burma, China. Malaya Archipelago and Philippines.

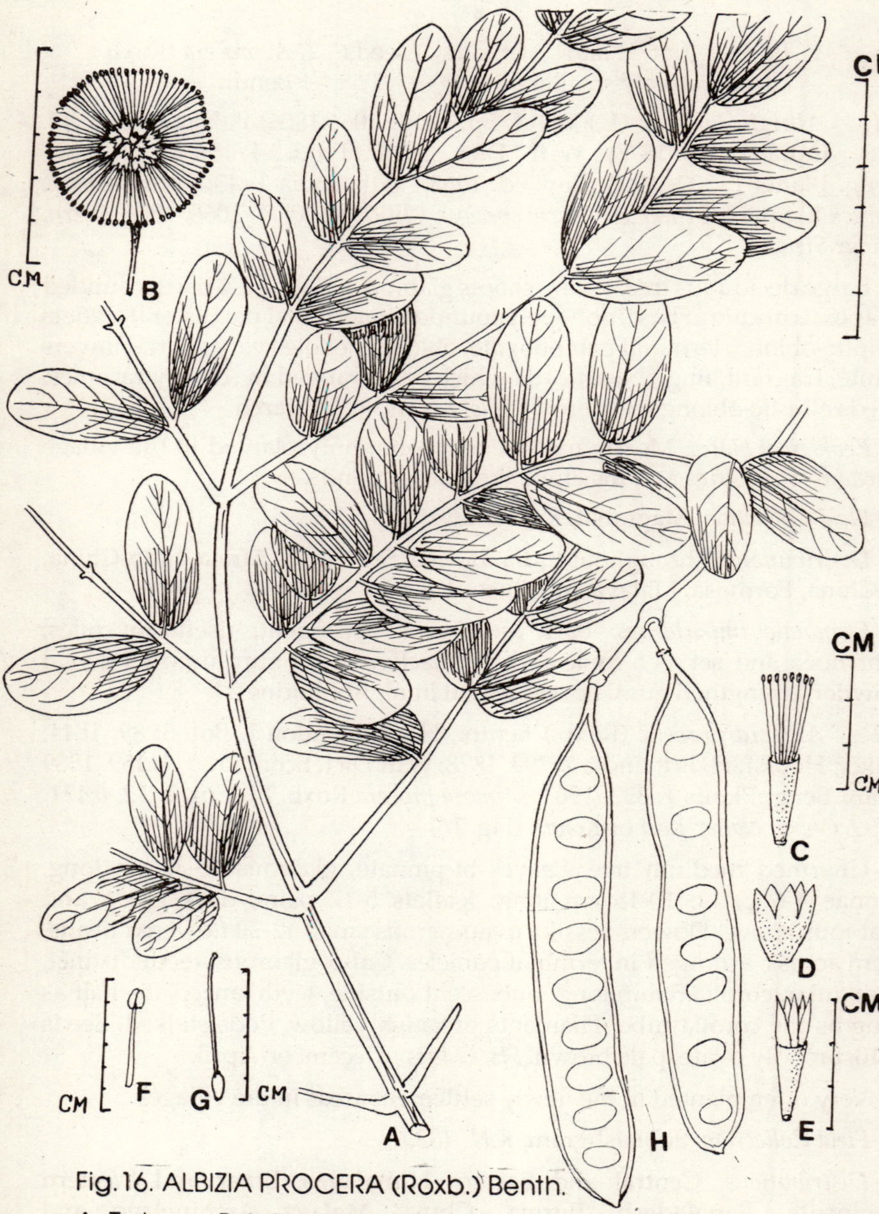

Fig. 16. ALBIZIA PROCERA (Roxb.) Benth.

A. Twig, B. Inflorescence, C. Flower, D. Calyx, E. Androecium, F. Anther & Filament, G. Ovary & Stigma.

Economic Importances: Quick growing timber yielding plants and sometimes used as avenue tree. The leaves have the insecticidal properties; leaves poultice are useful for ulcers.

3. *CALLIANDRA* Benth. *nom. cons.*

3.1. ***Calliandra brevipes*** Benth. Hook. J. Bot. 2: 140.1840.

Unarmed shrub. Leaves bipinnate, leaflets dark green; 20-40 pairs, sub-equal, sessile, linear, overlapping. Flowers purplish red, in loose hemi-spherical heads, stamens many, pinkish-purple with white base. Pods straight or faintly curved, flat with thickened sutures, gradually broadened from the base upwards; 2- valved; valves separating from their top to the base, recurving elastically. *Fls. & Frts.:* Almost throughout the year.

Planted in the ornamental gardens.

Field Collection: Ramnagar/Metia Bruz. *K.N.-1481.*

Distributions: Planted almost throughout India.

Economic Importance: Beautiful ornamental garden shrubs.

4. *LEUCAENA* Benth.

4.1. ***Leucaena latisiliqua*** (L.) Gills. Taxon 28: 190.1974. *Mimosa latisiliqua* L. Sp. Pl. 519.1753. *M. leucocephala* Lamk. Encycl. 1: 12. 1783. *Leucaena glauca* Benth., Hook. J. Bot. 4: 416.1842; Baker, Hook.f. Fl. Brit. India 2: 290. 1878; Watt Dict. Econ. Prod. 4: 632.1890: Prain Beng. Plants 1: 320. 1903 (Rep. ed. 1963). *L. leucocephala* (Lamk.) de Wit., Taxon 10: 54.1961. *(Vern. Name: Subabul* or *Khilkadam).*

Unarmed shrub or small tree. Branchlets glabrous. Leaves pinnae 3-7 pairs, main rachis grooved, acute at apex, obliquely rounded truncate or obtuse at base. Flowers in solitary or dense on globose heads. Pods upto 12 cm long 15-20 seeded, glabrous flat, straight, with about 1 cm long stalk. *Fls. & Frts.:* Throughout the year.

Commonly grown on the roadsides, river banks and hedges or waste lands and village shrubberies. Now commonly planted in the afforestation programmes.

Field Collection: Canning, *K.N.-202.*

Economic Importances: Quick growing plants, used as fire-wood and for some agricultural equipments. Pods and leaves contain highly protein fodder for the domestic cattles and useful organic manure or paper pulp.

5. *MIMOSA* L.

Key to the Species

1a	Spreading herb; pinnae digitately arranged; stamens 4	1. *M. pudica* L.
1b	Erect or straggling shrub; pinnae pinnately arranged, stamens 8	2. *M. rubicaulis* Lamk.

5.1. ***Mimosa pudica*** L. Sp. Pl. 518. 1753; Baker, Hook.f. Fl. Brit. India 2: 291. 1878; Watt Dict. Econ. Prod. 5: 248.1892; Prain Beng. Plants 1: 329. 1903 (Rep. ed. 1963). *(Vern. Name: Lajjabati* or *Najak).*

Prostrate or erect spreading herb. Stem armed with scattered recurved brownish coloured prickles and bristles; petiole, leaf- rachie, peduncle and stipules covered with bristles. Leaves 2-3 pinnae, leaflets 12-22 pairs, linear oblong, acute at apex, brightly-ciliate base obliquely rounded, linear lanceolate, acuminate. Heads in fascicles, globose. Flowers pink, bracteoles. Pods flat, slightly curved, 3-5 jointed, bristly along the sutures. *Fls. & Frts.:* October-March.

Commonly grow on the road-sides, railway tracks, waste lands, hedges and on the side of the village gardens.

Field Collection: Madarhat/Baruipur, *K.N.-75.*

Distributions: Throughout India; but introduced from Tropical America. Native of Tropical America.

5.2. ***Mimosa rubicaulis*** Lamk. Encycl. Meth. 1: 20.1783; Naskar, Aq. & Semi. Plants 124. 1990. *(Vern. Name: Shaih Kanta).*

Straggling, semi-aquatic shrub; Naskar (1.c.) described this taxon in respects of its taxonomy, ecological notes and distributions.

6. *NEPTUNIA* Lour.

6.1. ***Neptunia oleracea*** Lour. Fl. Cochinch. 654. 1790; Naskar Aq. & Semi Plants 124. 1990. *(Vern. Name: Pani Najak).*

Aquatic herbs. Very occasional in the village ponds and jheels. Naskar (1.c.) described this taxon in respect of its taxonomy and ecology.

7. *PITHOCELLOBIUM* Mart. *nom.cons.*

7.1. ***Pithocellobium dulci*** (Roxb.) Benth, Hook. London. J. Bot. 3: 199. 1844; Baker, Hook.f. Fl. Brit. India 2: 302.1878; Watt Dict. Econ. Prod. 6(1): 281.1893; Prain Beng. Plants 1: 333. 1903 (Rep. ed. 1963). *Mimosa dulcis*

Roxb. Pl. Cor. 1: 67.1795. *Inga dulcis* Willd. Sp. Pl. 4: 1005. 1806. *(Vern. Name: Belati Amlı).*

Medium tree with stipular spines on the bark; leaves branchlets, stipular throns with a pair or pinnae; leaflets 1-pair, sub-sessile, sub-coriaceous, glabrous, dark green and bright above, oblique, obovate-oblong, obtuse at both ends, entire. Peduncle short terminal penicles or axillary recemes; seeds black, half covered with white pulpy aril. *Fls. & Frts.:* March-September.

Abundant on the hedges, waste lands on the side of the railway tracks. Occasionally planted.

Field Collection: Canning, *K.N.-130.*

Distributions: Throughout India; early introduced into India. Native of Mexico.

Economic Importances: Pods edible; produce fire-wood and as hedge plants. Gum is also useful.

8. *PROSOPIS* L.

8.1. ***Prosopis chilensis*** (Molina) Stuntz, U.S. Dept. Agric. Bur. Pl. Invent. Seeds and Pl. Import. 31: 85.1914. *Ceratonia chilensis* Molina Sagg. Chil. 172.1782. *Prosopis juliflora* (Swartz) DC. Prod. 2: 447. 1825. *Mimosa juliflora* Swartz, Prodr. Veg. Ind. Occ. 85. 1788. *(Vern. Name: Bera Babla).*

Large shrub, with straight spines, branches long, sermentose. Leaves 1-3 at each nodes, bi-pinnate. Inflorescence in axillary spikes, 6-8 cm long, single or germinate at a node; stalks minutely hairy; calyx 5- toothed, capanulate, 1 cm long; corolla-petal 5, free tomentose on the inner surface, towards the apex. Stamens 10, slightly exserted. Pods linear compressed 10-15 cm X 1 cm. Seeds 20-23. *Fls. & Frts.:* Almost throughout the year.

Planted in the hedges or largely for the ornamental purposes.

Field Collection: Canning/on the riverbanks, *K.N.-205.*

Distributions: Arid zones and coastal areas; U.P., Bihar and Rajasthan. Native of Arid Regions of Mexico and Central America: introduced into India from Kew, England in the year 1877 for afforestation purposes.

Economic Importances: A huge plantation of this species check soil erosion and enriched the soil with nitrogen fixing bacteria: pods are also fodder for goats and sheeps. Add green manure in the soil.

9. *SAMANEA* Merr.

9.1. ***Samanea saman*** (Jacq.) Merrill. J. Wash. Acad. Sci. 6: 47.1916. *Mimosa saman* Jacq. Fragm. 15. t. 9. 1800. *Pithocellobium saman* Benth. Hook. Lond. J. Bot. 3: 101. 1864. *Enterolobium saman* Prain ex King. J. As. Soc. Bengal 66(2): 352. 1897: Prain Beng. Plants 1: 334. 1903 (Rep. ed. 1963): Wlh. India 3: 175. 1952. *(Vern. Name: Khirish).*

Large tree, branches spreading crown. Leaves pinnately compound: pinnae 5-7 pairs, leaflets 5-10 pairs, glabrous above, tomentose, pubescent beneath, obtuse or rounded at apex, acute at base, lateral one, rhomboid-oblong, apical pair obovate. Peduncle 1-3 nate, upper axillary or terminally, sub-racemed. Flowers pedicelled, showy, pinkish white, subtended by a bracts. Pods 10- 20 cm long. *Fls. & Frts.:* March- October.

Abundantly, elsewhere, sometimes planted as avenue tree, often naturalized.

Field Collection: Gosaba. *K.N.-674.*

Distributions: Introduced into the vicinity of Calcutta as an avenue tree and rapidly spread during the first half of the 18th century. Pantropic and cultivated indigenous to Mexico.

Economic Importance: Wood used as carpentry works and as fire woods.

Remarks: Some other species of the genus *Acacia* have also been collected by the different collectors from the lower Ganga delta and deposited in the Calcutta University Herbarium: but I have made no collection or I could not study in detail of all these taxa. Therefore. I have only mentioned the name of all these taxa:

1) *Acacia moniliformis* Griseb., Goett. Abh. 19: 136.1874. (Ballygunge/ Science College Campus. January. 1945. P.Pal- *s.n.*,C.U.H.)
2) *Acacia recurva* Benth., Hook. London J. Bot. 1: 519.1842 (Calcutta. October. 1938. J.C. Banerjee-s.n. C.U.H.)
3) *Acacia sphaerocephala* Cham. & Schlecht. Linnaea 5: 594. 1830 (Ballygunge/Science College Campus. August-1939. P. Pal. s.n. C.U.H)

Notes: The family Fabaceae, including Papilionaceae, Caesalpiniaceae and Mimosaceae is the largest family among the flowering plants in the lower Ganga Delta. The members of this family Fabaceae are very important as the pulses, vegetables and ornamental plants are mostly restricted within this family. The nitrogen fixing nodules and *Rhizobium* association of the family Fabaceae have also immense value for the intensive afforestation programmes and for producing fodder crops for the domestic cattles.

L. ROSACEAE

Key to the Genera

1a	Ovary inferior	1. *ERIOBOTRYA* Lindl.
1b	Ovary superior	(2)
2a (1)	Herb, not spiny	2. *POTENTILLA* L.
2b (1)	Erect or climbing, spiny shrub	3. *ROSA* L.

1. *ERIOBOTRYA* Lindl.

1.1. ***Eriobotrya japonica*** (Thunb.) Lind. Trans. Lind. Soc. 13: 102. 1822; Clarke, Hook.f. Fl. Brit. India 2: 372. 1878; Watt Dict. Econ. Prod. 3: 257. 1890; Prain Beng. Plants 1: 338. 1903 (Rep. ed. 1963). *Mespilus japonicum* Thunb. Fl. Jap. 206. 1784. *(Vern. Name: Loquat).*

Small, evergreen tree. Leaves large, 15-30 cm long, lanceolate or oblanceolate, acuminate, dentate or serrate, clustered at the ends, wooly beneath. Flowers dull white, fragrant, in terminal panicles. Fruits priform or globose, baccate, yellow or orange. *Fls. & Frts.:* November-April.

Field Collection: Subuddhipur/Baruipur, *K.N.-2300.*

Frequently planted in the orchards for its edible fruits.

Distributions: Occasionally cultivated throughout India. Native of China and Japan; but now cultivated in the warm countries.

Economic Importances: Fruits edible; the leaves are said to be useful in diarrhoea and in alcoholic intoxication; flowers considered as an expectorant.

2. *POTENTILLA* L.

2.1. ***Potentilla supina*** L. Sp. Pl. 497. 1753; Clarke, Hook.f. Fl. Brit. India 2: 359. 1878; Watt Dict. Econ. Prod. 6(1): 333. 1893; Prain Beng. Plants 1: 336. 1903 (Rep. ed. 1963); Naskar, Aq. & Semi. Plants 125. 1990.

Prostrate much branched leafy aquatic herb. Naskar (1.c.) described this taxon in respect of its taxonomy, key characters and economic importances.

3. *ROSA* L.

Key to the Species

1a	Erect shrub with prickles	3.1. *R. centifolia* L.
1b	Small spreading shrub with several prickles	3.2. *R. indica* L.

3.1. ***Rosa*** **X** ***centifolia*** L. Sp. Pl. 491. 1753; Clarke, Hook.f. Fl. Brit. India 2: 364. 1878; Prain Beng. Plants 1: 337.1903. *(Vern. Name: Basaraiya gulab).*

Erect shrub with prickles. Leaves pinnately compound, glandular ciliate. Flowers nodding, stalked often double, not bracteate, delicately scented, rose or purple coloured. Fruits without tomentum. *Fls. & Frts.:* Throughout the year.

Planted in the public gardens.

Field Collection: Shyamnagar, *K.N.-2091.*

Distributions: Most ancient cultivated Roses in India; introduced into Calcutta Botanic Garden during the year 1841. Native of Caucasus.

Economic Importances: Ornamental garden shrubs: root is astringent; flower decoction is used in ulcers of intestine. Floral wax is useful for soap manufacturing; petal oil is used in perfumery and cosmetics.

3.2. ***Rosa indica*** L. Sp. Pl. 492. 1753: Clarke, Hook.f. Fl. Brit. India 2: 364. 1878: Prain Beng. Plants 1: 337. 1903 (Rep. ed. 1963). *(Vern. Name: Kanta Golab).*

Small spreading shrub with a number of prickles, prickles equal in length. Leaves pinnately compound, leaflets serrate, stipules adnate. Flowers rose with strong scent; calyx tube obovate; sepals reflexed to flowers. *Fls. & Frts.:* Almost throughout the year, specially in the winter months.

Cultivated in the ornamental gardens and lawns.

Field Collection: Joynagar, *K.N.-871.*

Distributions: Planted almost throughout India.

Economic Importances: Ornamental garden shrubs.

Remarks: There are several other species and horticultural varieties of Rose, planted in the ornamental gardens in lower Ganga delta. *e.g. Rosa alba* L., *Rosa banksiae* R. Br., *Rosa bourboniana* Desportes. *Rosa brunonii* Lindl., *Rosa chinensis* Jacq., *Rosa damascena* Mill., *Rosa foetida* Herrm., *Rosa gallica* L., *Rosa gigantea* Collett., *Rosa macrophylla* Lindl., *Rosa multiflora* Thunb., *Rosa sericea* Lindl and *Rosa webbiana* Royle.

LI. CRASSULACEAE

1. *BRYOPHYLLUM* Salisb.

1.1. ***Bryophyllum pinnatum*** (Lamk.) Oken. Allg. Naturgesch. 3(3). 1966.1841: Ohba, Hara et al. Enum. Flw. Pl. Nepal 2: 159.1979. *Cotyledon pinnata* Lamk. Dict. 2: 141.1786. *Kalanchoe pinnata* (Lamk.) Pers. Syn. 1: 446 1805: Baker. Fl. Malesiana I. 1: 199.1953. *Bryophyllum calycium* Salisb. Parad. Lond. t. 3. 1805 & DC. Prodr. 3: 396. 1828: Clarke. Hook.f. Fl. Brit. India 2: 413. 1878: Prain Beng. Plants 1: 340. 1903 (Rep. ed. 1963). *(Vern. Name: Pathar Kuchi* or *Kop-pata).*

Glabrous fleshy herb. Leaves trifoliolate, oblong elliptic, crenate or very coarsely crenate leaflets. Flowers in panicles, greenish to purvish; calyx campanulate with 4-lobes: corolla sub-globose at the base and constricted in middle. Fruit enclosed in persistent perianth. Seeds small, oblong-ellipsoid, smooth. *Fls. & Frts.:* December-May.

Common on the dry waste lands, roadside thickets, railway tracks and clevices of the old rejected houses.

Field Collection: Jugdia-Benipur. *K.N.-230.*

Distributions: Tropical planes of India and very common on the lower Bengal. It was introduced in the Calcutta Botanic Garden by the East India Company and hence it spread all over Bengal and India. Native of Africa and distributed throughout the Tropics.

Economic Importances: Sometime cultivated as ornamental plant for its fleshy shinning foliages. Leaves are also used an efficient remedy for cholera and also for healing of wounds.

LII. DROSERACEAE

1. *ALDROVANDA* L.

1.1. ***Aldrovanda vesiculosa*** L. Sp. Pl. 381.1753; Naskar Aq. Semi. Plants 126. 1990. *(Vern. Name: Malacca Jhangi).*

Insectivorous aquatic succulent glabrous herb. Naskar (1.c.) described this taxa in respect of its taxonomy, key characters and economic importances etc.

The other species under this family, *Drosera burmanni* Vahl, Symb. 3 King J. Asiat. Soc. Bengal 66: 2.1983 is also present in the lower Ganga delta regions.

LIII. HALORAGACEAE

1. *MYRIOPHYLLUM* L.

Key to the Species

1a	Flowers white; stigma green, fruits with round ridges along the carpel backs	1. *M. indicum* Willd.
1b	Flowers pink: stigma pink: fruits acutely ridged along carpel backs	2. *M. tuberculatum* Roxb.

1.1. ***Myriophyllum indicum*** Willd. Sp. Pl. 4: 407.1805: Naskar Aq. Semi. Plants 127.1990. *(Vern. Name: Jhangi).*

Aquatic plants in the marshy places and ditches: Naskar (1.c.) described this plant in respects of its taxonomy, ecology and economic uses.

1.2. ***Myriophyllum tuberculatum*** Roxb. Fl. Ind. 1: 451: 1820: Naskar Aq. Semi. Plants 127.1990. Naskar (1.c.) described this plant.

LIV. RHIZOPHORACEAE

(The Mangrove Family)

Key to the Genera

1a	Moderate trees with extensive stilt roots or looping bow-roots: flowers tetramerous	4. *RHIZOPHORA* L.
1b	Small to moderate trees, without extensively aerial stilt like or bow like roots; flowers pentamerous or with more floral segments	(2)
2a (1)	Moderate tree without any aerial root like pneumatophores or stilt roots; stamens numerous	3. *KANDELIA* W. & A.
2b (1)	Small to moderate trees or shrubs: aerial roots knee-like, rounded or knobby apex of the aerial roots; stamens twice as many as the calyx lobes petals with marginal hairs	(3)
3a (2)	Moderate trees; leaves usually more than 12 cm long, apex usually acute; calyx 8-16 lobed; hypocotyle with persistent thin calyx teeth, scarcely ridged	1. *BRUGUIERA* Lamk.

3b (2)	Shrub or small trees, leaves usually less than 10 cm; rounded apex; calyx 5-6 lobed; hypocotyle ridged, persistent calyx dried and less prominant	2. *CERIOPS* Arn.

1. *BRUGUIERA* Lamk.

Key to the Species

1a	Flowers usually larger, 3-4 cm long, reddish; petals upto 16mm long	(2)
1b	Flowers usually small, 1-2 cm long or smaller, not reddish: petals 1.5-4mm long	(3)
2a (1)	Leaves elliptic oblong; petal lobe tip acute, 3-filamentous appendages	2. *B. gymnorrhiza* (L.) Lamk.
2b (1)	Leaves elliptic oblong: petal lobe tip blunt; without filamentous appendages	4. *B. sexangula* (Lour.) Poir.
3a (1)	Calyx lobes slender, short upto 3mm: spreading in fruits; petals 1.5-2.0mm long	3. *B. parviflora* W. & A.
3b (1)	Calyx lobes stout, long upto 5 mm, reflexed in fruits; petals 3-4mm long	1. *B. cylindrica* (L.) Bl.

1.1. ***Bruguiera cylindrica*** (L.) Bl. En Pl. Jav. 1: 93.1827; Naskar & Guha Bakshi, Mang. Swams Sundarbans 152.1987. *(Vern. Name: Bakul Kakra).*

A true mangrove species. Tall tree, upto 25m high, bark smooth, grey with few lenticels. Leaves oblanceolate, rarely elliptical acute or bluntly pointed apex, 15cm long and 5cm broad. Flowers 0.8-1.2 cm long, greenish: calyx tube very small smooth, 8-10 lobed: petals 0.3- 0.4 cm long. Fruits upto 1.5cm, obconical; fall with the hypocotyle seedling. Seed germination epigeal, viviparous. Hypocotyle 15cm long, diameter narrow. *Flowering* during winter and summer and *fruiting* during monsoon. (Fig. 17).

Grow abundantly on the inter-tidal mangrove forests in the Sundarbans mangrove forests.

Field Collection: Sajnakhali, *Mukherjee-5252* (CAL).

Distributions: In the coastal mangrove forests of Sundarbans, Mahanadi delta and Andaman & Nicobar Islands. In the Tropical Mangroves.

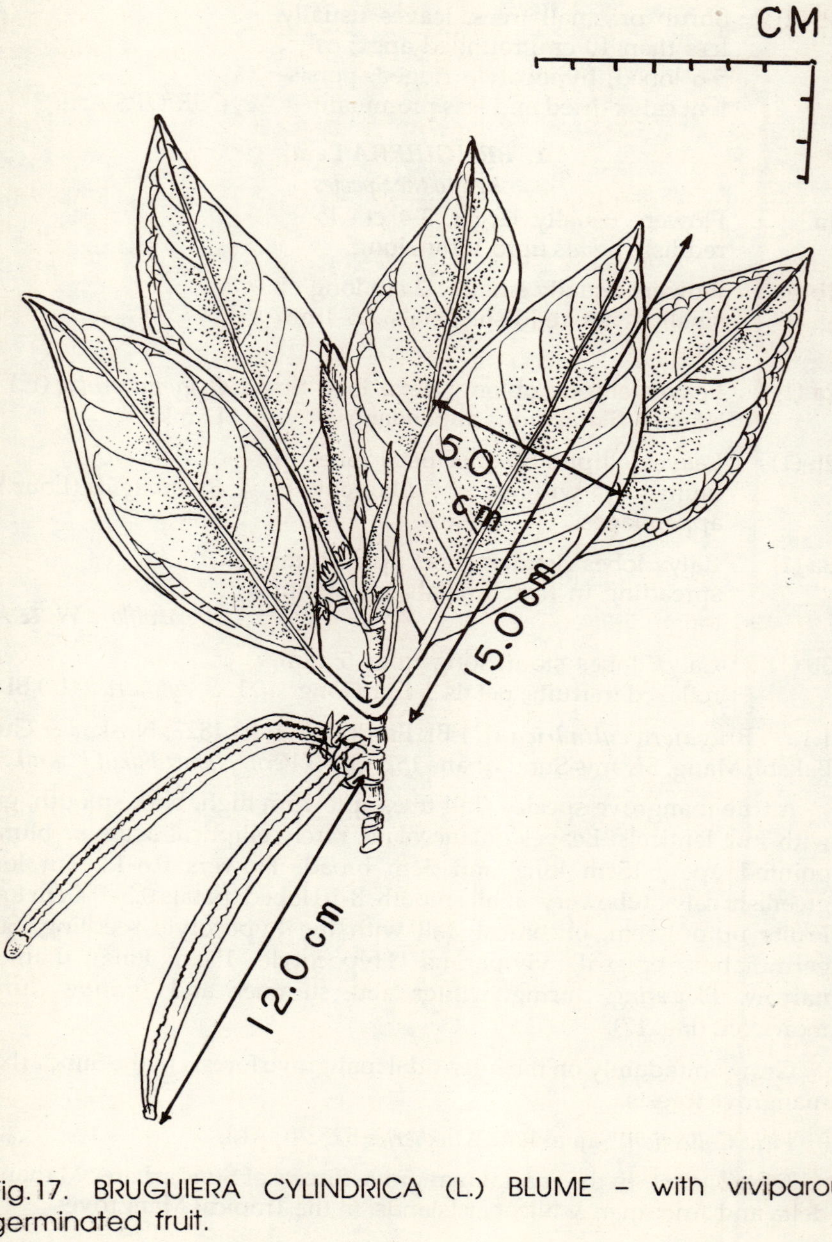

Fig. 17. BRUGUIERA CYLINDRICA (L.) BLUME – with viviparous germinated fruit.

Economic Importances: Produce fire-woods and timbers; bark tannin percentage-17.8%, useful for tann fisherman's nets and leather industries.

1.2. ***Bruguiera gymnorrhiza*** (L.) Lamk. Encycl. Meth. Bot. 4: 696.1997: Naskar & Guha Bakshi, Mang. Swamp Sundarbans 152. 1987 *(Vern. Name: Kankra).* (Fig. 18).

Medium height, 15-20m tall tree, much dense branched spreading, branches and stems marked with leaf-scars and stipules. Short or shallow buttressed like aerial roots, thicken trunks base. Leaves elliptic oblong, acute apex, or bluntly pointed, coriaceous 7.5-15.0 cm long and 5.0-7.5 cm broad, dark green. Flowers solitary axillary, calyx deep orange red/yellow, ribbed, campanulate, large (3-4 cm) flowered: calyx tip acute, 11-13 numbers. Petals 1.2-1.6 cm long, tips acute. Fruit fall on the ground along with the seedling; germination epigeal, viviparous, hypocotyle upto 25 cm long, blunt apically, slightly angular. *Flowers* during winter *fruits* hanging upto the next monsoon.

Commonly grow on the river banks, in the inter-tidal mangrove forests in the Sundarbans.

Field Collection: Harinbari/Sagar Islands, K.N.-490.

Distributions: Throughout the coastal mangrove areas in India, both east and west coasts; in Andaman and Nicobar Islands. In the Old World Tropical mangrove forests.

Economic Importances: High calorific valued fuel wood and timber. Tannin percentage is very high upto 36%.

1.3. ***Bruguiera parviflora*** Wight & Arnold ex Griffith Trans. Med. Phys. Soc. Cal. 8: 10.1936; Naskar & Guha Bakshi Mang. Swamps Sundarbans 154.1987. *(Vern. Name: Bakul Kankra).* (Fig. 19).

Medium tree upto 16m high; branches not spreading much, smooth grey bark. Shallow buttressed from the trunk base, root-knees occasionally blunt end, pneumatophores present. Leaves simple, 7.5-10.0cm long and 2.0-4.0cm broad, oblong lanceolate or elliptic lanceolate, bluntly shining above, dull beneath, glabrous. Flowers in lax, 2-5 flowered erect cymes, 2cm long, calyx-tube ridged. Petals 1.5 mm long, flowers yellow-green. Fruit upto 2cm, obconical, fall with the seedling. Fruit germination epigeal, viviparous, hypocotyle 15cm long. *Flowerings* during winter and *fruitings* during summer and monsoon months.

Grow gregariously in the inter-tidal mangrove forests.

Field Collection: Kulpi. *K.N.-1792.*

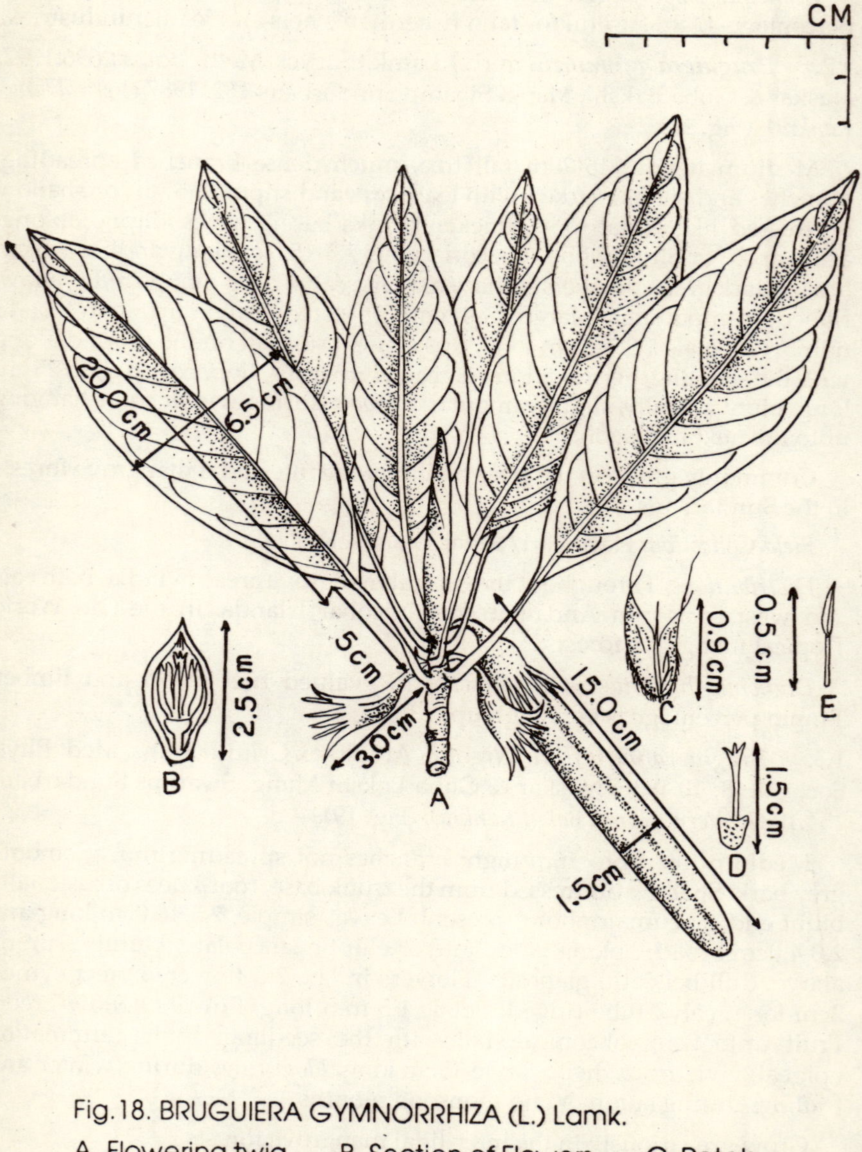

Fig. 18. BRUGUIERA GYMNORRHIZA (L.) Lamk.

A. Flowering twig, B. Section of Flowers, C. Petals, D. Gynoecium with ovary, E. Androecium.

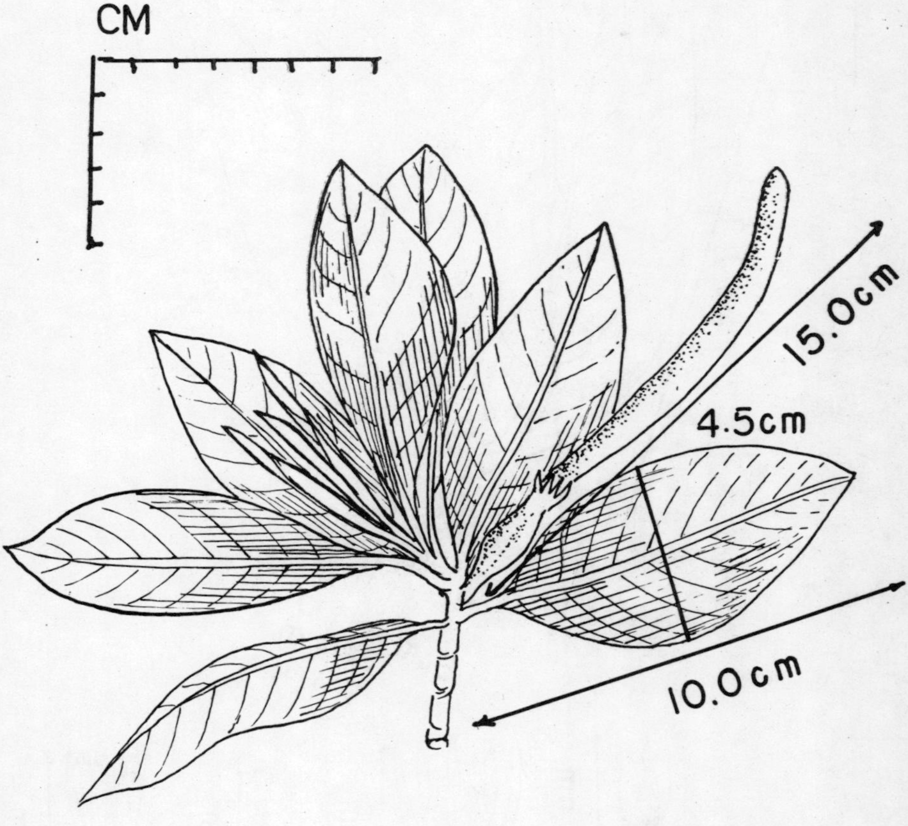

Fig. 19. BRUGUIERA PARVIFLORA (Roxb.) Wt. & Arn. ex Griff. – with viviparous germinated fruit.

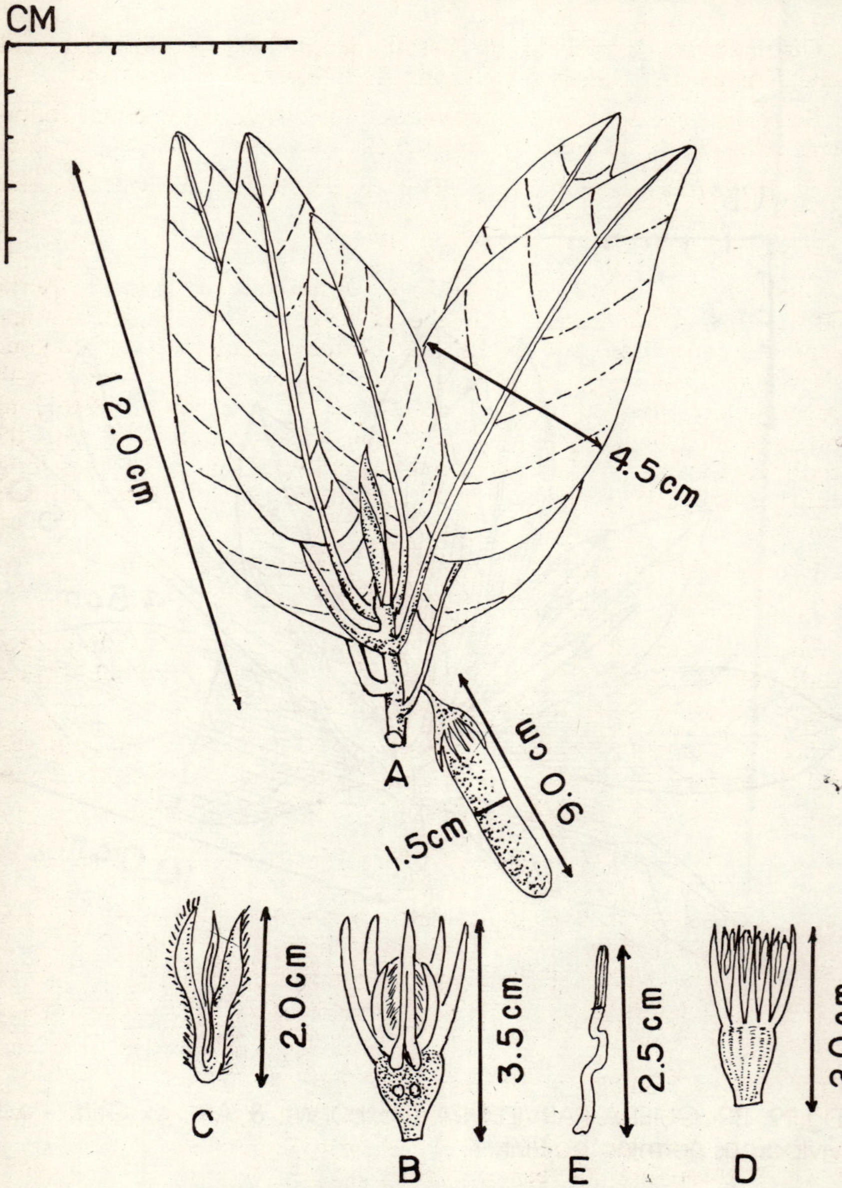

Fig.20. BRUGUIERA SEXANGULA (Lour.) Poir.

A. Fruiting twig, B. Section of Flower, C. Sepals,

D. Calyx tube, E. Androecium.

Distributions: Distributed in coastal mangrove forests of Mahanadi delta, Krishna and Godabari Delta and Andaman & Nicobar Islands

Economic Importances: Produce hard timber and fuel woods. Tannin percentange in the stem bark is upto 25.4%.

1.4. ***Bruguiera sexangula*** (Lour.) Poir. Lamk. Encycl. 4: 262.1816; Naskar & Guha Bakshi, Mang. Swamps Sundarbans 154.1987. *(Vern. Name: Kankra).* (Fig. 20).

Medium tree, upto 20m high, branches not spreading much. Aerial roots with shallow buttressed in the base of the trunk; root-knees and plank like roots. Leaves elliptic oblong or oblanceolate, 5-10cm long and 4-5cm broad, thinner than *B. gymnorrhiza.* Flowers 3-4cm long, sepals green initially and then turn yellow, whitish inside. Petals 1.2-1.5cm long and petal tips blunt. Fruit 2.0-2.5cm long obconic. Fruit fall with the seedlings. Germination epigeal, viviparous; hypocotyle upto 25cm long, blunt apically. *Flowering*-during winter and *fruiting*-during summer months.

Grow on the river slopes and inside the inter-tidal ridge forests.

Field Collection: Sajnakhali. *K.N.-1590.*

Distributions: Throughout the coastal mangrove areas of India and Andaman & Nicobar Islands. Throughout the tropical and sub-tropical mangrove forests in the Indo-West Pacific Regions.

Economic Importances: Produce good quality timber for boat-building and carpentary works and yield high calorific valued fuelwoods for the rural people of the Sundarbans. Tannin contain is considerable high.

2. CERIOPS Arn.

Key to the Species

1a	Shrubs upto 4 m high; stamens with a short filaments; hypocotyle upto 16cm long, warty toward apex, sharply ridged hypocotyl; fruit blunt apex, 10-15 mm long	1. *C. decandra* (Griff.) Ding Hou
1b	Shrub-small tree, upto 6m high; stamens with long slender filaments; hypocotyle upto 25cm long, worty throughout, terete or slightly ridged hypocotyle; fruit pointed, apex 15-20mm long	2. *C. tagal* (Perr.) Robinson

2.1. ***Ceriops decandra*** (Griff.) Ding Hou, Fl. Males (Ser. 1) vol. V (4): 471. 1958; Naskar & Guha Bakshi, Mang. Swamps Sundarbans 155. 1987. *Bruguiera decandra* Griff. Trans. Med. Phys. Soc. Calc. 8: 10. 1836. *(Vern. Name: Goran).* (Fig.21).

Shrub or small tree, upto 4 m high, erect branches formed its canopy conical, leaf-scar marks present on the branches. Aerial roots forming shallow buttress or root knees in the trunk base, leaves simple, opposite, obovate, caducous, rounded apex, dark green, glábrous. Flowers 2-3 chotomous, in condensed axillary cymes. Fruit length 1.0-1.5cm, blunt apically and hypocotyle length upto 16cm; germination epigeal, viviparous. Hypocotyle with sharply ridged and warted towards apex. *Flowering* during winter and *fruiting* during summer and monsoon months.

Grow abundantly in the inter-tidal ridge forests and river slopes in the Sundarbans mangrove forests.

Field Collection: Gosaba, *K.N.-182.*

Distributions: Throughout the coastal, deltaic and esturine tracts of India; Andaman and Nicobar Island. Throughout the Indo-West Pacific Ocean mangrove ecosystems.

Economic Importances: Cottage poles, pillers, pile and fuel woods. Tannin percentage in the stem bark is 19.0%, which is largely used by the fishermen for tanning their fishing nets.

2.2. ***Ceriops tagal*** (Perr.) Robinson, Philipp. J. Sci. Bot. 3: 306. 1908; Naskar & Guha Bakshi, Mang. Sawmps Sundarbans 156.1987. *(Vern. Name: Mat-Goran).* (Fig. 22).

Evergreen small or moderate tall tree, upto 6 m high: canopy conical. Aerial roots formed shallow buttressed and root-knees or flange like plank roots in the base of the trunks. Leaves simple, 8-10 cm long and 4-5 cm broad, coriaceous, obovate-oblong, apex rounded, glabrous. Flowers in axillary cyme, sepals 5 lobed, petals 5, white, shorter than the calyx. Fruits slightly conical, 1.5-2.0 cm long, pointed apically. Germination epigeal, viviparous, hypocotyle length 15-20 cm. Hypocotyle terete or slightly ridged, upto 25cm long warty throughout. *Flowering* period is winter and viviparous germination start soon after the maturation of the fruits; hypocotyle remain in the mother plant till the end of the monsoon.

Grow abundantly on the inter-tidal mangrove thickets and ridge forests.

Field Collection: Kulpi, *K.N.-1794.*

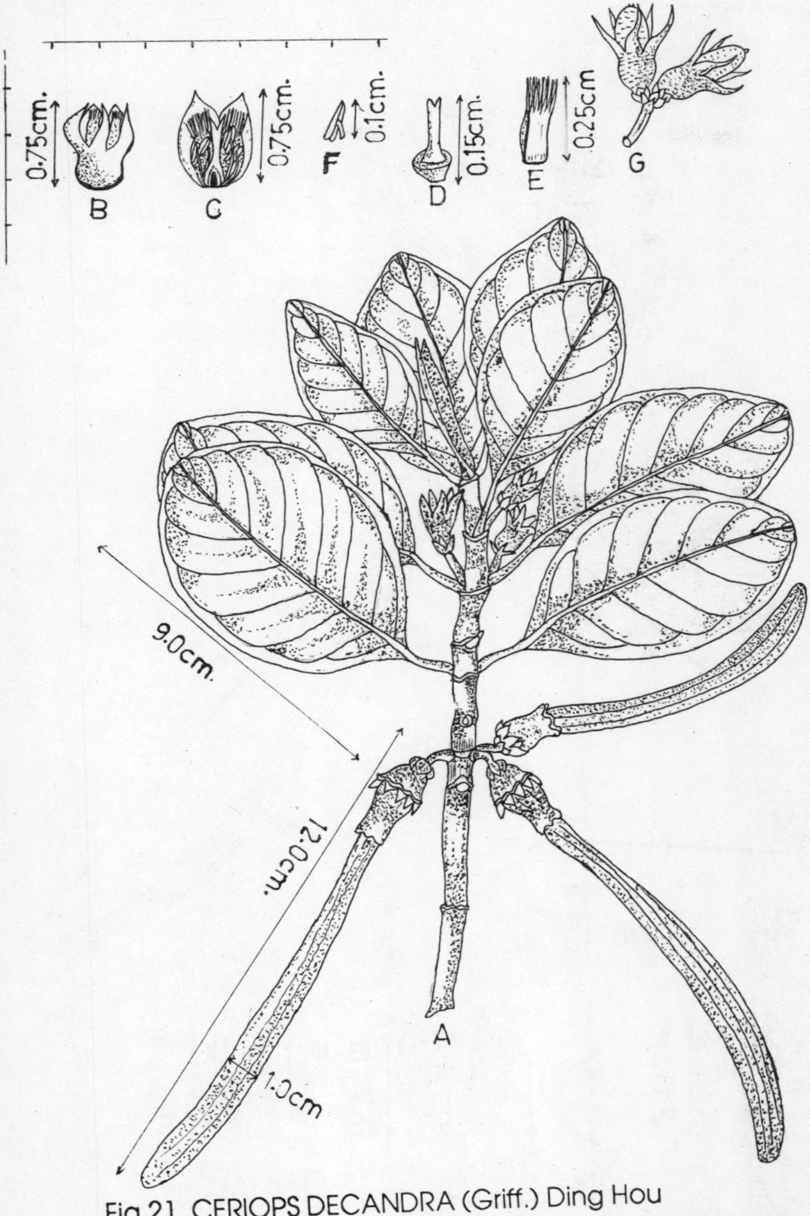

Fig.21. CERIOPS DECANDRA (Griff.) Ding Hou

A. Flowering twig, B. Single flower, C. T.S. of flower, D. Gynoecium, E. Calyx, F. Anther lobe, G. Two fruits on a stalk, starting viviparous germination.

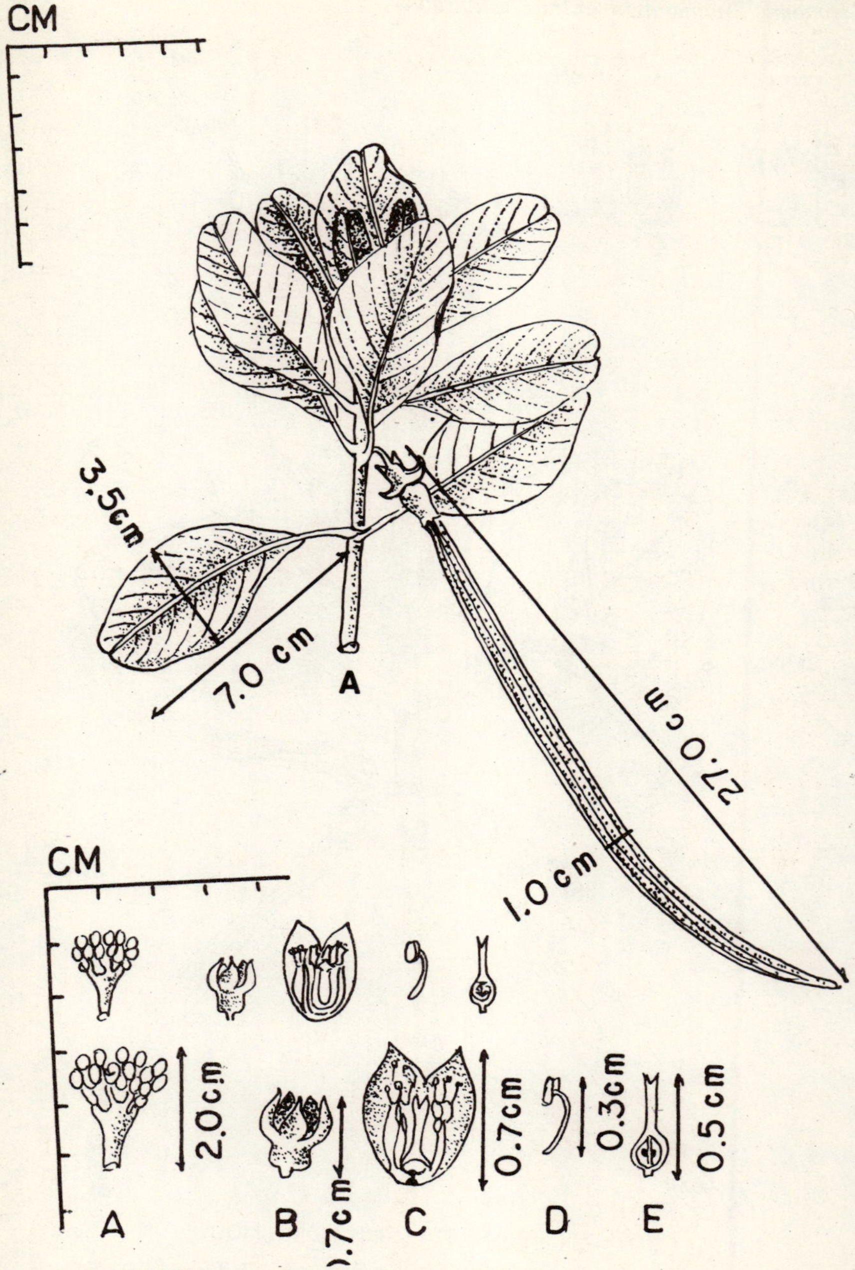

Fig.22. CERIOPS TAGAL (Perr.) Robins

A. Flowring twig, B. Single flower, C. Section of Flower, D. Androecium, E. Gynoecium with ovary, F. Inflorescence.

Distributions: Throughout the tidal estuaries and coastal areas in India and also in the Andaman & Nicobar Islands.

Economic Importances: Produce high calorific valued fuel-wood, timbers, pole for domestic cottage, piles and pillers. The wood is very hard and durable. Tannin content of the stem bark is 29.0%

3. *KANDELIA* Wight and Arnold

3.1. ***Kandelia candel*** (L.) Druce, Rep. Bot. Exch. Club. Brit. Isles 1913 (3): 420. 1914; Naskar & Guha Bakshi, Mang. Swamps Sundarbans 159.1987. *Rhizophora candel* L. Sp. Pl. 443. 1753. *(Vern. Name: Goria).* (Fig. 23).

Small to medium tree, upto 7m high; bark smooth, greyish or reddish brown; without any aerial root, but occasionally aerial roots developed in some cases. Leaves opposite, oblong-elliptic apex obtuse, 6.5 - 10 cm long and 2.5-4 cm broad, dark green and polished above. Flowers white 1.5- 2.0 cm long, in axillary cymes, 4-9 flowered, calyx lobes linear acute, petals bifid. Fruits 1.5-2.5 cm long, obclavate, viviparous. Hypocotyle of the seedling slender and tapered at each end, cylindrical acuminate, upto 40cm length at maturity. *Flowering* during winter and late winter and *fruiting* during summer.

Very occasional in the mangrove forests of the Sundarbans.

Field Collection: Pirkhali, *K.N.- 192.*

Distributions: Very rare in the other mangrove areas in India. Also occasional in the old World Tropics, in the mangrove ecosystems.

Economic Importances: Timber is very hard and durable; and good quality fuel-woods. Tannin content is also very high, upto 15.2%.

4. *RHIZOPHORA* L.

Key to the Species

1a	Leaves dark green, tips acute; inflorescence 2-flowered cymes; petals glabrous; stamens 12; stigma almost sessile; fruits and seedling maturing well below twig	1. *R. apiculata* Bl.
1b	Leaves dark green, tips pointed, mucronate; inflorescence more than 2-flowerd, cymes; petals hairy; stamens 8; stigma sessile; hypocotyle warted	2. *R. mucronata* Lamk.

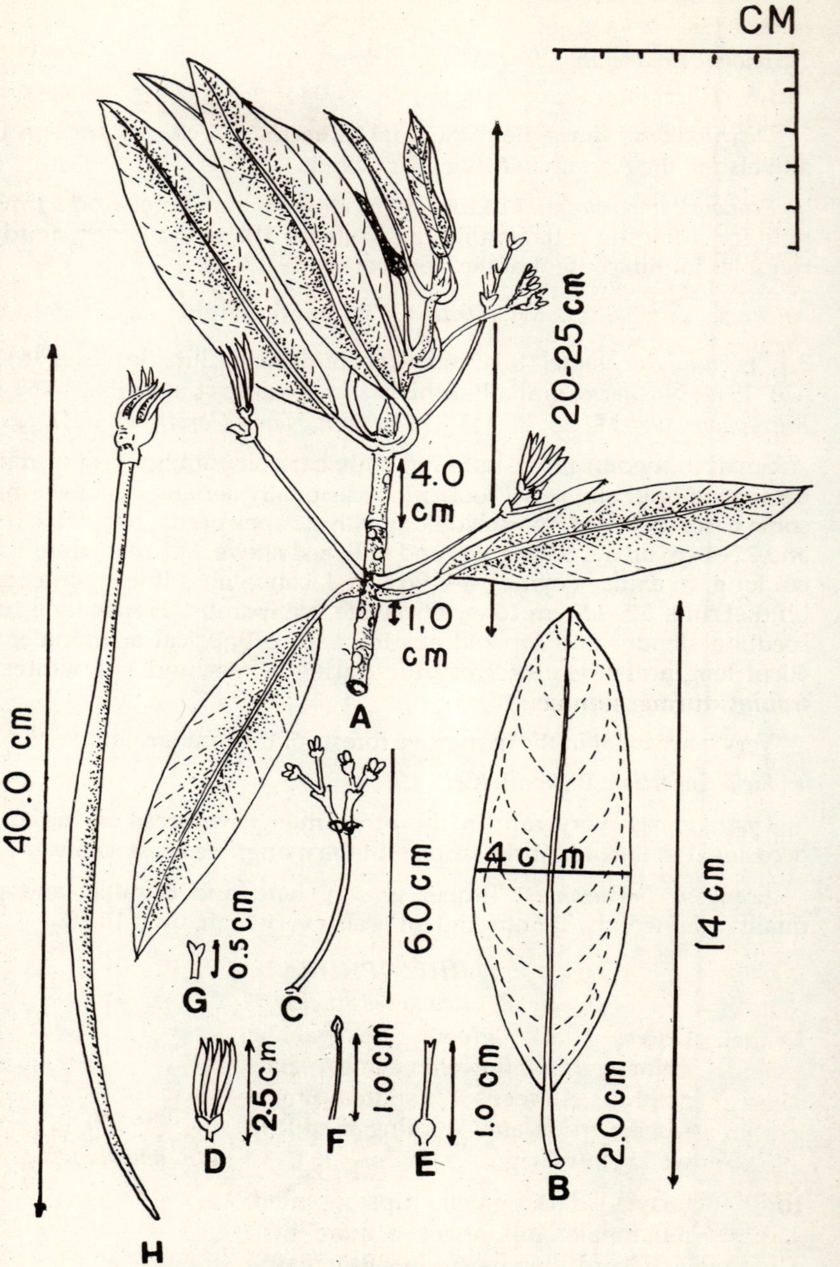

Fig.23. KANDELIA CANDEL (L.) Druce

A. Single flowering twig, B. Single leaf, C. Inflorescence, D. Single flower, E. Gynoecium with ovary, F. Androecium, G. Flower bud, H. Viviparous germinatd fruit.

4.1. ***Rhizophora apiculata*** Bl., En. Pl. Java 1: 91. 1827; Naskar & Guha Bakshi Mang. Swamps Sundarbans 159.1987. *(Vern. Name: Garjan).* (Fig.24).

Medium tree, upto 20 m high, co-existed with the other members of the Rhizophoraceae; branches much spreading horizontally and erect with prominant fallen leaf-scar marks. Looping stilt roots, upto 3-5 m high, occasionally bending like bow or pendulous/hanging roots; 5-10 cm diam. of these aerial stilt roots. Leaves dark green glabrous, oblong lanceolate, acute, 14 cm broad, petiolate. Flowers in pairs on the stubby axis, white stigma almost sessile. Fruit 2.5-3.0 cm long, seedling radicle length upto 50 cm, germination epigeal, viviparous: radicle surface smooth, cotyledonary collar present; seedling get detached from the fruit. *Flowering* during winter months but its hanging viviparous germinated radical remain for a longer period, *i.e..*, upto the end of the monsoon or next winter.

Common on the mid-estuary of the Sundarbans mangrove forests. Grow on the inter-tidal river banks and river flats on the newly formed sedimented soil.

Field Collection: Sanjnakhali/Sundarbans, *K.N.-1425.*

Distributions: Common almost throughout the tidal mangrove areas in the coastal India; Andaman & Nicobar Islands. In the Tropical and Sub-Tropical mangrove areas of the South East Asian regions.

Economic Importances: Produce high quality timber and fuel-woods. Tannin contents of the barks is very high, 33.0%.

4.2. ***Rhizophora mucronata*** Lamark, Encycl. 6: 189.1804; Naskar & Guha Bakshi Mangs. Swamps Sundarbans 162.1987. *(Vern. Name: Garjan).*

Medium tree, upto 20 m high, co-existed with *Ceriops* spp.; and *Bruguiera* spp.; branches spreading horizontally and erect marked with fallen leaves-scars. Stilt roots looping from the lower trunks and horizontal branches, formed very gregarious dense growth of corky, fleshy, cylindrical 5-6 m long, 5-6 cm diam. Sometime hanging (pendulous) from the trunks or bow roots. Leaves dark glossy green, thick and cuticularised, coriaceous elliptic, 15 cm long and 8 cm broad, apex pointed, mucronate. Flowers white, large in few flowered axillary cymes, 2-3 chotomously branched; stigma sessile. Fruit coriaceous 2.5-3.5 cm long. Seedling germination modified epigeal–viviparous, seedling length attain upto 70 cm or more. Cotyledonary collar absent in the radicles: radicle with warted surface and more pointed. Seedling detached from the fruit.

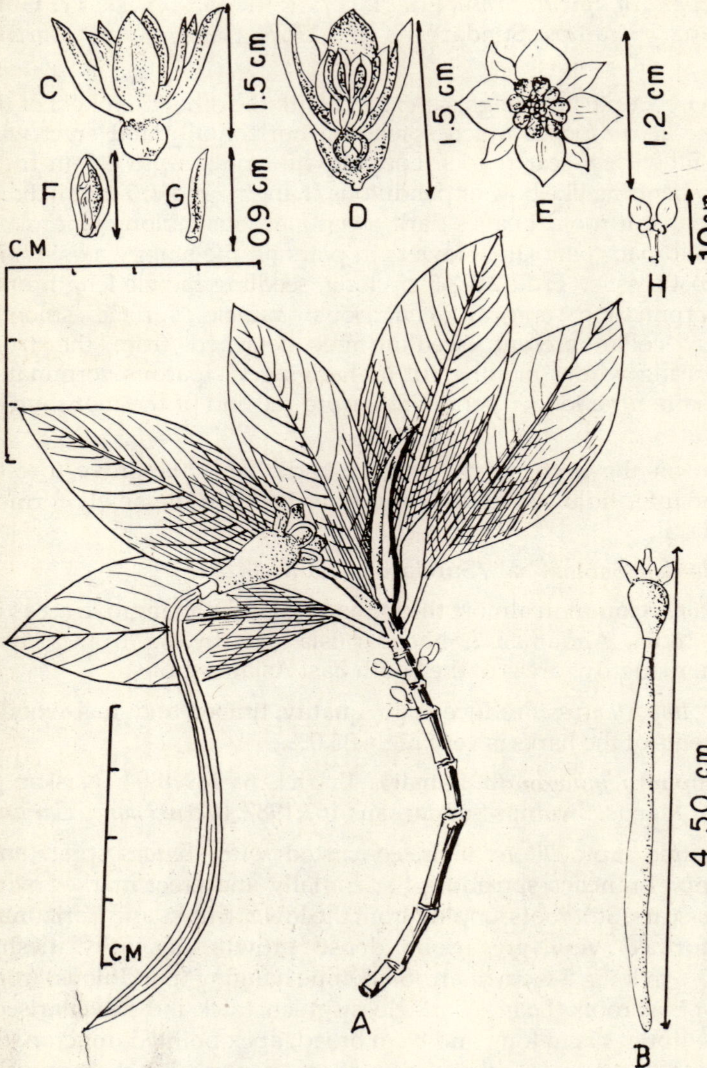

Fig.24. RHIZOPHORA APICULATA Blume

A. Flowering twig, B. Viviparous germinated fruit, C. Single flower, D. Section of flower (side view), E. Top view of flower, F. Single petal, G. Single sepal, H. Two flower buds with peduncles.

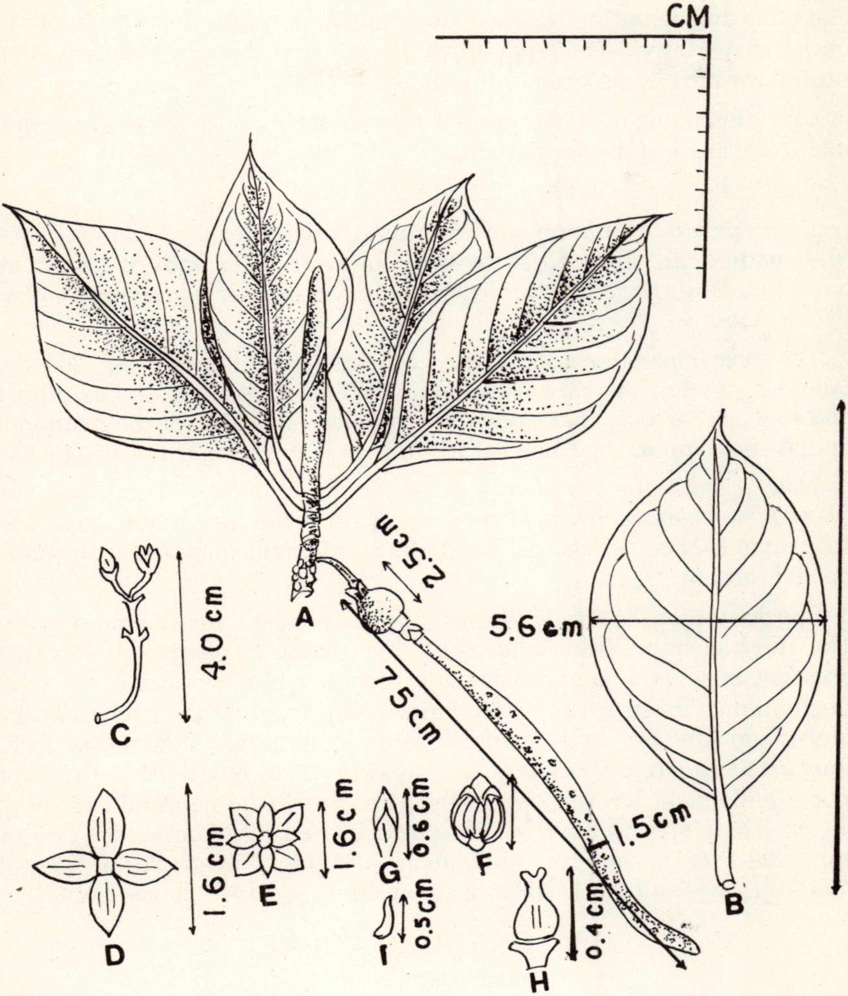

Fig.25. RHIZOPHORA MUCRONATA Lamk.

A. Twig with viviparous germinated fruit, B. Single flower,

C. Part of Inflorescence, D. Sepals arrangement, E. Top view of flower,

F. Section of flower, G. Single petal, G. Gynoecium, I. Sepal.

Flowering during the late monsoon to winter and *fruiting* during the month of January-March. The viviparous germinated hypocotyles are seen during the next monsoon months. (Fig. 25).

Common in the tidal mangrove forests, in the river slopes and swampy inter-tidal areas of the Sundarbans.

Field Collection: Sajnakhali, K.N.-1501.

Distributions: Common on the mangrove areas of Sundarbans, Mahanadi delta, Goa, Krishna and Godavari delta and Andaman & Nicobars. Throughout the tropical mangrove ecosystems in the Indo-West Pacific regions.

Economic Importances: Stem bark contains useful tannin materials; the tannin content of this taxa is estimated 20%. Woods are useful as timber and fuel or charcoals.. Bark powder is astringent; useful for the treatment of diabetes, hemorrhage, haematuria and angina. Young fruits are edible.

Notes: The other Indian species of the family Rhizophoraceae, *Rhizophora stylosa* Griff. is also reported from the Mahanadi delta and Andaman & Nicobar Islands. But in the Sundarbans mangrove forest this taxon is absent.

Species under the family Rhizophoraceae is called the true mangroves. The most characteristic features of this group of plants are that the presence of several halophytic adaptations and the tolerance of the tidal fluctuations. Its structural peculiarities help to adapt themselves in the physiologically dry and optimum saline conditions. *Rhizophora* spp., *Bruguiera* spp. are generally grow and very much restricted to the river slopes and brackish water zones, which are directly influenced by the tidal thrust. These species have failed to invade the ridge forests; but *Ceriops* spp., *Kandelia* sp. can grow in the inter-tidal ridge forests. The aerial stilt roots of *Rhizophora* check soil erosion in the river slopes, considerably.

LV. COMBRETACEAE

Key to the Genera

1a	Climbing shrub	2 *QUISQUALIS* L.
1b	Medium to large tree	(2)
2a (1)	Petals absent; calyx limb deciduous; generally grow in all types of soils	3. *TERMINALIA* L. *nom. cons.*
2b (1)	Petals 5: calyx limb persistent; restricted on the mangrove forest	1. *LUMNITZERA* Willd.

1. *LUMNITZERA* Willd.

1.1. ***Lumnitzera racemosa*** Willd., Neue Schr. Ges. Naturf. Fr. Berl. 4: 187.1803; Naskar & Guha Bakshi Mangr. Swamps Sundarbans 225.1987. *(Vern. Name: Kripal)*. (Fig. 26).

Evergreen tree, 20m tall, fibrous fissured bark, with spreading branches. Pneumatophores occasionally developed as looping lateral roots. Leaves spirally arranged, leaf blades hairy, when young, cordate at the top of the branches, oblong obovate or oblanceolate, fleshy. Flowers in dense terminal and axillary spikes, 1-2 cm, white, sessile actinomorphic. Fruit 1 seeded oblong ovoid, 1-2 cm long with hard coat. Seed germination hypogeal or modified epigeal. *Flowering* and *fruiting* during the months of May to November.

Grow occasionally on the above tidal forests, as the back mangal species.

Field Collection: Namkhana, *K.N. 1498.*

Distributions: In the coastal and estuarine saline environment and also in the Tropical and Sub-Tropical areas of the Old World zones.

Economic Importances: Stem bark contain 19% tannin; stem juice is said to be useful for herpes and itches.

2. *QUISQUALIS* L.

2.1. ***Quisqualis indica*** L. Sp. Pl. 556.1762; Clarke. Hook.f. Fl. Brit. India 2: 459.1878; Watt Dict. Econ. Prod. 6(1): 388.1892; Prain Beng. Plants 1: 350.1903 (Rep. ed. 1963). *Q. villosa* Roxb. Fl. Ind. 2: 426. 1832. *(Vern. Name: Rangoon-ke-bel* (Hindi)).

Rapid growing deciduous climbing shrub. Leaves oblong lanceolate or elliptic, glabrous, 6-15 cm long, acuminate, the lower parts of the petiole persistent and becoming hard and thorny. Flowers in terminal, drooping, spike, whitish, changing deep pink to red, fragrant. Fruit with very sharp angles, scarcely winged, glabrous, black, top very acute. *Fls. & Frts.:* Almost throughout the year; mainly from April -June .

Commonly planted as an ornamental plant, sometime also as an escape near gardens and waste lands of the village thickets.

Field Collections: Sarberia /Sandeshkhali, *K.N.-787.*

Distributions: Planted almost throughout India; now naturalized in India. It is a native of Java and Malay Peninsula and distributd in Africa.

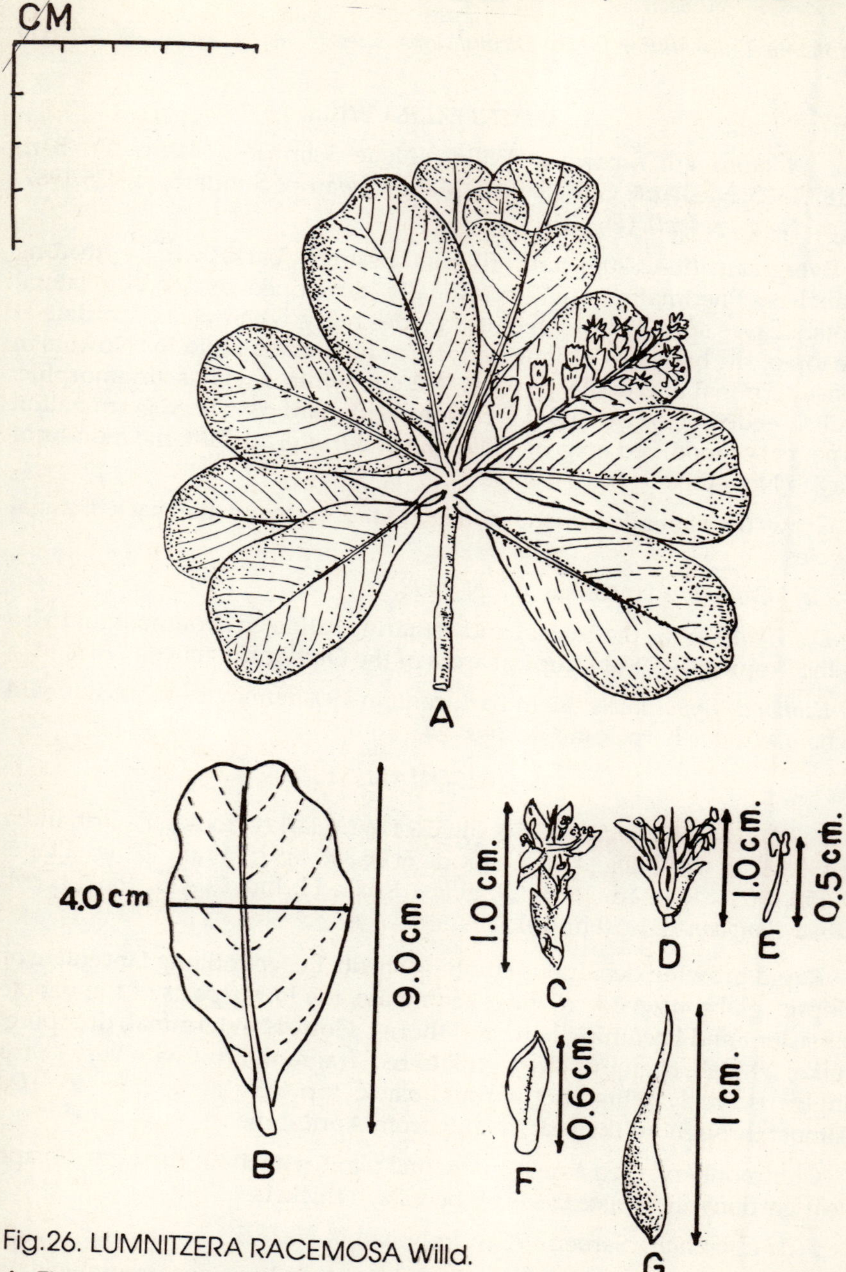

Fig.26. LUMNITZERA RACEMOSA Willd.

A. Flowering twig, B. Single leaf, C. Single flower, D. T.S. of flower, E. Androecium, F. Petals, G. Gynoecium.

Economic Importances: Ornamental garden plants; its flowers are bright, showy. Ripe roasted seeds are given in diarrhoea and fever in China and also used for rickets of children in Indo-China. With oil the seeds are externally applied to parasitic skin diseases; and seed-oil has purgative properties. Root and stem extractions are effective as anthelmintic. In Malay, the leaves are used as lotion for boils and ulcers. Tender shoots are edible.

3. *TERMINALIA* L. *nom. cons.*

Key to the Species

1a	Fruits with distinct wings, generally 5 in number	1. *T. arjuna* W. & A.
1b	Fruits not winged, rather ovoid or sub-compressed	(2)
2a (1)	Leaves not clustered at the ends of branches: spikes panicled	4. *T. chebula* Retz.
2b (1)	Leaves clustered at the ends of the twigs and alternate: spikes simple	(3)
3a (2)	Petioles long; fruits tomentose, pyriform	2 *T. bellirica* Roxb.
3b (2)	Petioles very short, fruits ellipsoid, more or less compressed	3. *T. catappa* L.

3.1. ***Terminalia arjuna*** (Roxb. ex DC.) Wt. & Arn. Prodr. 314.1834; Clarke, Hook.f. Fl. Brit. India 2: 447.1878; Watt Dict. Econ. Prod. 6(4): 16. 1893; Prain Beng. Plants 2: 348.1903 (Rep. ed. 1963). *Pentaptera arjuna* Roxb. Fl. Ind. 2: 438.1832. *P. glabra* Roxb. Fl. Ind. 2: 440.1832. *(Vern. Name: Arjuna).*

Large, handsome tree, upto 30 m high. Leaves sub-opposite or opposite, oblong or elliptic, obtuse or shortly acute at the apex, narrowed towards the base and often cordate. Flowers in pendulous, terminal and axillary spikes, pale yellow. Fruit ovoid or obovoid-oblong, 5 to 7 hard coriaceous narrow wings. *Fls. & Frts.:* April-September.

Generally planted as avenue tree; largely as wild on the village shruberries.

Field Collection: Sagar Mela Premises/Sagar Islands. *K.N.-464.*

Distributions: Native of Central and South India and distributed throughout India; recently this species is planted under the social afforestation programmes in Sri Lanka and India.

Economic Importances: Wood is hard; used in house building. The bark is a source of tannin and dye. The fresh and dried bark is considered as excellent stomach remedy and heart diseases. The leaves juice is used for earach. Barks powder relieve hypertension, diuretic and toxic effect in liver cirrhosis.

3.2. ***Terminalia bellirica*** (Gaertn.) Roxb. ex Fleming, Cor. Pl. 2: 54. t. 198.1798; Clarke, Hook.f. Fl. Brit. India 2: 445.1878; Watt Dict. Econ. Prod. 6(4): 19.1893; Prain Beng. Plants 1: 348.1903 (Rep. ed. 1963). *Myrobalanus bellirica* Gaertn., Fruct. 2: 90.t. 97.1791 *(bellirina). (Vern. Name: Bohera).*

Medium to large deciduous tree. Leaves clustered towards the tips of the twigs, broadly elliptic or obovate, sub-acute or acute, long petioled, coriaceous. Flowers in interrupted spikes, greenish-yellow, foetid smelling. Drupes ovoid or globose, brown, obscurely angled after drying. *Fls. & Frts.:* November-May.

Planted as avenue tree or ornamental plants.

Field Collection: Malirdhar/Canning, *K.N.-142.*

Distributions: Wild in most part of the deciduous forests of India, but not in the dry and wet places. Sri Lanka, Malaya and all other Tropical countries.

Economic Importances: Wood is hard but not durable; normally used in packing boxes. Fruits are an article of commerce, used as tannin and dyeing materials: radially eaten by goats, sheeps, cattles and monkeys.

3.3. ***Terminalia catappa*** L. Syst. Nat. (ed. 12) 2: 674 (err-638). 1767; Clarke, Hook.f. Fl. Brit. India 2: 444.1878; Watt Dict. Econ. Prod.6 (4): 22.1893; Prain Beng. Plants 1: 348.1903 (Rep. ed. 1963). *(Vern. Name: Katbadam* or *Deshibadam).*

Large, handsome, deciduous tree with spreading crown and buttressed trunk, upto 30 m high. Leaves clustered towards the tip of the branchlets, ovate with narrowly cordate base, short petioled. Flowers solitary, in slender, spikes, greenish white. Fruits glabrous, ellipsoid, slightly compressed with very hard and fibrous endocarp, 2 marked ridges when mature. *Fls. & Frts.:* March - August.

Planted as avenue tree or in the garden edges but occasionally grow wild.

Field Collection: Dakshmin Barasat, *K.N.-900.*

Distributions: Planted in most parts of India, especially in the coastal regions. Native of Malaya. All Tropical Countries.

Economic Importances: The wood is hard but light; also used as timber and fire-woods. Kernels are edible; the nut yields an oil similar to almond oil, but it has little use. The oil-cake is a good food for pigs. The bark and leaves yield a black tannin/dye. The bark is astringent and employed as remedy for dysentery and biliousness.

3.4. ***Terminalia chebula*** (Gaertn.) Retz., Obs. Bot. fasc. 5: 31.1789; Clarke, Hook.f. Fl. Brit. India 2: 446.1878; Watt Dict. Econ. Prod. 6(4): 24.1893; Prain Beng. Plants 1: 348.1903 (Rep.ed.). *Myrobalanus chebula* Gaert. Fruit. 2: 91.t.97.1991. *(Vern. Name: Haritaki).*

Medium to large deciduous tree. Leaves 7-15 cm long, usually sub-opposite, ovate, or elliptic, acute, rounded or sub-acute at base; petiole with 2 or more glands at its apex on the upper side. Spike terminal, often panicled. Flowers dull-white; calyx limb deciduous. Drupes ellipsoid, glabrous, 5-ribbed, when dry. *Fls. & Frts.:* April-August.

Commonly planted as ornamental and avenue tree.

Field Collection: Dhopagachi/Baruipur, *K.N.-2139.*

Distributions: Throughout the greater parts of India, except some of the hotter areas. Sri Lanka and all other Tropical countries.

Economic Importances: Wood is durable and used in making the agricultural equipments. Fruits are one of the best source of tannin and an article of commerce. *'Haritaki'* is very familiar in the Hindu Medicine as alternative and tonic: unripe fruits cure dysentery and ripe fruit as a remedy for asthma, sore throat, anaemia, gout, bilioussness, heart diseases and many other disorders. Fruits are also useful in snake-bite. Fruits laxative, stomachic and purgative.

Notes:: *Apogeissus acuminata* Wall. and *A. pendula* Edgew are occasionally planted in the ornamental gardens in the lower Ganga delta.

LVI. MYRTACEAE

Key to the Genera

1a	Inflorescence resembling bottle brush; flowers in spikes	(2)
1b	Inflorescence not as above: flowers solitary, clustered or in umbels or heads	(3)

2a (1)	Flowers in axillary spikes; stamens in bundles, opposite the petals; flowers yellowish white	3. *MELALEUCA* L.
2b (1)	Flowers in proliferous spikes; stamens free, flowers crimson in colour	1. *CALLISTEMON* R. Br.
3a (1)	Fruits dry, loculicidally splitting at top	2. *EUCALYPTUS* L. 'Herit.
3b (1)	Fruits baccate	(4)
4a (3)	Ovary 4-5 celled	4. *PSIDIUM* L.
4b (3)	Ovary 2-celled	5. *SYZYGIUM* Gaertn.

1. *CALLISTEMON* R. Br.

1.1. ***Callistemon citrinus*** (Curt.) Stapf., Bot. Mag. 150: t. 9050. 1925. C. *lanceolatus* DC. Prodr. 3: 223.1828. *(Vern. Name: Botal Brass).*

Large evergreen shrub or small tree with twisted trunk and drooping branches. Leaves alternate, linear lanceolate or lanceolate. Flowers crimson in terminal spikes; sepals deciduous; stamens numerous, long exerted: style red. Capsule with a truncate apex. *Fls. & Frts.:* October-April.

Commonly planted in the gardens for ornamental purposes.

Field Collection: Madhyamgram, *K.N.-801.*

Distributions: Introduced into India about 1804; planted throughout India, as an ornamental plant. A Native of Australia.

Economic Importance: A beautiful ornamental plant.

Notes: Hook.f., Fl. Brit. India (1772-1797) and Prain (1903a, 1903b and 1905) have not mentioned this taxon.

2. EUCALYPTUS L 'Herit.

2.1. ***Eucalyptus citriodora*** Hook. Mitch. J. Exped. Trop. Austral. 235.1848; Benthall Trees Calc. 246.1946; Wlh. India 3: 210.1952. *(Vern. Name: Eucalyptus).*

Planted, tall evergreen tree, upto 40 m high, with clear smooth shining grey trunk and long obliquely linear-lanceolate leathery leaves. Flowers white in umbels. Fruits ovoid. *Fls. & Frts.:* April-August.

Purely planted as avenue tree and as an ornamental plant on the lawn: no wild or naturalized form has been recorded.

Field Collection: Jadavpur, *K.N.-2379.*

Distributions: Planted almost throughout India; introduced in India. A Native of Australia.

Economic Importances: Ornamental tree; leaves yield an essential oil; oil is used in perfumery, used in manufacturing citronellol and menthol. Wood is useful as paper pulp.

Notes:: Prain (1903a and 1905) has not mentioned this taxon.

Remarks: A good number of other species are now introduced in India. For their quick growth, these taxa intake a huge amount of water from the sub-soil layer; but the ratio between water requirement and wood production is favourable for the bio-mass production programmes.

3. *MELALEUCA* L. *nom. cons.*

3.1. ***Melaleuca leucadendron*** L. Mart. 105.1767; Duthie, Hook.f. Fl. Brit. India 2: 466.1879; Prain Beng. Plants 1: 352. 1903 (Rep. ed. 1963). *(Vern. Name: Cajaputi).*

Small to medium tree; branches mostly pendulous. Leaves 5-10 cm long, elliptic oblong, tapering at both ends, 3-7 parallel veins. Flowers creamy-white in terminal spikes. Capsules nearly himispherical. *Fls. & Frts.:* February-May.

Rarely planted in the gardens, parks and as avenue tree.

Field Collections: Specimen not collected and preserved.

Distributions: Planted almost throughout India. A Native of Burma, the Malay Islands and Australia; widely planted in the Tropics.

Economic Importances: Ornamental garden plants; wood is useful for boat building. Leaf oil (Cajeput oil) is used as insecticides or mosquitoes expelling.

4. *PSIDIUM* L.

4.1. ***Psidium guajava*** L. Sp. Pl. 470. 1753: Duthie, Hook.f. Fl. Brit. India 2: 468.1878; Watt Dict. Econ. Prod. 6(1): 361.1892; Prain Beng. Plants 1: 353.1903 (Rep. ed. 1963). *P. pyriferum* L. Sp. Pl. (ed.2) 672.1762. *(Vern. Name: Piyara).*

Small, much-branched tree. Leaves elliptic-oblong or oblong; pubescent beneath. Flowers 1-3 on axillary peduncles, white. Fruit sub-globose or pyriform, yellow when ripe. Many seeded without much juice. *Fls. & Frts.:-*

February-September; but now-a-days almost throughout the year; the flowering and fruiting is artificially induced by hormones application.

Widely planted in the orchards in all the regions of India.

Field Collection: Dhanyakuria, *K.N.-571*.

Distributions: Throughout India; introduced into India by the Portuguese. A Native of Tropical Central America, and now naturalized in the Old World Tropics.

Economic Importances: Fruits are delicious; both green and ripe fruits are edible. Contain a good amount of vitamins and iron properties.

Notes: Guava or *Piyara* is the most economic fruit and the farmers of the Baruipur area of the district South 24-Parganas largely planted this fruit plant in their orchards. Recently, by the application of hormones and adapting several methods these farmers are producing the fruits almost throughout year.

5. *SYZYGIUM* Gaertn. nom. cons.

Key to the Species

1a	Woody perennial shrub; rusty villous peduncles: berries orange red	5. *S. ruscifolium* Sant.& Wag.
1b	Medium to small tree. Flowers in laxly or axillary cymes, panicles or in terminal clusters; berries generally white, black or brown	(2)
2a (1)	Leaves sessile or very shortly petioled; flowers in lateral or terminal cymes, large white; fruits whitish, sometime reddish but very much juicy	6. *S. samarangense* Merr. & Perr.
2b (1)	Leaves not sessile; fruits not so much juicy, black or yellowish white	(3)
3a (2)	Flowers not very large, sessile in short compact panicles from the older branches	(4)
3b (2)	Flowers large, bright-crimson or white in terminal clusters or dense racemes	(5)
4a (3)	Leaves abruptly acuminate or acute; berries varies in size, broadly ellipsoid, shining black when ripe	1. *S. cumini* (L.) Skeels

4b (3)	Leaves shortly acuminate or acute; berries 6-9, globose	2. *S. fruticosum* DC.
5a (3)	Fruits round to oval, creem white when ripe; seeds round, brown, nut-like single	3. *S. jambos* (L.) Alston
5b (3)	Fruits oval, larger than the above: seeds more than one, not nut like	4. *S. malaccense* Merr. & Perr.

5.1. ***Syzygium cumini*** (L.) Skeels, U.S. Dept. Agri. Bur. Pl. Ind. Bull. 248: 25. 1912. *Myrtus cumini* L. Sp. Pl. 471.1753. *Eugenia jambolana* Lamk. Encycl. 3: 198.1789; Duthie, Hook.f. Fl. Brit. India 2: 499.1879; Watt Dict. Econ. Prod. 3: 284.1890; Prain Beng. Plants 1: 356.1903 (Rep. ed. 1963). *(Vern. Name: Kalo-jam).*

Medium sized tree, bark grey smooth, much-branched. Leaves coriaceous, elliptic oblong or ovate-oblong, abruptly acuminate or acute at apex, petiolate. Flowers fragrant, sessile in short compact or laxly paniculate axillary cymes on old branches. Berry varies much in size, broadly ellipsoid, shining black when ripe. *Fls. & Frts.:* March-July.

Rarely planted but generally self-grown on the village areas.

Field Collections: Amtala, *K.N.-368*; Dewanganj, *K.N.-631*.

Distributions: Almost throughout the plains of India. Widely distributed in the Indo-Malaya-Asian region. Introduced in the other tropical regions.

Economic Importances: Ripe fruits are sour test, but delicious. Leaves fodder. Bark decoction used in dysentery and diarrhoea.

5.2. ***Syzygium fruticosum*** DC. Prodr. 3: 260.1828. *Eugenia fruticosa* Roxb. Fl. Ind. (Ed.2) 2: 487.1824; Duthie, Hook.f. Fl. Brit. India 2: 499.1879; Prain Beng. 1: 356.1903 (Rep. ed. 1963). *(Vern. Name: Ban Jam).*

Medium tree, bark pale grey. Leaves coriaceous, elliptic or elliptic-oblong, shortly acuminate or acute at apex, entire, cuneate at base, smooth, dark-green lateral nerves arched into an intramarginal nerves. Flowers sessile, in panicles from old branches: panicle-branches divaricate acutely 4-angled. Berries 6-9, globose. *Fls. & Frts.:* March-September.

Generally wild or self-sown.

Filed Collection: Basanti/on the river bank, *K.N-2271*.

Distributions: Almost throughout India, Bangladesh and Burma.

Economic Importances: Produce timber and fire-wood: fruit edible but very poor quality.

5.3. ***Syzygium jambos*** (L.) Alston. Trimen, Nandb. Fl. Ceyl. 6.115.1931. *Eugenia jambos* L.Sp. Pl. 470.1753; Duthie, Hook.f. Fl. Brit. India 2:474. 1878; Watt Dict. Econ. Prod. 3:293.1891; Prain Beng. Plants 1: 355.1903 (Rep. ed. 1963). *Myrtus jambos* H.B.K. Nov. Gen. Sp. 6. 144.1823. *Jambosa vulgaris* DC. Prodr. 3: 286.1828. (*Vern. Name: Golap Jam*).

Small to medium, much branched tree. Leaves simple, tapering into petioles. Flowers in terminal clusters, white. Fruits dull yellow when ripe, oval to round. Seed brown, round nut-like, one. *Fls. & Frts.:* March-June.

Commonly planted in the orchard.

Field Collections: Nabagram, *K.N.-643*.

Distributions: Planted throughout India; occasionally wild. Probably an Indian origin. Distributed in Malaya and Burma.

Economic Importances: Ripe fruits edible; valuable as a tonic for brain and liver complaints. Stem barks are said to be useful in asthma, bronchitis and also improve the voice, leaves after boiling applied in the eye sores.

5.4. ***Syzygium malaccense*** (L.) Merr. & Perry, J. Arn. Arab. 19: 115.1938. *Eugenia malaccensis* L. Sp. Pl. 470.1753; Duthie, Hook.f. Fl. Brit . India 2: 471.1878; Watt Dict. Econ. Prod. 3: 288.1890; Prain Beng. Plants 1: 355.1903 (Rep.ed.1963). *Jambosa malaccensis* DC. Prodr. 3: 286.1828. *(Vern. Name: Malacca Jamrul)*.

Small to medium tree. Leaves simple, 20-30 cm long. Flowers dense racemes of large bright crimson; petals 4 sub-orbicular, glandular; stamens numerous about 2.5cm long. Calyx lobes incurved in fruit. *Fls. & Frts.:* April-June.

Planted as ornamental tree throughout the ornamental gardens.

Field Collection: Titagarh, *K.N.-2224*.

Distributions: Planted throughout India, in the ornamental gardens.

Economic Importances: Fruits edible but scarcely. Bark astringent, used for mouth wash for thrush; dried and powdered leaves useful against cracked tongue.

5.5. ***Syzygium ruscifolium***(Willd.) Santapau & Wagh, Bull. bot. Surv. India 5: 109.1963; Naskar & Guha Bakshi, Mang. Swamps Sundarbans 139.1987. *(Vern. Name: Hidjili Menadi)*.

Woody perennial shrub. Naskar & Guha Bakshi (l.c.) described this taxon in relation to its taxonomy, ecology and distributions.

5.6. ***Syzygium samarangense*** (Blume) Merrill & Perry, J. Ara. Arb. 19: 115.1938. *Myrtus samarangense* Blume, Bijdr. 1084.1826. *Eugenia javanica* Lamk. Ency. 3: 200.1789; Duthie, Hook.f. Fl. Brit. India 2: 474.1878; Watt Dict. Econ. Prod. 6(3): 284.1893; Prain, Rec. bot. Surv. India 3(2): 212.1905. *(Vern. Name: Jamrul).*

Medium sized tree. Leaves sessile or shortly petioled, glabrous broadly or sometimes narrowly elliptic oblong. Flowers in lateral and terminal cymes. Fruit whitish, very much juicy, rarely reddish, turbinate. *Fls. & Frts.:* April-June.

It is generally planted in the orchards, occasionally self sown.

Field Collection: Nilgunj/Barrackpore, *K.N-2198.*

Distributions: Plains of India, native of Andamans & Nicobar Islands and Malaya. Widely distributed in the old and tropical world.

Economic Importance: Fruits are juicy and edible.

LVII. LECYTHIDACEAE

(This family splitted out from Myrtaceae)

1. *BARRINGTONIA* J.R. & Forsst. *nom. cons.*

Key to the Species

1a	Calyx slightly imbricate; fruit oblong, fusiform, 4-angled	1. *B. acutangula* (L.) Gaertn.
1b	Calyx valvate, fruit ovoid, slightly 4-angled	2. *B. racemosa* Roxb.

1.1. ***Barringtonia acutangula*** (L.) Gaertn. Fruct. 2: 9. t.101. 1780; Naskar & Guha Bakshi, Mang. Swamps Sundarbsns 129.1987. *(Vern. Name: Hijal).*

Moderate tree on the banks of the rivers in Sundarbans.

Field Collections: Mollakhali, *K.N-1846.* Naskar & Guha Bakshi (l.c.) described this plant in relation to its taxonomy, ecology and economic importances.

1.2. ***Barringtonia racemosa*** Roxb. Fl. Ind. 2: 634.1824. *(Vern. Name: Sumudra).*

Tall tree, frequently grow on the rivr banks of the Sagar Islands of the Sundarbans. Naskar & Guha Bakshi (Mangr. Swamps Sundarbans 129.1987) described this plant in relations to its toxonomy, ecology and economic Importances.

Field Collections: Basirhat, *K.N.-937.*

Remarks:. Careya arborea Roxb. Cor. Pl. 3: 14, t. 218.1819. *(Vern. Name: Kumb* or *Kumbi)* - a native of the warmer parts of India, also grow in this lower Ganga delta. The other plantation trees of the family are– *Couroupita guianensis* Aubl. *(Vern. Name: Nagalingam)* - a native of Tropical America: *Gustavia augusta* L. - a native of Tropical America. I have seen all these three taxa in the ornamental gardens and parks like in Rabindrasarabar, Tollygunge but I could not collect these taxa.

LVIII. MELASTOMETACEAE

1. *MEMECYLON* L.

1.1. ***Memecylon umbellatum*** Burm. Fl. Ind. 87.1768. *M. edule* Roxb. Cor. Pl. 1: 59.t. 82.1795: Clarke, Hook.f. Fl. Brit. India 2: 563.1879: Watt Dict. Econ. Prod. 5: 226.1892; Prain Beng. Plants 1: 361.1903 (Rep. ed. 1963). *(Vern. Name: Anjan).*

Small tree, young branches terete, pale when dry. Leaves 4-7.5 cm x 1.5-4 cm, elliptic or ovate. Flowers numerous in umbellate cymes from the axils of old leaves. Berry globose. *Fls. & Frts.:* January-April.

Very rare on the village thickets.

Field Collection: Falta, *K.N.-1995.*

Distributions: Throughout India; Bangladesh, Sri Lanka.

Economic Importances: Fruits are edible; leaves useful in leucorrhoea, gonorrhoea. Root decoction useful for excessive menstrual discharges.

LIX. LYTHRACEAE

Key to the Genera

1a	Herbs with many minute flowers, calyx membranous	(2)
1b	Trees or shrubs; flowers often large; calyx moderately thick	(3)
2a (1)	Flowers axillary, sessile, solitary or in spikes; capsules 2-4 valved	4. *ROTALA* L.
2b (1)	Flowers whorls; cymes or clusters, capsule circumsciss, bursting	1. *AMMANNIA* L.
3a (2)	Flowers 6-merous; stamens many; seeds winged	2. *LAGERSTROEMIA* L.
3b (2)	Flowers 4-merous; stamens 8; seeds pyramidal	3. *LAWSONIA* L.

1. AMMANNIA L.

Key to the Species

1a	Leaves tapering to the base, lanceolate, more or less distinctly petioled; capsule globose	1. *A. baccifera* L.
1b	Leaves rounded or sub-auricled at the base, lanceolate or elongated	(2)
2a (1)	Flowers in peduncled cymes; calyx striate with 8 vertical lines; capsule not covered by calyx-teeth, hardly exceeding calyx-tube	2. *A. multiflora* Roxb.
2b (1)	Flowers sessile in axillary cymes, clusters, many flowered; calyx not striate with vertical lines; capsule not covered by calyx teeth	3. *A. salicifolia* Monti

1.1. ***Ammannia baccifera*** L. Sp. Pl. 120. 1753; Naskar & Guha Bakshi, Mang. Swamps Sundarbsns 132. 1987 & Naskar, Aq. Semi. Plants 129. 1990. *(Vern. Name: Dadmari).*

Annual erect marsh herb in the Sundabans area. Naskar & Guha Bakshi (l.c.) described this taxon from the Sundarbans area.

Economic Importances: Leaves useful for raise blisters, in rheumatism pains and fever.

1.2. ***Ammannia multiflora*** Roxb. Fl. Ind. 1: 447. 1820; Naskar & Guha Bakshi, Mang. Swamps Sundarbans 133. 1987; Naskar, Aq. Semi. Plants 129.1990.

Erect branched marshy herb, common in the marshy areas. Naskar (l.c.) described this taxon, and collected from Amtala/Diamond area (K.N.-376).

1.3. ***Ammannia salicifolia*** Monti, Consm. Bonon 1: 112.1767; Naskar & Guha Bakshi Mang. Swamps Sundarbans 132.1987 & Naskar Aq. Semi. Plants 130.1990.

Erect annual herb, grow on the marshy places and river banks. Naskar & Guha Bakshi (l.c.) and Naskar (l.c.) described this taxon after collecting from Lakshmikantapur *(K.N.-1775).*

2. *LAGERSTROEMIA* L.

Key to the Species

1a	Erect shrub; leaves elliptic; flowers pink, white or purple in large panicle; capsule woody	1. *L. indica* L.
1b	Medium to large deciduous tree; leaves lanceolate: flowers violet in long panicle; fruits broadly ovoid capsule	2. *L. speciosa* (L.) Pers.

2.1. ***Lagerstroemia indica*** L. Nat. (ed. 10) 2: 1076.1759; Clarke, Hook.f. Fl. Brit. India 2: 575. 1879; Watt Dict. Econ. Prod. 4: 583.1890; Prain Beng. Plants 1: 367.1903 (Rep. ed. 1963). *(Vern. Name: Telinga China).*

Erect shrub. Leaves elliptic, sessile, acute, opposite. Flowers pink, white, purple or rosy in large panicles. Capsule woody in nature. *Fls. & Frts.:* June-August.

Commonly planted in the ornamental gardens.

Field Collection: Joynagar/Mazilpur, *K.N.-860.*

Distributions: Throughout India, planted in the ornamental gardens. Throughout the warmer parts of Asia, originally from China.

2.3. ***Lagerstroemia speciosa*** (L.) Pers. 2: 72.1806. *Munchausia speciosa* L. Muench. Hans. 1: 357.t. 2. 1770. *Lagerstroemia flos-reginae* Rtez. Obs. 5: 25.1788; Clarke, Hook.f. Fl. Brit. India 2: 577.1879; Watt Dict. Econ. Prod. 5: 582.1892; Prain Beng. Plants 1: 366. 1903 (Rep. ed. 1963). *(Vern. Name: Jarul).*

Medium to large deciduous tree. Leaves opposite lanceolate. Flower violet in long terminal panicles. Calyx tube ribbed, tomentose. Fruit broadly ovoid capsule. *Fls. & Frts.:* April-August.

Planted as avenue tree or in the ornamental gardens and parks.

Filed Collection: Gosaba, *K.N.-644.*

Distributions: Throughout India; native of China and Malaya.

Economic Importances: Beautiful ornamental tree. Leaves and dried fruits prepared like tea – is useful for diabetes in Philippines: leaves also has the purgative properties. Fruits and leaves yield tannin.

3. *LAWSONIA* L.

3.1. ***Lawsonia inermis*** L. Sp. Pl. 349.1753. *L. alba* Lamk. Encycl. 3: 106.1789; Clarke, Hook.f. Fl. Brit. India 2: 573.1879; Watt Dict. Econ.

Prod. 4: 597.1890; Prain Beng. Plants 1: 365.1903 (Rep. ed. 1963). *(Vern. Name: Mehudi* or *Mehedi).*

Spiny hedge shrub. Leaves opposite, elliptic. Flowers tetramerous, cream coloured, fragrant, in large terminal panicles. Capsule 4-celled, dehiscing irregularly. Seeds angular brown. *Fls. & Frts.:* June-September.

Extensively planted as a hedge plant and sometime as an escape.

Field Collection: Falta, *K.N.-1832.*

Distributions: Planted almost throughout India. East Africa, Australia, Egypt, Sudan, Persia, Madagascar and Pakisthan.

Economic Importances: The leaves are usd to dye the nails, skin and beard; various parts of the plant are used medicinally. Leaves are antibiotic; leaves poultice or paste or decoction is used for boils, burns, bruises and skin inflammations.

4. *ROTALA* L.

Key to the Species

1a	Capsule 3-4 valved	(2)
1b	Capsule 2-valved: seeds pink, narrowly oblong, angular	2. *R. indica* Koehne
2a (1)	Flowers small in axillary and solitary and seeds black	1. *R. densiflora* Koehne
2b (1)	Flowers pinkish, closely packed in terminal simple or panicled spikes: seeds elliptic, peltate	3. *R. rotundifolia* Koehne

4.1. ***Rotala densiflora*** (Roth ex Roem. & Schult.) Koehne, Bot. Jahrd. 1: 164.1880; Naskar, Aq. Semi. Plants 131.1990.

Spreading herb with divarcating branches. Naskar (l.c.) described this taxon after collecting from Bamanghata/Salt Lake area *(K.N.-1732).*

4.2. ***Rotala indica*** (Willd.) Koehne, Bot. Jahrd. 1: 172.1880; Naskar Aq. Sem. Plants 131.1990.

Decumbent herb. Naskar (l.c.) described this taxon and collected from Baruipur *(K.N.-16).*

4.3. ***Rotala rotundifolia*** (Buch.Ham.) Koehn. Bot. Jard. 1: 175.1881; Naskar Aq. Semi. Plants 132.1990.

Naskar (l.c.) described this taxon and collected from Herobhanga *(K.N.-186).*

LX. SONNERATIACEAE

(This family was previously treated under Lythraceae)

1. *SONNERATIA* Linn. f.

Key to the Species

1a Medium tree with conspicuous pneumatophores; foliages and branches are drooping or pendulous; leaves elliptic, narrow, 7.5-12.5 cm long and 4.0-5.0cm broad, gradually tapering towards the apex. Flowers upto 2cm diam.; calyx 4-6, not exceeding 2cm, initially cup-shaped but flat during maturity of the fruits; petals absent, ovary 5-8 celled, with expanded stigma. Fruits upto 2cm diam., round or globular, green smooth with style tip 1. *S. apetala* Buch. Ham.

1b Small to medium tree with pneumatophores in the intertidal mangrove forests or inner-estuaries; foliages and branches not much pendulous or drooping like *S. apetala*; leaves broad 5cm or more, 8-10cm long, not tapering towards the apex, apex round or emarginate. Flowers above 2cm diam. or more, solitary, terminal or axillary, calyx 6-8 lobed, exceeding 2.5cm, initially cup-shaped or tubular but flat latter, petals present (but absent in *S. griffithii*), ovary 2.5cm or more in diam., stigma capitat. Fruit 3-7cm diam., globose, berry, green leathery with long style tip (2)

2a (1) Plants of the inner estuary, twigs diffused branched, slender, occasionally slightly pendulous, leaves glabrous opposite, elliptic oblong or oval obovate with short

	petiole, mid-vein often red at base, inconspicuous without prominant veins; petals slender reddish, alternate with calyx, best seen in flower buds	2. *S. caseolaris* (L.) Engler
2b (1)	Plants of outer estuaries or shore lines or true estuaries; twigs not much diffused or not pendulous; leaves glabrous; opposite, obovate-sub-orbicular, petiole scarcely developed, mid-vein green throughout, veins conspicuous, petals absent, filaments white	3. *S.griffithii* Kurz.

1.1. ***Sonneratia apetala*** Buch. Ham. Symes Embassy Ava 3: 477.1800; Naskar & Guha Bakshi Mangr. Swamps Sundarbans 164.1987. *(Vern. Name: Keora)*. (Fig. 27).

Naskar & Guha Bakshi (l.c.) described this taxon in relation to its nomenclature, taxonomy and economic importance. Collections were made from Rangabelia/Sundarbans, *K.N.-1508*.

1.2. ***Sonneratia caseolaris*** (L.) Engler, Engler & Nachtr. 261.1897; Naskar & Guha Bakshi, Mangr. Swamps Sundarbans 166. 1987. *(Vern. Name: Chak-Keora)*. (Fig. 28).

Naskar & Guha Bakshi (l.c.) described this taxon in relation to its nomenclature, taxonomy and economic importance. Collections were made from Kulti estuary *(K.N.-1760)*.

1.3. ***Sonneratia griffithii*** Kurz. J. Asiat. Soc. Bengal 2: 56.1871; Naskar & Guha Bakshi Mangr. Swamps Sundarbans 71.1987. *(Vern. Name: Ora)*. (Fig. 29).

Naskar & Guha Bakshi (l.c.) mentioned this taxon from Sundarbans.

Discussion: Most of these species of *Sonneratia* Linn. f. are very occasional in the Sundarbans and other Indian mangrove ecosystems. In the Indian parts of the Sundarbans mangrove exosystems, *Sonneratia* spp. are estimated only 5%; the most dominant flora here *Avicennia* L.- 40%. The estimated total biomass of the *Sonneratia* spp. in the Indian Sundarbans is 220000 metric tones (Naskar, 1990); among them the *S. apetala* is the most dominant species of the Sundarbans in West Bengal. *S. caseolaris* is very much restricted in a minor estuary, Sakha Bidyadhari (Kulti estuary) and

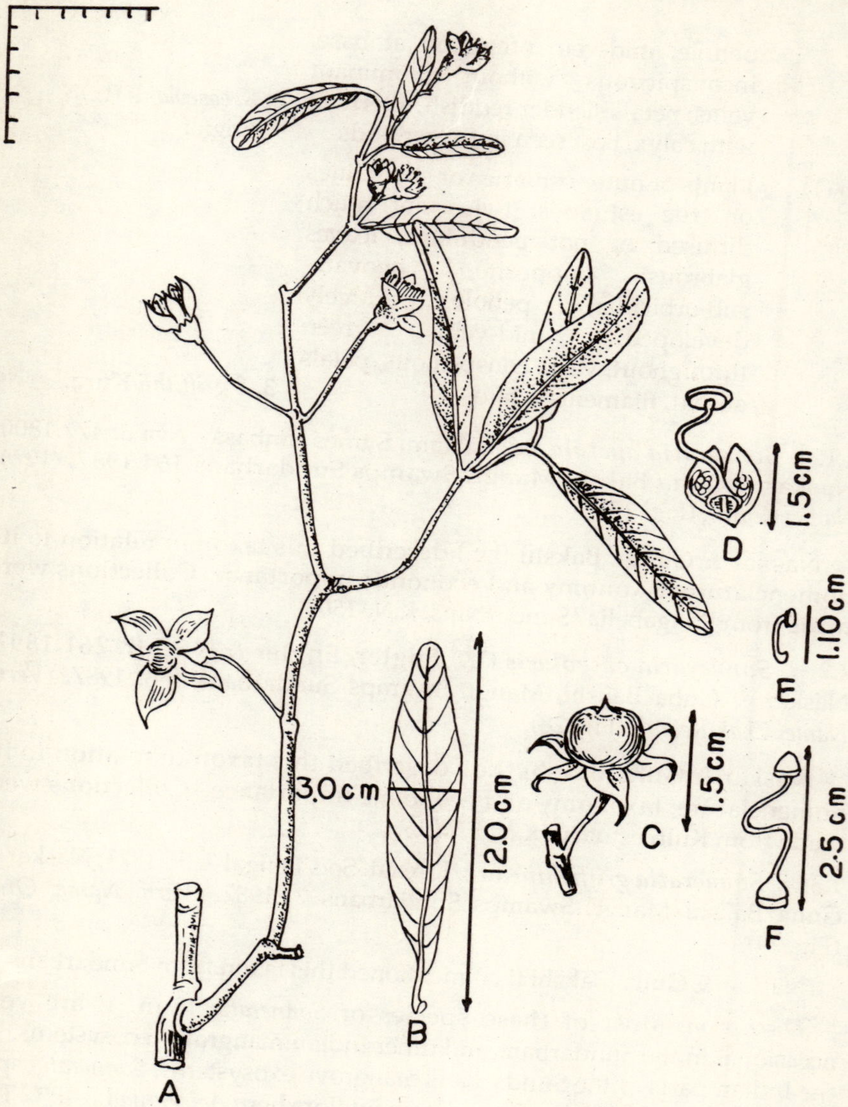

Fig.27. SONNERATIA APETALA Buch. Ham.
A. Flowering twig, B. Single flower, C. Fruit with persistent calyx, D. Section of gynoecium, E. Gynoecium, F. Androecium.

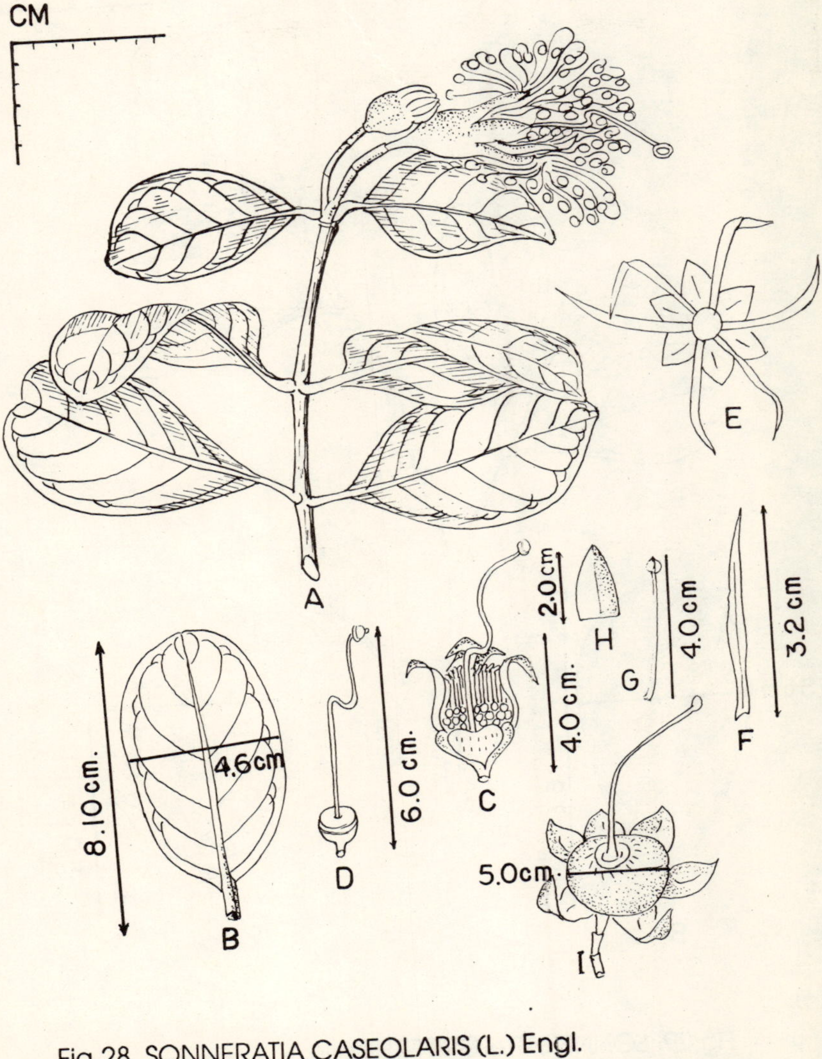

Fig.28. SONNERATIA CASEOLARIS (L.) Engl.
A. Flowering twig, B. Single leaf, C. T.S. of flower, D. Gynoecium, E. Top view of flower, F. Single petal, G. Androecium, H. Sepals, I. Single mature fruit.

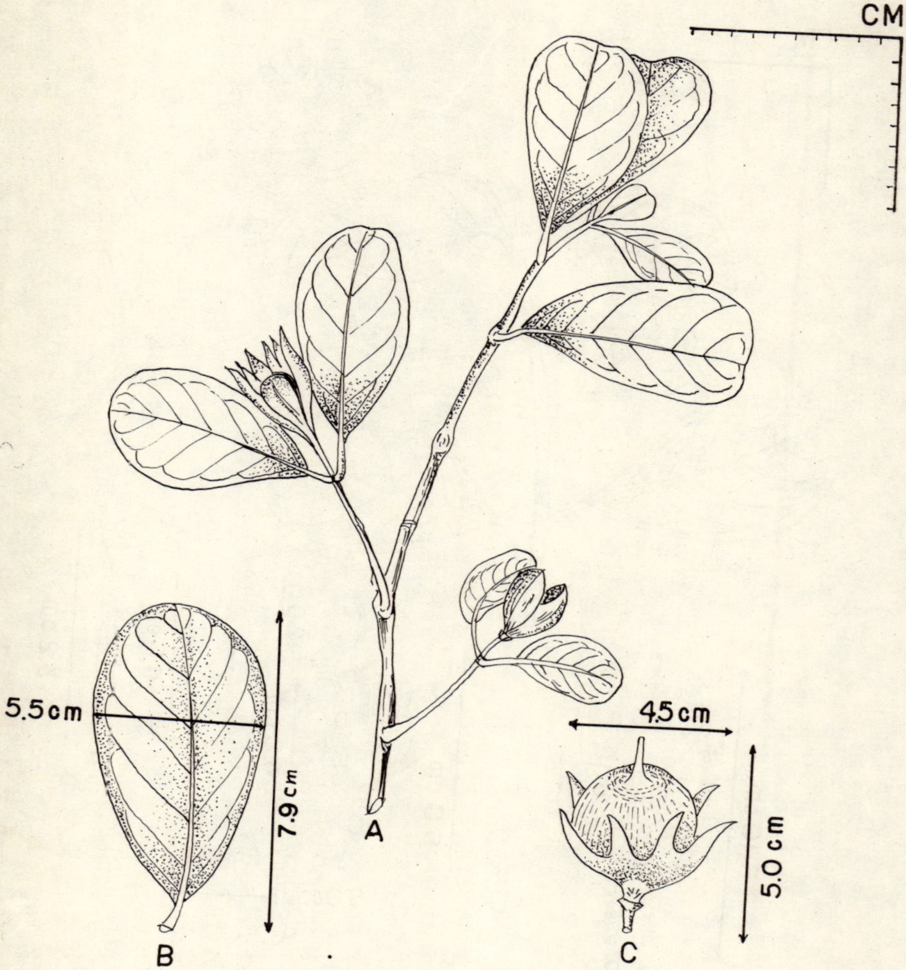

Fig.29. SONNERATIA GRIFFITHII Kurz.
A. Flowering twig, B. Single leaf, Single fruit.

also present in some mangrove reclaimed rice fields of the Sundarbans. The average leaf litter from the *Sonneratia* spp. is estimated 220 metric tons per annum, which are most important through biodegradation mechanism and enriching the soil and water of the mangrove ecosystems. *Sonneratia* spp. leaves are less thick and less cutacularised, in comparison to the other true mangrove species, as such these require less time for the biodegradation process. The aerial roots, pneumatophores of the *Sonneratia* spp. are also studied as very good substrata for holding different benthic and periphytic algal forms. These benthic and periphytic algal flora are the very encouraging and nutritious natural food for the estuarine fish and prawn species.

The speies of the *Sonneratia* Linn. f. are noted suitable for the natural growth, but in the changed ecological conditions and gradual degradation of the mangrove ecosystems, it require to undertake the artificial afforestation programmes along with the other mangrove flora.

LXI. PUNICACEAE

(This family is derived from Lythraceae)

1. *PUNICA* L.

1.1 ***Punica granatum*** L. Sp. Pl. 472.1753; Hook.f. Fl. Brit. India 2: 581.1879; Watt Dict. Econ. Prod. 6(1): 368. 1892; Prain Beng. Plants 1: 367.1903 (Rep. ed. 1963). *(Vern. Name: Dalim).*

Evergreen shrub or small tree. Leaves oblong or obovate, narrowed to the short petiole. Flowers bright red on terminal, solitary or in 3- flowered cymes. Sepals 5-7 persistent; petals crumpled: stamens many; ovary with superposed locules. Fruits a globose, reddish, crowded by the hardened persistent sepals, with leathery epicarp. Seeds numerous covered by juicy, acidic aril. *Fls. & Frts.:* January-October.

Rarely planted in the old orchards for its fruits and beautiful flowers; but it does not flourish well in this area.

Field Collection: Canning/Bhangankhali, *K.N. - 153.*

Distributions: Culitvated almost throughout India. Native of Iran and now naturalized in the Mediterranean area and South Asia.

Economic Importance: The acidic and juicy aril of the seeds are eaten on ripening. The bark and rind of the fruit are used for tannin.

LXII. ONAGRACEAE

1. *LUDWIGIA* L.

Key to the Species

1a	Stamens twice as many as the calyx-lobes	(2)
1b	Stamens as many as the calyx lobes	(3)
2a (1)	Plants creeping or floating; petals 5	1. *L. adscendens* (L.) Hara
2b (1)	Plants erect, petals 4	2. *L. octovalvis* sub sp. *sessiliflora* Raven
3a (1)	Capsules oblong, seeds in several rows in each cell	3. *L. perennis* L.
3b (1)	Capsules narrowly linear; seeds in single row in each cell	4. *L. prostrata* Roxb.

1.1. ***Ludwigia adscendens*** (L.) Hara, J. Jap. Bot. 28: 291. 1953; Naskar Aq. Semi. Plants 133.1990. *(Vern. Name: Kesara- dam).*

1.2. ***Ludwigia octovalvis*** sub. sp. ***sessiliflora*** (micheli) Raven, Kew Bull. 15: 476. 1962; Naskar, Aq. Semi. Plants 134. 1990. *(Vern. Name: Lal Banlanga).*

1.3. ***Ludwigia perennis*** L. Sp. Pl. 119.1753; Naskar Aq. Semi. Plants 135. 1990. *(Vern. Name: Banlanga).*

1.4. ***Ludwigia prostrata*** Roxb. Fl. Ind. 1: 420.1820; Naskar Aq. Semi. Plants 135. 1990.

All these four species of *Ludwigia* L. were described by Naskar (1.c.) in relations to their taxonomy, key notes and ecology and economic importances.

LXIII. TRAPACEAE

This family is often included under the Onagraceae; Raimann (1898) created a family Hydrocaryaceae for accommodating the genus *Trapa* L. Benthum & Hooker (1883), Rendle (1925) and Hutchinson (1959) included *Trapa* L. in the Onagraceae also. But Ram (1956) removed the taxon from the Onagraceae to Trapaceae on the basis of embryological studies; here the present author followed Ram (1956) and ranked the taxon under the distinct family—Trapaceae.

1. *TRAPA* L.

1.1. ***Trapa bispinosa*** Roxb. Pl. Cor. 3: 29, t. 234.1798; Naskar, Aq. Semi. Plants 135. 1990. *(Vern. Name: Pani Phal* or *Singara Phal).*

Naskar (1.c.) described this taxon in relation to its taxonomy, ecology and economic importances.

LXIV. SAMYDACEAE

1. *CASEARIA* Jacq.

1.1. ***Casearia elliptica*** Willd. Sp. Pl. 2: 628. 1800; *C. tomentosa* Roxb. Fl. Ind. 2: 421.1832; Hook.f. Fl. Brit. India 2: 593.1879; Watt Dict. Econ. Prod. 2: 209.1889; Prain Beng. Plants 1: 371.1903 (Rep. ed. 1963).

Small tree. Leaves shortly petioled, oblong or lanceolate from an oblique rounded or acute base, crenate, serrate or almost entire, sub-coriaceous, sparsely gland dotted. Flowers numerous, small, fascicled in the axis of the leaves, greenish yellow; stamens 8, alternating with hairs; scals like staminodes. Fruits 2-valved, broadly ellipsoid, fleshy, yellow when ripe; seeds embedded in red pulp. *Fls. & Frts.:* March-April.

Occasionally planted on the road-sides as avenue tree.

Field Collections: Shyamnagar, *K.N. - 2095.*

Distributions: Throughout India; Sri Lanka and Indo-Malaysia.

Economic Importances: Bark is used for adultering the red dye, 'Kamela' and also employed medicinally as an external application in dropsy.

Notes: Saldanha *et al.* (1976) treated this taxon under the Flacourtiaceae.

LXV. PASSIFLORACEAE

1. *PASSIFLORA* L.

Key to the Species

1a	Climbing shrub with foetid smell; leaves cordate; flowers large, petals present; fruit green	1. *P. foetida* L.
1b	Climbing or prostrate herb, without foetid smell; leaves polymorphic; flowers small, petals absent; fruits black	2. *P. suberosa* L.

1.1. ***Passiflora foetida*** L. Sp. Pl. 959.1753; Masters, Hook.f. Fl. Brit. India 2: 599.1879; Prain Beng. Plants 1: 373. 1903 (Rep. ed. 1963).

Slender climbing shrub with foetid smell. Leaves 4-7cm long, ciliate and denticulate with gland-tipped, setaceous hairs, similar hairs also beneath and simple hairs above; stipules cleft into gland tipped segments. Flowers greenish white mostly solitary, axillary, with an involucre of 3 pinnatisect bracteoles with capillary glandular segments; segments 5, on a gynophore. Fruit like a small green globose berry. *Fls. & Frts.:* August-January.

Frequently found, climbing on the bushes as a wild species.

Field Collections: Canning, *K.N. - 124.;* Amtala, *K.N. - 372.*

Distributions: Almost throughout India; introduced into India during the 19 Century. An American Species, in the Tropics.

1.2. ***Passiflora suberosa*** L. Sp. 958.1753; Masters, Hook.f. Fl. Brit. India 2: 599. 1879; Prain Beng. Plants 1: 373.1903 (Rep. ed. 1963). *(Vern. Name: Jhumko).*

Climbing or prostrate herb, glabrous or pubescent; tendril slender. Leaves membranous, polymorphous, entire to deeply 3 lobed, rarely subpeltate, apex acute or obtuse and apiculate, base rounded. Flowers solitary or few in the axils or rarely in axillary racemes: bracts minute; sepals ovate-lanceolate. Fruit 8-12cm diam., globose, black. Seeds compressed, acute at apex, base tapering. *Fls. & Frts.:* July-September.

In the hedges and thickets, throughout the lower Ganga delta.

Field Collection: Sashan/near Baruipur, *K.N. - 1175.*

Distributions: Sometime in the Eastern India. Native of Tropical America.

Notes: Its leaves shapes and sizes are largely variable.

Remarks: Another cultivated species of this genus is *Passiflora edulis* Sims.

LXVI. CARICACEAE

The family Caricaceae was treated under Passifloraceae by Benthum & Hooker system of Classification, but Hutchinson placed it under the order Cucurbitales and treated as separate distinct family.

1. *CARICA* L.

1.1. ***Carica papaya*** L. Sp. Pl. 1036. 1753; Masters, Hook.f. Fl. Brit. India 2: 599. 1879; Watt Dict. Econ. Prod. 2: 158.1889; Prain Beng. Plants 1: 374.1903 (Rep. ed. 1963). *(Vern. Name: Papaya).*

Rapid growing, branched or unbranched, herbaceous, tree like; with week, succulent trunk and milky sap. Leaves palmately lobed, lobes 7, each again divided into a number of lobes, nearly orbicular in outline, pale or glaucous beneath; petioles yellow; male flowers in long drooping panicles and female in short clusters. Fruits 1-celled; seeds many with an outer sappy and inner hard testa, albuminous.*Fls. & Frts.*: Almost throughout the year, mainly during May to August.

Extensively planted or cultivated, throughout Bengal.

Field Collection: Bamanghata/Tapsia, *K.N.-2301.*

Distributions: Papaya was not cultivated in India before the arrival of the Portuguese, who introduced this species in India in the 16th century. Now it is commonly planted almost throughout India. A native of Tropical America.

Economic Importances: Both green and ripe fruits are edible. The milky juice contain an excellent digestive action on meat and milk curdling ferment and it also regarded as vegetable pepsin. The juice have an extraordinary property of hastening the digestion of musclar (meat) fibres. Both papin and pepsin possess the power of converting the content of stamach into peptone. The ferment present in the milky juice possesses extraordinary energetic action upon nitrogenous substances. Milky juice of the fruits (unripe) is useful as a cosmetic to remove freckles and other blemishes from the skin; anthelmintic, particularly effective in the expulsion of lumbrici.

LXVII. CUCURBITACEAE

Key to the Genera

1a	Flowers mostly white and grow wild in the village thickets and hedges; fruits comparatively smaller in size	(2)
1b	Flowers mostly yellow, rarely white; commonly cultivated for its edible fruits, very frequently as garden escape; fruits generally larger in size	(5)
2a (1)	Tendrils 2-3 fid; fruits mostly round and beautifully ornamented or deep-red	12. *TRICHOSANTHES*

2b (1)	Tendrils mostly simple or rarely bi-fid; fruits mostly ovoid, not much; ornamented and greenish	(3)
3a (2)	Fruits with fine pointed spines throughout the body	9. *MOMORDICA* L.
3b (2)	Fruits with smooth surface throughout	(4)
4a (3)	Corolla campanulate; fruits with blunt ends	3. *COCCINIA* Wt. & Arn.
4b (3)	Corolla rotate, fruits with acute tips	7. *GYMNOPETALUM* Arnold
5a (1)	Fruit generally large to very large, round or ovoid; seeds flat, oval, whitish or yellowish, 1-2 cm long; 0.5cm broad with blunt ends and pointed tips	5. *CUCURBITA* L.
5b (1)	Fuit comparatively smaller than the above, normally ovoid or spherical	(6)
6a (5)	Anther cells floded together or sigmoidly curved	(7)
6b (5)	Anther cells straight or if curved not conduplicate nor sygmoid	(10)
7a (6)	Stamens inserted at the mouth of the calyx; filaments exserted, recurved; anther free	(8)
7b (6)	Stamens inserted below the mouth of the calyx; filaments hardly exerted, not recurved; anther coherent	(9)
8a (7)	Male flowers racemose: furits dry; endocarp fibrous	8. *LUFFA* Cav.
8b (7)	Male flowers solitary; furits soft	1. *BENINCASA* Savi.
9a (7)	Tendrils simple, connectives produced beyond anther cells	4. *CUCUMIS* L.
9b (7)	Tendrils bi- or tri-fid; connectives not produced	2. *CITRULLUS Sch. nom cons.*

10a (6)	Male and female pedicles alike, 1-flowered, clustered	(11)
10b (6)	Male flowers in corymbose, umbels or racemes; connectives not produced; tendril simple or bi-fid	11. *SOLENA* Lour.
11a (10)	Tendril simple	10. *MUKIA* Arnott.
11b (10)	Tendril bi-fid	6. *DIPLOCYCLOS* van Post & Kt.

1. *BENINCASA* Savi.

1.1. ***Benincasa hispida*** (Thunb.) Cogn. DC. Mon. Phan. 3: 513. 1888; Wlh. India 2: 173.1950. *Cucurbita hispida* Thunb., Fl. Jap. 322. 1784. *Benincasa cerifera* Savi., Bibl. Ital. 9: 158.1818; Clarke, Hook.f. Fl. Brit. India 2: 616. 1879; Prain Beng. Plants 1: 379.1903 (Rep. ed. 1963). *(Vern. Name: Chal-Kumra).*

Large trailing or climbing annual herb. Tendril 2-fid. Leaves 10-15cm diam., 5-7 lobed, hispid beneath, cordate, petiole without glands. Flowers large yellow, solitary. Fruits large 30-40cm long, with a waxy blooms: flesh white. *Fls. & Frts.:* August-January.

Commonly cultivated in the village gardens.

Field Collection: Titagarh, *K.N. - 2221.*

Distributions: Cultivated throughout India. A native of Japan and Java.

Economic Importances: Fruits are cooked as vegetable and various preparations by the confectioners.

2. *CITRULLUS* Schrad. *nom. cons.*

2.1. ***Citrullus vulgaris*** Schard. ex. Eckl., Zeyh. Enum. 2: 279. 1836; Clarke, Hook.f. Fl. Brit. India 2: 621.1879; Watt Dict. Econ. Prod. 2: 331. 1889; Prain Beng. Plants 1: 381.1903 (Rep. ed. 1963). *C. citrullus* L. Sp. Pl. 1010. 1753. *Colocynthis citrullus* (L.) O. Kunte, Rev. Gen. 1: 256. 1891. *(Vern. Name: Tarmuz).*

Annual, climbing or trailing herb. Stem angular. Tendril 2 fid. Leaves deeply divided or lobed. Flowers large, corolla yellow within; greenish outside and villous. Fruits large, sub-globose or ellipsoid, smooth, often with glaucous waxy coating; fleshy juicy, red or yellowish white; seeds usually margined, black. *Fls. & Frts:* March-July.

Cultivated on the sandy soils for its edible fruits: mainly in the rabi season.

Field Collection: Malancha/near Gobindapur, *K.N. - 2153.*

Distributions: Almost throughout India. In the Tropical countries; indigenous in Tropical Africa.

Economic Importance: Ripe fleshy/juicy fruits are edible, but low quality.

3. *COCCINIA* Wt. & Arn.

3.1.1. ***Coccinia grandia*** (L.) Voigt., Hort. Suburb. Calcut. 59.1845. *Bryonia grandis* L. Mant. Pl. 1: 126. 1767. *Coccinia indica* Wight & Arn. Prodr. 347. 1834. *Cephalandra indica* Naud. Ann. Sci. Nat. (ser. 5) 5: 116. 1866; Watt Dict. Econ. Prod. 2.: 252. 1889; Clarke, Hook.f. Fl. Brit. India 2: 621. 1879; Prain Beng. Plants 1: 381. 1903 (Rep. ed. 1963). *Coccinia cordifolia* Cogn. DC. Nonogr. Phan. 3: 529. 1881. *Bryonia cordifolia* L. Sp. Pl. 1012.1753. *(Vern. Name: Tela-Kucha).*

Climbing much branched shrubs, stem angular, glabrous. Tendrils simple, slender, striate. Leaves bluntly angled to palmately 5-lobed, with a cordate base, bright green and punctate above, paler with raised mark of the punctation beneath. Male-flower solitary; peduncle slender. Female peduncle shorter. Fruit fusiform ellipsoid, pulp-red. Seeds oblong obovoid, much compressed, greyish yellow. *Fls. & Frts.:* May-February.

Common on the hedges and thickets, everywhere.

Field Collection: Diamond Harbour, *K.N. - 391.*

Distributions: Throughout India, Sri Lanka, China, Malaya, Tropical Africa and Central America.

Economic Importances: Occasionally, fruits are cooked as vegetable; leaves juice reduce blood-sugar and urine sugar; useful in glycosuria.

4. *CUCUMIS* L.

1a	Leaves more or less harshly scarbrid; fruits not tuberculate	(2)
1b	Mostly cultivated, often as garden escape; leaves hispidulous and sometime also softly hairy but not scabrid; fruits almost cylindric, faintly muricate	3. *C. sativus* L.

2a (1)	Plants widely grow on the hedges; leaves deeply 5-lobed, hardly 5-angled; fruits faintly or decidedly 3-sided, markedly 10-striped, ellipsoid	1. *C. callosus* Cogn.
2b (1)	Commonly cultivated; leaves faintly 5-lobed fruits not at all 3-sided; variously spherical, ovoid	2. *C. melo* L.

4.1. ***Cucumis callosus*** (Rottlar) Cogn., Engl. Pflanzenr. 88: 129. 1924; Chakraborty, Rec. bot. Surv. India 17 (1): 100. 1959. *C. trigonum* Roxb. Fl. Ind. 3: 722.1832; Clarke, Hook.f. Fl. Brit. India 2: 619. 1879; Prain Beng. Plants 1: 380.1903 (Rep. ed. 1963).

Annual or perennial climber or creeper. Leaves more or less harshly scabrid; rather deeply 5-lobed; hardly 5-angled, never with soft hairs. Fruits faintly or decidedly 3-sides, usually markedly 10-striped, ellipsoid. *Fls. & Frts.:* November-March.

Common on the railway tracks, waste lands and roadsides.

Field Collection: Bamanpukur, *K.N.-1085.*

Distributions: Plains of India. Tropical countries of the World.

4.2. ***Cucumis melo*** L. Sp. Pl. 1011.1753; Clarke, Hook.f. Fl. Brit. India 2: 620.1879; Watt Dict. Econ. Prod. 2: 627.1889; Prain Beng. Plants 1: 381.1903 (Rep. ed. 1963); Chakraborty, Rec. bot. Surv. India 17: 102.1959. *(Vern. Name: Karbuz.).*

Annual creeping climber. Leaves faintly 5-lobed, softly hairy and scabrid. Flowers yellow in short peduncled; calyx densely hairy. Fruits variously spherical, ovoid, elongated or contorted. Seeds numerous oblong smooth. *Fls. & Frts.:* January-May.

Commonly cultivated, but often meet as a garden escape.

Field Collection: Sonarpur, *K.N.-350.*

Distributions: Throughout India; all the tropical countries.

Economic Importances: Fruits edible either cooked or raw.

4.3. ***Cucumis sativus*** L. Sp. Pl. 1: 1012.1753; Clarke, Hook.f. Fl. Brit. India 2: 620.1879; Watt Dict. Econ. Prod. 2: 632.1889; Prain Beng. Plants 1: 381.1903 (Rep. ed. 1963). *(Vern. Name: Sasa).*

Medium, annual creeper or climber; tendril simple. Leaves hispidulous and sometime also softly hairy but not scabrid; 5-angular or slightly

5-lobed, lobes acute. Flowers monoecious, yellow, short peduncled, axillary; the male fascicled and the female solitary. Fruit cylindric, faintly muricate. *Fls. & Frts.:* Almost throughout the year.

Commonly cultivated, but often found as an escape in the waste lands, roadsides and railway tracks.

Field Collection: Dhopagachhi, *K.N.-1227.*

Distributions: Widely cultivated throughout India; native of North India. Tropical and sub-tropical countries in the world.

Economic Importances: Edible as vegetable or as green fruit. Very helpful as it act as enzyme for digestion.

5. *CUCURBITA* L.

5.1. ***Cucurbita maxima*** Buch. Ham. Encycl. 2: 151.1786; Clarke, Hook.f.Fl. Brit. India 2: 622.1879; Watt Dict. Econ. Prod. 2: 638.1889; Prain Beng. Plants 1: 382.1903 (Rep. ed. 1963) *(Vern. Name: Mitha Kumra).*

Large, annual climber or creeper, hispid or hairy; tendril 2-3 cleft. Leaves 20-30 cm diam., not deeply lobed, leaf-stalks with hairs on both surface, not rigid and pungent. Flowers large, yellow; calyx lobes narrow-subulate. Fruits very large, fleshy berry. Seeds ovoid or oblong, compressed-*Fls. & Frts.:* Throughout the year.

On the roadside garbages; but commonly cultivated for its fruits.

Field Collection: Nilgunj/Barrackpore, *K.N.-2197.*

Distributions: Cultivated throughout India. In the most warmer regions of the World; native of Peru.

Economic Importance: Fruits cooked as vegetable.

6. *DIPLOCYCLOS* (Endl.) van Post & Ktze.

6.1. ***Diplocyclos palmatus*** (L.) Jeffrey, Kew Bull. 15: 352.1962. *Bryonia palmata* L. Sp. Pl. 1012.1753 (Excl. Syn.). *B. laciniosa* L. Sp. Pl. 1013.1753; Clarke, Hook.f. Fl. Brit. India 2: 622.1879; Watt Dict. Econ. Prod.1: 542.1889; Prains Beng. Plants 1: 383 (Rep. ed. 1963) & Rec. bot. Surv. India 3: 218.1905. *Bryonopsis laciniosa* Naud., Ann. Sci. nat. (Ser.4) 12: 141.1859. *(Vern. Name: Mala).*

Stem slender, much branched, glabrous. Leaves scabrous above and glabrous beneath, margin undulate, crenulate or denticulate; male flowers 3-6 in fascicles; female flowers fasciculate in solitary, ovary globose. Fruits

14 thick, spherical, yellowish green; 6-stripped. *Fls. & Frts.:* April-December.

Commonly found on the hedges and thickets.

Field Collections: Birati. *K.N.-3* (CAL).

Distributions: Major parts of India; Sri Lanka, Malaya, Australia and Africa.

7. *GYMNOPETALUM* Arn.

7.1. ***Gymnopetalum cochinchinense*** (Lour.) Kurz. J. As. Soc. Bengal 40: 57.1871: Clarke, Hook.f. Fl. Brit. India 2: 611.1879: Prain Beng. Plants 1: 377. 1903 (Rep. ed. 1963) & Rec. bot. Surv. India 3 (2): 216.1905; Chakraverty, Rec. bot. Surv. India 17 (1): 62. 1959. *Bryonia cochinchinensis* Lour., Fl. Cochin. 595.1790.

Climber, stem angular, sulcate, hirsute-scabrous. Tendril usually simple. Leaves membranous, ovate, angular or shortly lobed, often triangular, acute, margin undulate, or distantly denticulate, intermediate lobes larger and longer. Male flowers solitary on 3-6 cm long peduncle. Female peduncle 3-4 cm long. Fruits 3-5 cm long, ovoid or ovoid oblong, red, 10-ribbed. shortly villous. Seeds brown, marginate compressed. *Fls. & Frts.:* May-February.

Rarely found on the waste lands of the village shrubberies.

Field Collection: Narendrapur/near Kamalgazi, *K.N.-1985.*

Distributions: Throughout India; Malaya and South China.

8. *LUFFA* Cav.

8.1. ***Luffa cylindrica*** (L.) Roem. Fam. 2: 64.1846; Chakraverty, Rec.bot. Surv. India 17 (1) : 75.1959. *Momordica cylindrica* L. Sp. Pl. 1009.1753. *Luffa aegyptica* Mill. Dict. (ed. 8) 4: 500. 1785; Clarke, Hook.f. Fl. Brit. India 2: 614.1979; Watt Dict. Econ.Prod. 5: 94.1892; Prain Beng. Plants 1: 379. 1903 (Rep.ed.1963) & Rec. Bot. Surv. India 3 (2): 216. 1965. *L.pentandra* Roxb. Hort. Beng. 70. 1814. *(Vern. Name: Dhundal).*

Large climber; stem stout, 5 angled, twisted, glabrous; tendrils 3-fid. Leaves orbicular, reniform in outline, 10-20 cm long, often broader than long, palmately 5-lobed, glabrous. Male flowers in the axils 4-20 together, racemes usually crowded towards the ends; petals spreading yellow with green veins. Female flowers–such as male. Staminodes 5. Fruits 12.5-35 cm long, cylindric, blunt at the ends, marked with longitudinal lines. *Fls. & Frts.:* August-January.

Commonly cultivated, but sometimes grow wild on the village areas.

Field Collection: Subuddhipur/Baruipur, *K.N.-340.*

Distributions: Cultivated throughout India. In the Old World Tropics.

Economic Importances: Edible as vegetable fruits; seeds have some medicinal values. Fibrous network in the ripe fruits are employed as flesh-brush.

9. *MOMORDICA* L.

Key to the Species

1a	Flowers monoecious; bracts of male flowers about the middle or below the middle of the peduncle	1. *M. charantia* L.
1b	Flowers dioecious; bracts near top of male peduncle and embracing the flower	(2)
2a (1)	Flowers white; fruits muricated	2. *M. cochinchinensis* Spreng.
2b (1)	Flowers yellow; fruits densely covered with soft spines	3. *M. dioica* Roxb. ex Willd.

9.1. ***Momordica charantia*** L. Sp. Pl. 1009. 1753; Clarke, Hook.f. Fl. Brit.India 2: 616.1879; Watt Dict. Econ. Prod. 5: 256.1891; Prain Beng. Plants1: 380.1903 (Rep. ed. 1963). *(Vern. Name: Karala).*

Annual much branched, slender climbing, angled and grooved herb. Tendrils simple, slender, elongated, pubescent, leaves orbicular in outline, 5-10 cm diam., pubescent, deeply divided into 5-7 lobed. Flowers yellow, monoecious; male flowers solitary, peduncles 5-10 cm long, glabrous or pubescent. Ovary fusiform, muricate, stigma 3, bi-fid. Fruits greenish white when young and bright orange yellow at maturity, 5-15 cm long, 5.0-7.5 cm diam., pendulous, fusiform. Seeds many, flat, ovoid, red on ripening. *Fls. & Frts.:* May-August or throughout the year.

Commonly cultivated in the village gardens; occasionally as garden escape.

Field Collection: Kadampur, *K.N.-629.*

Distributions: Widely cultivated throughout India; Sri Lanka, Malaya, China and Tropical Africa.

Economic Importances: Fruits bitter in taste; but a common vegetable; also used medicinally for blood sugar.

9.2. ***Momordica cochinchinensis*** (Lour.) Spreng. Syst. Veg. 3: 14. 1826; Clarke, Hook.f. Fl. Brit. India 2: 618. 1879; Watt Dict. Econ. Prod. 5: 257.1891; Prain Beng. Plants 1: 380.1903 (Rep. ed. 1963); Wlh. Ind.6: 410.1962. *Muricia cochinchinensis* Lour. Fl. Cochinch. 2: 596. 1790. *(Vern. Name: Kakrol).*

Strong climbing shrub; roots tuberous, perennial; stem robust, angular, glabrous. Tendrils simple, stout, angled, glabrous. Leaves sub-orbicular in out line, cordate at the base, glabrous in both the surface. Male flowers-peduncle 1-flowered, 5-15 cm long, angularly furrrowed, more or less pubescent; corolla white, tinged with yellow. Female flowers-peduncle 2.5-5.0 cm long, with a small bract about the middle. Fruit 10-15 cm long, ovoid, pointed, red, fleshy, terete. *Fls. & Frts.:* June-January.

Occasionally cultivated, but very often grow as wild in orchard.

Field Collection: Habra/Ashokenagar, *K.N.-1433.*

Distributions: Throughout India, Malaya and Philippines.

Economic Importance: Green fruits are cooked as vegetable.

9.3. ***Momordica dioica*** Roxb. ex Willd. Sp. Pl. 4: 604.1805; Clarke, Hook.f. Fl. Brit. India 2: 617.1879; Watt Dict. Econ. Prod. 5: 258.1891; Prain Beng. Plants 1: 380.1903 (Rep. ed. 1963); Chakraverty, Rec. bot. Surv. India 17 (1): 91.1959; Wlh. India 6: 411. 1962. *(Vern. Name: Ban-Karala).*

Perennial, slender, much branched climber, furrowed, glabrous herb; tendril simple, elongated, striate, glabrous. Leaves membranous, deeply 3-5 lobed. Flowers dioecious, yellow; male flowers solitary peduncled, slender, angled; bract cucullate. Female flowers peduncled nearly as long as those of the male; ovary clothed with long soft papillae. Fruit ellipsoid 2.5-8.0 cm long, shortly beaked, densely echinate with soft spines. Seeds many broadly ellipsoid, slightly compressed. *Fls. & Frts.:* May-August.

Common throughout the lower Ganga delta, both as cultivated and as garden escape.

Field Collection: Taki/Sankchura, *K.N.-984.*

Distributions: Cultivated throughout India; Sri Lanka and Malaya.

Economic Importances: Both the plant and young fruits are cooked as vegetable.

10. *MUKIA* Arnott.

Formerly this genus was included under *Melothria* L. *(sensu lato)*; but now it is treated as a distinct genus, according to the recent classification proposed by Jeffrey, Kew Bull. 15: 345.1962.

10.1. ***Mukia maderaspatana*** (L.) Roem. Fam. Syn. Rep. 2: 47.1846. *Cucumis maderaspatanus* L. Sp. Pl. 1012.1753. *Bryonia scabrella* L. f. Suppl. 424.1781. *Mukia scabrella* Arn. J. Bot. 3: 276.1841; Clarke, Hook.f. Fl. Brit. India 2: 623.1879; Watt Dict. Econ. Prod. 5: 287.1891; Prain Beng. Plants 1: 382.1903 (Rep. ed. 1963). *Melothria maderaspatanus* (L.) Cogn. Dc. Mon. Phan. 3: 623. 1881; Chakraverty, Rec. bot. Surv. India 17 (1): 141.1959. *(Vern. Name: Bilasi).*

Monoecious, hispid or scabrid annual, climbing or prostrate; stem slender. Tendrils slender. Leaves dark-green above and pale green beneath, shallowly 3-7 lobed, deeply pedicelled or sessile. Female flowers solitary or 3-5 in fascicle; pedicels very short or absent. Fruit globose, when ripe orange red, smooth and shining. Seeds turgid oblong, grey. *Fls. & Frts.:* August-December.

Common in the waste lands and vacant places.

Field Collection: Herobhanga, *K.N.-1861.*

Distributions: Almost throughout India. Tropical Africa and Asia through Malaya to Tropical Australia.

Economic Importances: Vegetative parts, occasionally cooked as vegetable.

11. *SOLENA* Lour.

11.1. ***Solena amplexiculis*** (Lamk.) Gandhi, Saldhana & Nicolson, Fl. Hassan 179. 1976. *S. heterophylla* Lour. Fl. Coch. 1: 514.1790. *Bryonia umbellata* Klein, Willd. Sp. Pl. 4: 618. 1805. *Zehneria umbrellata* Thw. Enum. 125. 1858; Clarke,Hook.f. Fl. Brit. India 2: 625. 1879 (pro parte); Watt Dict. Econ. Prod. 6 (4): 355. 1893; Prain Beng Plants (Rep. ed. 1963). *Melothrica heterophylla* (Lour) Cong. DC. Monog. Phan. 3: 618.1881 *(Vern. Name: Kudri).*

Dioecious climber, root-stock perennial; stem slender, furrowed, glabrous. Tendrils simple, slender, glabrous. Leaves polymorphous, coriaceous, ovate, sub-orbicular, oblong or narrow-lanceolate, acuminate or acute at apex, deeply or shallowly emarginate at base. Male flowers few to many in sub-usual or racemes or facicle; peduncle slender, glabrous or sub-glabrous. Female flowers solitary; fruits oblong-ovoid, sub-angular, bright red when ripe, purpled. Seeds many, scarcely compressed. *Fls. & Frts.:* May-January.

Grows in the village thickets and hedges; occasionally cultivated.

Field Collection: Ramnagar/Jugdia, *K.N.-220.*

Distributions: Throughout India; Sri Lanka, Malaya, China.

12. *TRICHOSANTHES* L.

Key to the Species

1a	Leaves lobed or angled	(2)
1b	Leaves entire; male racemes with small bracts	3. *T. nervifolia* L.
2a (1)	Male racemes with large bracts; leaves palmately out into 5	1. *T. bracteata* Voight
2b (1)	Male racemes without bracts; leaves sub-reniform, cordate usually 5, sometimes 3-7 lobed or angled	2. *T. cucumerina* L.

12.1. ***Trichosanthes bracteata*** (Lamk.) Voight, Cat. Hort. Sub. Calc. 58.1845; Chakraverty, Rec. bot. Surv. India 17 (1): 44. 1959. *Modecca bracteata* Lamk. Encycl. 4: 210. 1796. *T. palmata* Roxb. Fl. Ind. 3: 704. 1832; Clarke, Hook.f. Fl. Brit. India 2: 696. 1879; Watt Dict. Econ. Prod. 6 (4): 84. 1839; Prain Beng. Plants 1: 377. 1903 (Rep. ed. 1963). *(Vern. Name: Makhal).*

Dioecious climber, stem robust, scabrous, paberulous or sub-glabrous, woody below. Tendrils 2-3 fid. Leaves scabrous, minutely punctate, sub-entire, deeply or shallowly 3-5 lobed with scattered dark-green. Male flowers usually in 5-12 flowered, axillary, solitary or rarely paired racemes; pedicels very short, stout, fimbriate on outer surface, near base a few dark green glands scatterded. Female flowers axillary, solitary or rarely in racemes. Fruits globose, white when green but ripened fruits are red with 10 longitudinal orange steaks, very hard, seeds many ovate oblong, smooth. *Fls. & Frts.:* August-March.

Occasionally found in the village thickets, climbing in trees.

Field Collection: Bamanpukur, *K.N.-1145.*

Distributions: Throughout India. Sumatra, Java, Borneo and Timor.

Economic Importances: Roots and fruits are used as a medicine for headache; and said to be useful in certain diseases of cattle.

12.2. ***Trichosanthes cucumerina*** L. Sp. Pl. 1008. 1753; Clarke, Hook.f. Fl. Brit. India 2: 609. 1879; Watt Dict. Econ. Prod. 6 (4): 82. 1893; Prain Beng. Plants. 1: 377.1903 (Rep. ed. 1963). *(Vern. Name: Ban Chichinga).*

Annual climber; stem slender, striate, hairy. Tendrils slender, 2-3 branched. Leaves sparcely hairy, orbicular-reniform, lobed deeply or shallowly into usually 5 or rarely 3 or 7, deeply cordate at base; lobes acute at apex, wavy or irregularly toothed. Male flower peduncles in pairs, female flowers axillary, solitary. Fruits ovate fusiform. Seeds ovate-oblong, compressed, margin undulate, truncate and slightly emerginate at apex, attenuate at base. *Fls. & Frts.:* July-December.

Common on the village scrub-jungles.

Field Collection: Basanti, *K.N.-1281.*

Distributions: Throughout India; Malaya to Tropical Australia.

12.3. ***Trichosanthes nervifolia*** L. Sp. Pl. 1008. 1753; Clarke, Hook.f. Fl. Brit. India 2: 609. 1879; Watt Dict. Econ. Prod. 6(4): 84.1893. *(Vern. Name: Ban-Chichinga).*

Perennial, stem woody below, much-branched slender, striate glabrous shrub; tendril bi-fid, glabrous. Leaves 5-10 cm X 2.5 -6.0cm, ovate oblong, acute, mucronate, margin denticulate, glabrous. Male flowers in axillary 4-10 flowered corymbose racemes; peduncles slender, solitary, on short peduncles. Fruits 4-8 cm long, ellipsoid, shortly beaked, tapering to both the ends. Seeds semi-ellipsoid, compressed, thickets at the margins *Fls.& Frts.:* June-October.

Occasionally found on the waste lands.

Field Collection: Taki/Sankchura, *K.N.-992.*

Distributions: Mostly in the South India; Sri Lanka.

Notes: Prain (1903) has not mentioned this taxon from Bengal.

Remarks: Other purely cultivated species of this family are as follows:

(1) *Cucurbita pepo* DC., *(Vern. Name: Kumra);*

(2) *Lagenaria siceraria* (Mol.) Standl., *(Vern. Name: Lau);*

(3) *Luffa acutangula* Roxb., *(Vern. Name: Jhinga);*

(4) *Trichosanthes anguina* L. *(Vern. Name: Chichinga);*

(5) *Trichosanthes dioica* Roxb. *(Vern. Name: Patal).*

LXVIII. CACTACEAE

1. *OPUNTIA* L.

1.1. ***Opuntia dillenii*** (Ker-Gawler) Haw. Supp. Succ. 79.1819; Naskar & Guha Bakshi, Mang. Swamps Sundarbans 100. 1987. *(Vern. Name: Nag Phana).*

Naskar & Guha Bakshi (l.c.) described this plants regarding its toxonomy, ecology, economic importances and the distribution patterns.

This is the only member of the Cactaceae in the lower Ganga delta and very common on the coastal areas, on the sandy sea shores.

LXIX. AIZOACEAE (nom. alt. Ficoidae)

Key to the Genera

1a	Halophytic herbs; stem reddish, shining and glabrous; flowers yellowish green; ovary and capsule 3-5 celled; capsule dry scarious	1. *SESUVIUM* L.
1b	Mesophytic herbs; stems tinged with purple shining; flowers pinkish or dull white; ovary and capsule 1-2 celled; capsule breaking transversely	2. *TRIANTHEMA* L.

1. *SESUVIUM* L.

1.1. ***Sesuvium portulacastrum*** L. Syst. (ed. 10) 1058.1753); Naskar and Guha Bakshi Mang. Swamps Sundarbans 81. 1987. *(Vern. Name: Gada Bani).*

Naskar & Guha Bakshi (l.c.) described this plant, which are most common in the Sundarbans mangrove forest areas as mangrove associates.

2. *TRIANTHEMA* L.

Key to the Species

1a	Prostrate glabrous or decumbent, much-branched herb; flowers solitary, peduncled or sessile, pinkish or dull white	1. *T. portulacastrum* L.

1b	Small, diffuse, much-branched herb; flowers sessile, several in an axil; fruit enclosed in the calyx tube	2. *T. triquetra* Rottl. & Willd.

2.1. ***Trianthema portulacastrum*** L. Sp. Pl. 233.1753; Naskar & Guha Bakshi Mang. Swamps Sundarbans 82. 1987. *(Vern. Name: Sabuni).*

2.2. ***Trianthema triquetra*** Rottl. & Willd., Ges Naturf. Neus. Schr. 4: 191.1803; Naskar & Guha Bakshi, Mang. Swamps Sundarbans 82. 1987.

Naskar & Guha Bakshi (l.c.) described both these taxa from the lower Ganga delta in relations to their taxonomy, ecology and economic uses.

LXX. MOI LUGINACEAE

(Splitted out from Azoiaceae)

Key to the Genera

1a	Inflorescences in axillary; petaloids, staminodes absent; stamens 12; stigma-3; capsule opening by 5 valves	1. *GLINUS* L.
1b	Inflorescence in compound terminal cymes; petaloid staminodes present, stamens-5; stigma 3; capsule opening by 3 valves	2. *MOLLUGO* L.

1. *GLINUS* L.

Key to the Species

1a	Plants densely stellate hairy; flowers sessile; seeds membranous scale	1. *G. lotoides* L.
1b	Plants glabrous; flowers pedicelled; seeds with a short, subulate bristles at hilium	2. *G. oppositifolius* DC.

1.1. ***Glinus lotoides*** L. Sp. Pl. 463.1753; Majumdar, Ind. Agr. 6: 110.1962. *Mollugo hirta* Thunb. Prodr. Pl. Cap. 24.1794; Clarke, Hook.f.Fl. Brit. India 2: 662.1879; Watt Dict. Econ. Prod. 5: 255. 1892; Prain Beng. Plants 1: 389.1903 (Rep. ed. 1963). *M. glinus* Rich. Tent. Fl. Abyss. 1: 48.1847. *M. lotoides* Ktz. Rev. Gen. 264.1891.

Densely hairy, prostrate annual herb, with deep root. Stem spreading from the centre. Leaves 1-3 cm long, opposite or falsely whorled, orbicular, obovate or rarely narrowly lanceolate, obtuse, usually downy on both surfaces. Flowers in axillary clusters, unequally pedicelled. Fruit oblong, seeds dark brown, many. *Fls. & Frts.:* February-June.

Everywhere in the waste places and dry fields associated with other weed flora.

Field Collection: Dewanganj, *K.N.-638.*

Distributions: Common throughout India; Bangladesh, Sri Lanka and the warmer regions of the World.

1.2. ***Glinus oppositifolia*** (L.) DC. Bull. Herb. Boiss. 2, 1: 559.1901. *Mollugo oppositifolia* L. Sp. Pl. 89. 1753.*M. spergula* L. Syst. (ed.10) 881.1759; Clarke. Hook.f. Fl. Brit.India 2: 662.1879; Prain Beng. Plants 1: 389.1903 (Rep. ed. 1963). *(Vern. Name: Gima Sak).*

Glabrous or slightly hairy annual herb. Stem branched, diffuse or prostrate, leafy with long internodes. Leaves 1.25-2.5 cm long, usually in whorls, spathulate, lanceolate or elliptic, apiculate. Flowers axillary. Capsule 3-celled, oblong, ovoid; seeds many, covered with raised tubercular points. *Fls. & Frts.:* May-August.

Commonly found in waste lands, open pastures and along the roadsides.

Field Collection: Madarhat/Baruipur, *K.N.-73.*

Distributions: Almost throughout India; Burma, Sri Lanka, Tropical Africa and Australia.

Economic Importances: Entire plant is edible as vegetable; said to contain valuable medicinal properties.

2. *MOLLUGU* L.

2.1. *Mollugo pentaphylla* L. Sp. Pl. 89.1753; Naskar, Aq. Semi-Aq. Plants 136.1990. *(Vern. Name: Khet Papra).*

Naskar (1.c.) described this taxon from the lower Ganga delta. The other species of this Genus is *Mollugo disticha* (L.) Ser. was also reported by Banerjee (Sci. & Cult. 35: 63.1969) , which was collected from Salt Lake on 3.3.1966, *Banerjee, 5466* (CAL).

LXXI. APIACEAE

(*nom. alt.* Umbelliferae)

Key to the Genera

1a	Umbels inflorescence simple type	2. *CENTELLA* L.
1b	Umbels inflorescence compound types	(2)

2a (1)	Flowers yellow	(3)
2b (1)	Flowers white or pinkish	(4)
3a (2)	Fruits winged	1. *ANETHUM* L.
3b (2)	Fruits not winged	6. *FOENICULUM* Mill.
4a (2)	Leaflets not with the linear or filiform segments	(5)
4b (2)	Leaflets not with the linear or filiform segments	(8)
5a (4)	Fruit glabrous	(6)
5b (4)	Fruit not glabrous	(7)
6a (5)	Fruit globose; bracts absent	3. *CORIANDRUM* L.
6b (5)	Fruit narrowly oblong; bracts linear	4. *CUMINUM* L.
7a (5)	Root tuberous; fruit elliptic	5. *DAUCUS* L.
7b (5)	Roots not tuberous, fruit ovoid	9. *TRACHYOSPERMUM* Link. *nom. cons.*
8a (4)	Plant erect	7. *OENANTHE* L.
8b (4)	Plant erect or diffused, hairy	8. *SESELI* L.

1. *ANETHUM* L.

1.1. ***Anethum graveolens*** L. Sp. Pl. 263.1753. *Peucedanum sowa* Kurz.; Prain Beng. Plants 1: 394.1963 (Rep. ed. 1963). *P. graveolens* Hiern. Oliver Fl. Trop. Africa 3: 19.1871; Benth. & Hook.f. Gen. Pl. 1: 919; Clarke, Hook.f. Fl. Brit. India 2: 700. 1879; Watt Dict. Econ. Prod. 6 (1): 181. 1893. *(Vern. Name: Salpha).*

Glabrous, branched, herb. Leaves 2-3 pinnate, finely dissected. Bracts and bracteoles absent. Flowers yellow in umbels. Fruit narrowly winged. *Fls. & Frts.:* January-March.

Cultivated for its foliages, which are used for flavouring curries. It is also found as an garden escape in the villages.

Field Collection: Basanti/Sundarbans, *K.N.-2272.*

*Distributions:*Cultivated almost throughout India. A cosmopolitan species.

Economic Importances: The leaves and yellow flowers are used for culinary and medicinal purposes.

2. *CENTELLA* L.

2.1. ***Centella asiatica*** (L.) Urban, Mart. Fl. Brass. 11: 287.t.78.1897; Naskar, Aq. & Semi-Aq. Plants 138. 1990. *(Vern. Name: Thankuni).*

Naskar (1.c.) described this taxon from the lower Ganga delta.

3. *CORIANDRUM* L.

3.1. ***Coriandrum sativum*** L. Sp. Pl. 256. 1753; Clarke, Hook.f. Fl. Brit. India 2: 717. 1879; Watt Dict. Econ. Prod. 2: 567.1889; Prain Beng. Plants 1: 395. 1903 (Rep. ed. 1963). *(Vern. Name: Dhane).*

Annual, slender branched, glabrous, strongly scented herb. Leaves pinnately decompound. Bracts absent, bracteoles linear about 2 cm long. Flowers white or pinkish in compound umbels; outer one zygomorphic. Fruits globose with the 4 secondary ridges on each coccus. *Fls. & Frts.:* December-April.

Winter annuals, widely cultivated in the village gardens.

*Field Collections:*Garia, *K.N.-320; Budge Budge, K.N.-750.*

Distributions: Cultivated almost throughout India. Most of the Tropical countries.

Economic Importances: The foliages and fruits are much used for flavouring curries.

4. *CUMINUM* L.

4.1. ***Cuminum cuminum*** L. Sp. Pl. 254. 1753: Clarke, Hook.f. Fl. Brit. India 2: 718.1879; Watt Dict. Econ. Prod. 2: 642. 1889. *(Vern. Name: Zira).*

Slender, glabrous, annual herb. Leaves dissected into filiform segments. Bracts and bracteoles with scarious membranous basal wings, filiform. Flower white, in compound umbels. Fruits cylindric, ridged, aromatic with pointed-ends. *Fls. & Frts.:* January-March.

Widely cultivated for its aromatic fruits,which are used as spices; very often found in the open fields as escape or wild.

Field Collection: Bamanghata, *K.N. - 2302.*

Distributions: Cultivated throughout India.

Economic Importance: Important spices for the common Indian curries.

5. *DAUCUS* L.

5.1. ***Daucus carota*** L. Sp. Pl. 442. 1753; Clarke, Hook.f. Fl. Brit. India 2: 718. 1879; Watt Dict. Econ. Prod. 3: 43.1890; Prain Beng. Plants 1: 395. 1903 (Rep. ed. 1963). *(Vern. Name: Gajar).*

Hispid, much branched, erect herbs. Leaves pinnately decompound, segments narrow-lanceolate or nearly linear and acute. Flowers white, numerous, in showy compound umbels. Bracteols many, 3-fid and simple. Fruit bristly along the secondary ridges. *Fls. & Frts.:* January-March.

Extensively grown in the gardens for its tuberous roots.

Field Collections: Titagarh/Bhanagar, *K.N.-2212.*

*Distributions:*Cultivated almost throughout India; Native of Europe.

Economic Importances: Conical roots are a common vegetable, eaten both cooked and raw.

Notes: The cultivated carrot is *D. carota* L. subsp. *sativus* (Hoffm.) Archangeli.

6. *FOENICULUM* Mill.

6.1. ***Foeniculum vulgare*** Mill., Gard. Dict. N.1.1768; Clarke, Hook.f. Fl. Brit. India 2: 695.1879; Watt Dict. Econ. Prod. 6(1): 405.1890; Prain Beng. Plants 1: 392.1903 (Rep. ed. 1963). *(Vern. Name: Pan-Muhuri).*

Glabrous annual herb, 30-50 cm high, with aromatic smell. Stems fistular in the lower region, but slender and solid in upper part. Leaves divided into filiform segments, 15-20 cm long. Flowers yellow, in compound umbels, stamens much longer than the petals. Bracts, bracteoles and calyx absent. Fruits oblong or lipsoid, ribbed. *Fls. & Frts.:* January-April.

Rarely cultivated in the gardens; occasionally as garden escape.

Field Collection: Sarberia/Sundarbans, *K.N.-784.*

*Distributions:*Cultivated throughout India. Native of Mediterranean region.

Economic Importances: Root has purgative properties.

7. *OENANTHE* L.

7.1. ***Oenanthe javanica*** (Bl.) DC. Prodr. 4: 132. 1830; Naskar Aq. & Semi-Aq. Plants 139. 1990. *(Vern. Name: Pan-Turasi).*

Naskar (1.c.) described this taxon from the lower Ganga Delta.

8. *SESELI* L.

8.1. ***Seseli diffusum*** (Roxb. ex Sm.) Sant. & Wagh., Bull. bot. Surv. India 5(2): 108.1963; Naskar, Aq. & Semi-Aq. Plants 139.1990. *(Vern. Name: Ban Jowan).*

Naskar (l.c.) described this taxon from the lower Ganga delta.

9. *TRACHYOSPERMUM* Link. *nom cons.*

9.1. ***Trachyospermum amnia*** (L.) Sprague, Kew Bull. 228. 1929. *Sison amnia* Sp. Pl. 252. 1753 (in part). *Ammi copticum* L. Mant. 1: 56.1767. *Carum conticum* Hiern., Oliver. Fl. Trop. Africa 3: 12. 1871; Clarke, Hook.f. Fl. Brit. India 2: 682.1879; Watt Dict. Econ. Prod. 2: 198.1890; Prain Beng. Plants 1: 392. 1903 (Rep. ed. 1963). *(Vern. Name: Jurani).*

Erect, annual, glabrous or minutely pubescent, herb. Leaves 2-3 pinnate, the ultimate segments very slender, linear or filiform. Flowers white, in compound umbels; calyx teeth obscure. Bracts and bracteoles linear, scarious at the base. Fruits ovoid, muricate, sub-bispid. *Fls. & Frts.:* November-March.

Cultivated winter annuals.

Field Collection: Kulpi, *K.N. - 1794.*

Distributions: Cultivated almost throughout India. North Africa and North Asia.

Economic Importances: Fruits are medicinally important.

LXXII. ALANGIACEAE

1. *ALANGIUM* Lamk. *nom.cons.*

1.1. ***Alangium salvifolium*** (Linn.f.) Wangerin., Engl. Pflanger, 41: 9. 1910 (Incl. spp. *hexapetalum* & spp. *decapetalum*); Naskar, Aq. & Semi-Aq. Plants 140. *(Vern. Name: Ankura).*

Economic Importances: Fruits are edible; barks and roots are medicinally used for jaundice and seeds are used for the cure of boils; root-bark is used as purgative.

Naskar (1.c.) described this taxon after collecting from lower Ganga Delta.

LXXIII. RUBIACEAE

Key to the Genera

1a Herbs or undershrubs (2)

1b	Shrubs or trees	(5)
2a (1)	Ovules solitary in each cell	(3)
2b (1)	Ovules many in each cell	(4)
3a (2)	Fruits indehiscent carpels	9. *KNOXIA* L.
3b (2)	Fruits of 2-separable mericarps	3. *BORRERIA* Mey
4a (2)	Corolla-lobes toothed; fruits indehiscent	5. *DENTELLA* Forsst
4b (2)	Corolla-lobes entire; fruits usually 2-valved capsule	14. *OLDENLANDIA* L.
5a (1)	Flowers in dense globose heads	(6)
5b (1)	Flowers not in globose heads	(10)
6a (5)	Ovules numerous in each cell	(7)
6b (5)	Ovules solitary in each cells; hypanthia confluent	11. *MORINDA* L.
7a (6)	Corolla-lobes imbricate: flowers ebracteolate	(8)
7b (6)	Corolla-lobes valvate, flowers bracteolate	(9)
8a (7)	Hypanthia confluent or connate	2. *ANTHOCEPHALUS* A. Rich.
8b (7)	Hypanthia not connate	13. *NAUCLEA* L.
9a (7)	Stigma clavate; calyx-tube 5-lobed	1. *ADINA* Salisb.
9b (7)	Stigma mitriform; calyx-tube truncate	10. *MITRAGYNA* Korth. *noms. cons.*
10a (5)	Calyx tube expanded in a coloured leaf	12. *MUSSAENDA* L.
10b (5)	Not as above	(11)
11a (10)	Ovule solitary in each cell	(12)
11b (10)	Ovule many in each cell	(15)
12a (11)	Fruit baccate or drupaceous	(13)
12b (11)	Fruit capsular; turning shrub, foetid when bruished; style-2	15. *PAEDERIA* L.

13a (12)	Flowers in axillary fascicles at the leafless branches	4. *COFFEA* L.
13b (12)	Flowers in terminal corymbose panicles	(14)
14a (13)	Style twice as long as tube	16. *PAVETTA* L.
14b (13)	Style not as above	8. *IXORA* L.
15a (11)	Inflorescence in a terminal cymes	7. *HAMELIA* Jacq.
15b (11)	Inflorescence not as above	(16)
16a (15)	Ovary 1-celled; placentae 2-5	6. *GARDENIA* Ellis *noms. cons.*
16b (15)	Ovary 2-celled	17. *XEROMPHIS* Rafin.

1. *ADINA* Salisb.

1.1. ***Adina cordifolia*** (Roxb.) Benth. & Hook.f. Gen. Pl. 2: 30.1873; Hook.f. Fl. Brit. India 3: 24. 1880; Watt Dict. Econ. Prod. 1: 114. 1879; Prain Beng. Plants 1: 403. 1903 (Rep. ed. 1963). *Nauclea cordifolia* Roxb. Pl. Cor. 1: 40 t. 53. *N. cordifolia* Willd., Roxb. Fl. Ind. 1: 514. 1820. *(Vern. Name: Kali Kadam* or *Haldu).*

Medium tree, wood hard. Leaves orbicular, stipules deltoid, large, caducous. Flowers many, yellow, in dense globular head, intermixed with paleaceous bracts. Calyx 5-lobed. Corolla-valvate in aestivation; funnel-shaped. Fruit a cluster of capsules each with 2-folicular cocci and many seeded. *Fls. & Frts.:* February- April.

Frequently found in the village shrubberies.

Field Collections: Shyamnagar, *K.N. - 2087.*

Distributions: Native of India and distributed throughout Sri Lanka and Burma.

Economic Importances: Wood is used for making toys. It provides good shade due to its horizontal branches.

2. *ANTHOCEPHALUS* A. Rich.

2.1. ***Anthocephalus chinensis*** (Lam.) Rich. ex Walp., Rep. 2: 491. 1843. *A. cadamba* (Roxb.) Miq. Fl. Ind. bot. 2: 135. 1856; Hook.f. Fl. Brit. India 3: 23.1880; Watt Dict. Econ. Prod. 1: 266. 1879: Prain Beng. Plants 1: 403.1903 (Rep. ed. 1963). *Nauclea cadamba* Roxb. Fl. Ind. 2: 12. 1824. *(Vern. Name: Kadam).*

Large avenue tree with spreading branches. Leaves coriaceous, elliptic oblong or ovate. Stipules, lanceolate caducous. Flowers small in dense and large globose heads, orange red; corolla yellowish. Fruits globose, solid with confluent ovaries. *Fls. & Frts.:* May-August.

Frequently planted as avenue tree or on the lawn.

Field Collection: Bamanpukuria/Minakhan. *K.N.-1074.*

Distributions: Throughout the hotter parts of India. Native of China and Malaya.

Economic Importances: The wood is used for packing and preparing boxes. Fruits are sometimes edible.

3. *BORRERIA* G.F.W. Mey. *nom. cons.*

Key to the Species

1a	Procumbent herb; leaves ovate-oblong or elliptic; flowers pale blue or pale violet	1. *B. articularis* Will.
1b	Usually erect; leaves linear to linear lanceolate; flowers white	2. *B. stricta* Schum.

3.1. ***Borreria articularis*** (Linn.f.) Williams, Bull. Herb. Boiss. (ser.2) 5: 956. 1905. *Spermacoce articularis* Linn. f. Suppl. 114.1781. *S. hispida* L. Sp. Pl. 102.1753; Hook.f. Fl. Brit. India 3: 200.1880; Watt Dict. Econ. Prod. 6(3): 320. 1893; Prain Beng. Plants 1: 425. 1903 (Rep. ed. 1963). *Borreria hispida* (L.) K. Schum., Engl. & Prantl. Nat. 4(4): 144.1891. *(Vern. Name: Madanghanti).*

Procumbent, scabrid or hispid, deep-rooted, much branched, annual herb. Leaves deccusate, ovate, ovate-oblong or elliptic-oblong. Flowers pale violet or pale blue, in nodal whorls, tetramerous. Capsule oblong, hairy; seeds oval dark brown. *Fls. & Frts.:* July- January.

Common weed of open waste lands and pastures.

Field Collection: Brace Bridge/Jingira Bazar, *K.N.-1220.*

Distributions: Almost throughout India. Old World Tropics.

3.2. ***Borreria stricta*** (Linn.f.) Schum., Pfam. 4(4): 143.1891. *Spermacoce stricta* Linn.f. Suppl. 120. 1781; Hook.f. Fl. Brit. India 3: 200.1880; Prain Beng. Plants 1: 425.1903 (Rep. ed. 1963).

Erect to sub-erect slender herb. Stems angled scabrid with recurved scabrous hairs. Leaves sub-sessile, linear to linear lanceolate, acute or acuminate. Flowers white, in dense axillary and terminal clusters.

Capsules hispid scabrous above, glabrous below, both mericarp dehiscing vertically, seeds ellipsoid, dark brown. *Fls. & Frts.:* September-December.

Very common in the open places on the dry sandy soil.

Field Collection: Herobhanga, *K.N. - 1862.*

Distributions: Throughout India. Sri Lanka, Tropical Asia and Africa.

Notes: D. Prain (1903) described this taxon from the upper parts of Bengal.

4. *COFFEA* L.

4.1. ***Coffea bengalensis*** Roxb. Fl. Ind. 1: 540 1832; Hook.f. Fl. Brit. India 3: 153.1880; Prain Beng. Plants 1: 419.1903 (Rep. ed. 1963).

Shrub with slender branches. Leaves opposite dull with broad stipules. Flowers axillary appear when the plant is leafless; corolla salver-shaped. Fruits small drupe. Seeds plano-convex, albumen horny. *Fls. & Frts.:* April-November.

Occasionally on the waste lands, roadsides or on railway tracks.

Field Collection: Budge Budge, *K.N. - 728.*

Distributions: Almost througout India. Burma, Siam and Java.

5. *DENTELLA* Forsst.

5.1. ***Dentella repens***(L.) Forsst. Char.Gen. 26,t.13.1776. *Oldenlandia repens* L. Mant. 1: 40.1767. *(Vern. Name: Bhumipat).*

Naskar, (Aq. & Semi-Aq. Plants 141.1990) described this taxon after collecting from the lower Ganga Delta.

6. *GARDENIA* Ellis *nom. cons.*

Key to the Species

1a	Leaves petiolate, 10-15cm; flowers peduncled, lobes longer than the tube	1. *G. jasminoides* Ellis.
1b	Leaves sessile 15-25cm; flowers peduncled, lobes shorter than the tube	2. *G. latifolia* Aiton.

6.1. ***Gardenia jasminoides*** Ellis., Phil. Trans. 51 (2): 935.1781; Wlh. India 4: 110.1956. *Gardenia florida* L. Sp. Pl. (ed-2.) 305.1762; Hook.f. Fl. Brit. India 3: 115. 1880; Watt Dict. Econ. Prod. 3: 480.1890; Prain Beng. Plants 1: 413.1903 (Rep. ed. 1963). *(Vern. Name: Gandharaj).*

Cultivated horticultural shrub. Leaves opposite, obovate-lanceolate. Flowers terminal solitary, medium, white, very fragrant. *Fls.* March-October.

Commonly planted in the ornamental gardens.

Field Collection: Gobardanga, *K.N. - 532.*

Distributions: Native of China and Japan. But now planted throughout India: introduced in India from China in the 17th Century.

6.2. ***Gardenia latifolia*** Aiton., Hort. Kew (ed.1) 1: 294. 1789: Hook.f. Fl. Brit. India 3: 116.1880; Watt Dict. Econ. Prod. 3: 482.1890; Prain Beng. Plants 1: 413. 1903 (Rep. ed. 1963).

Handsome, deciduous tree. Leaves opposite or trinate, broadly elliptic or orbicular or obovate when young. Flowers large, solitary, white, calyx lobes 7-9, pubescent, recurved, corolla- tube long, funnel shaped lobes 5-9. Fruits globose. *Fls. & Frts.:* May-June.

Occasionally on the village thickets.

Field Collections: Palta, *K.N. - 1831.*

Distribution: Dry hilly districts of the Western Central and South India, Bihar and W. Bengal. Bangladesh and Sri Lanka.

Economic Importance: Timber yielding plant.

7. *HAMELIA* Jacq.

7.1. ***Hamelia nates*** Jacq., Enum. Pl. Syst. Carib. 2: 15. 1760; Prain Beng. Plants 1: 412.1903 (Rep. ed. 1963) & Rec. bot. Surv. India 3(2): 222. 1905:

Exotic horticultural shrub, branches reddish, hairy, quadrangular. Leaves wide oblong, narrow at both ends, hairy membranous, opposite. Cylindrical, coloured flowers in terminal scropioid cymes. *Fls. & Frts.:* March-August.

Planted as hedge plants and also for the ornamental purposes.

Field Collection: Dhopagachhi, *K.N.-2145.*

Distributions: Almost throughout India; introduced into India in the early 18th Century. Native of Tropical America.

8. *IXORA* L.

Key to the Species

1a Plants mostly grow wild: flowers white (2)

1b	Plants mostly cultivated as ornamental garden shrubs: flowers bright-scarlet, pink or yellow	(3)
2a (1)	Leaves sub-sessile, rounded; cymes not jointed at the ramifications brachiate	1. *I. arborea* Roxb. ex Sm.
2b (1)	Leaves acuminate; cymes jointed at the ramifications, brachiate	4. *I. undulata* Roxb.
3a (1)	Corolla lobes acute	2. *I. coccinea* L.
3b (1)	Corolla lobes rounded	3. *I. stricta* Roxb.

8.1. ***Ixora arborea*** Roxb. ex. Sm. Rees. Cycl. 19: 5. 1811; Wlh. Ind. 5: 275.1959. *I. parviflora* Vahl Symb. 3: 11, t.52.1794; Hook.f. Fl. Brit. India 3: 142.1880; Watt Dict. Econ. Prod. 4: 533.1890; Prain Beng. Plants 1: 418.1903 (Rep. ed. 1963). *(Vern. Name: Rangan).*

Evergreen tree. Leaves coriaceous, petiole stout, stipules short, cups short. Flowers sub-sessile, white, odorous, sessile or pedicelled. Fruits depressed globose, shining black.*Fls. & Frts.* March-September.

Planted as ornamental hedge plants, but most common as wild.

Field Collection: Dakshin Chatra/Maslandapur, *K.N. - 524.*

Distributions: Throughout the hilly districts of India. Bangladesh and Sri Lanka.

8.2. ***Ixora coccinea*** L. Sp. Pl. 110.1753; Hook.f. Fl. Brit. India 3: 145. 1880; Watt Dict. Econ. Prod. 4: 533. 1890; Prain Beng. Plants 1: 418. 1903 (Rep. ed. 1963); Wlh. Ind. 5: 276.1959. *I. bandhuca* Roxb. Fl. Ind. 1. 376. 1820. *(Vern. Name: Rangan).*

Glabrous shrub, 0.66-1m high. Leaves coriaceous, pale at maturity, sessile, oblong, obtuse, apiculate, base rounded or subcordate; main nerves 8-12 pairs; slender, stipules with a long rigid cuspidate point. Flowers numerous bright scarlets; pedicels very short, glabrous or suberulous; bracts and bracteoles small. Fruits globose. Fls. & Frts.: Almost throughout the year.

Cultivated in the ornamental gardens edges and as hedge plant.

Field Collection: Gosaba, *K.N.-175.*

Distributions: Throughout India. Native of the W. Peninsula. Sri Lanka and throughout the tropics.

8.3. ***Ixora stricta*** Roxb. Fl. Ind. 1: 379.1820; Hook.f. Fl. Brit. India 3: 145. 1880; Prain Beng. Plants 1: 418.1903 (Rep. ed. 1963).

Shrub, glabrous. Leaves subsessile, obovate or obovate-oblong, obtuse. Flowers vary from scarlet and pink to white and yellowish. *Fls. & Frts.:* July-october.

Planted in the ornamental gardens.

Field Collection: Dhosarhat, *K.N. - 618.*

Distributions: Different parts of India, mostly cultivated. Native of the Moluccas and China.

Economic Importances: Important garden/ornamental plants.

8.4. ***Ixora undulata*** Roxb. Fl. Ind. 1: 385.1820; Hook.f. Fl. Brit. India 3: 147.1880; Prain Beng. Plants 1: 418.1903 (Rep. ed. 1963). *(Vern. Name: Palaka Jui).*

Shrub, 2-4m high. Leaves opposite glabrous, shining above, elliptic-oblong, lanceolate or oblanceolate, apex acute or acuminate, base rounded or obtuse, magin mostly undulate. Inflorescences 10-20 cm long and broad peduncled. Flowers small, bracts and bracteoles subulate. Fruits succulent, dull purple, globose. *Fls. & Frts.:* March-August.

In the village shrubberies and the village surroundings.

Field Collection: Joygram/near Minakhan. *K.N.-305.*

Distributions: Eastern Tropical Himalaya. Nepal, Bhutan and Sikkim.

Remarks: Other purely cultivated but less common species of *Ixora* L. are as follows:-

I. barbata Roxb. Fl. Ind. 1: 384.1820. Small shrub with white flowers.

I. brunnescens Kurz, J. As. Soc. 2: 317.1872. Glabrous shrub, white flowers.

I. subsessilis Wall. Cat. 6139.1828-1849. Shrub or small trees, flowers rosy.

9. *KNOXIA* L.

9.1. ***Knoxia sumatrensis*** (Retz.) DC. Prodr. 4: 569.1830. *Spermacoce sumatrensis* Retz. Obs. Bot. 4: 23.1786. *K. corymbosa* Willd. Sp.Pl.1: 582.1797; Hook.f. Fl. Brit. India 3: 128.1880: Prain Beng. Plants 1: 424.1903 (Rep. ed. 1963).

Erect annual 0.3-1.5 m high shrub; stem sparingly branched, somewhat stout with long internodes, often obtusely 4 angled, softly and more or less densely hairy. Leaves lanceolate, sessile or shortly petioled, acute, more or less hairy on both the sides, tapering much at the base into a short petiole, main nerves 8-12 pairs. Flowers small numerous, on the side of slender pubescent branches of large lax corymbose cymes; pedicels very short. Fruits sessile or shortly pedicelled, ellipsoid, 4 angled. *Fls. & Frts.:* August-November.

Occasionally on the village thickets.

Field Collection: Bamanghata/Salt Lake, *K.N. - 1733.*

Distributions: Throughout the Tropical India. Sri Lanka, Malaya, Archipelago and Tropical Australia.

10. *MITRAGYNA* Korth. nom. cons.

10.1. ***Mitragyna parviflora*** (Roxb.) Korth., Obs. Naucl. Ind. 10.1839. *Nauclea parviflora* Roxb. Pl. Cor. 1: 40.t.52.1795. *Stephegyne parviflora* (Korth.) Verh. Nat. Gen. Bot.161. 1840; Hook.f. Fl. Brit. India 3: 25.1880; Watt Dict. Econ. Prod. 6(3): 360. 1893; Prain Beng. Plants 1: 404.1903 (Rep. ed. 1963). *(Vern. Name: Kadam).*

Large deciduous tree, glabrous or pubescent. Leaves variable in shape and size, elliptic, suborbicular or obovate, rounded acute or bluntly acuminate. Flowers greenish-yellow, fragrant; peduncles solitary, terminal, short. Fruits capsules, oblong with blunt rounded tops and 10 blunt ribs. *Fls. & Frts.:* June-October.

Occasionally planted in the parks and roadsides as avenue tree.

Field Collection: Duttapukur, *K.N.-1709.*

Distributions: Throughout the drier parts of India and Sri Lanka.

Economic Importances: Produce fire wood and also as ornamental.

11. *MORINDA* L.

11.1. ***Morinda citrifolia*** L. Sp. Pl. 176.1753; Hook.f. Fl. Brit. India 3: 155.1880; Watt Dict. Econ. Prod. 5: 262.1891; Prain Beng. Plants 1: 420.1903 (Rep. ed. 1963). *(Vern. Name: Ach).*

Shrub or small tree. Leaves glabrous, shining, short-petioled. Flowers white in dense ovoid heads; peduncles solitary, usually leaf-opposed; corolla infundibuliform; stamens-5, filaments hairy. Fruit white when ripe, smooth. *Fls. & Frts.:* April- October.

Occasionally planted but also found as an escape in the villages.

Field Collection: Kestopur/near Baguihati, *K.N.-1894.*

Distributions: Throughout the hotter parts of India. Indo-malayasia.

Economic Importance: Ornamental plants.

12. *MUSSAENDA* L.

12.1. ***Mussaenda glabrata*** (Hook.f.) Hutch. ex Gamble Fl. Madr. 610.1921. *M. frondosa* var. *glabrata* Hook.f. Fl. Brit. India 3: 90.1880. *M. frondosa* L. Sp. Pl. 177.1753, *pro parte*: *M. corymbosa* Roxb. Fl. Ind. 1: 556.1820.

Shrub, leaves broadly elliptic, shortly acuminate, more or less pubescent. Flowers golden yellow to orange in terminal cymes; calyx-hairy. Berries sub-globose or obovoid. *Fls. & Frts.:* July- January.

Frequently planted in the gardens and orchards.

Field Collection: Basanti/public gardens, *K.N. - 2274.*

Distributions: Tropical Himalaya, Assam, Khasia Hills and Western Peninsula. Sri Lanka and Malaya Islands.

Economic Importances: Ornamental garden plants.

Remarks: Other most ornamental species is *M. philippica* A Rich., Mem. Soc. Hist. Nat. Par. 5: 248.1834 (*Field Collection* is made from Barrackpore, *K.N. - 632).*

13. *NAUCLEA* L.

13.1. ***Nauclea orientalis*** L. Sp. Pl. (ed.2) 243.1962; Benthall Trees Calc. 275.1949; Wlh. Ind. 7: 6.1966. *Sarcocephalus cordatus* Miq. Fl. Ind. Bat. 2: 133; Hook.f. Fl. Brit. India 3: 22. 1880; Watt Dict. Econ. Prod. 6(2): 467.1893. *Nauclea cordata* Roxb. Hort. Beng. 14. 1814 & Fl. Ind. 1: 509.1820.

Small tree with bushy crown; bark smooth greyish. Leaves large broadly ovate or cordate; flowers in globose heads. Small, yellow or orange, fragrant. Fruits globose, fleshy. Seeds small, albuminous. *Fls. & Frts.:* March-October.

Commonly planted as ornamental plants.

Field Collection: Basanti, *K.N.-2277.*

*Distributions:*Assam and Eastern Himalaya. Native of Malaya, Burma and Sri Lanka.

Economic Importances: Leaves are applied to boils, the bark is used as fish poison; fruits are edible.

14. *OLDENLANDIA* L.

Key to the Species

1a	Flowers in terminal or in upper axillary simple or paniculate cymes	1. *O. biflora* L.
1b	Flowers usually axillary, solitary or in 2-5 flowered corymbs	(2)
2a (1)	Leaves elliptic, about 2 times as long as broad	5.*O. pumila* Roxb.
2b (1)	Leaves usually linear, 4-10 times as long as broad	(3)
3a (2)	Calyx-teeth considerably shorter than the corolla tube	4. *O. herbacea* (Linn.f.) DC.
3b (2)	Calyx teeth not much shorter than the corolla tube	(4)
4a (3)	Flowers always solitary, axillary, subsessile or on peduncles shorter than the flower	3. *O. diffusa* (Willd.) Roxb.
4b (3)	Flowers usually in corymbs, rarely solitary on peduncles, longer than the flowers	2. *O. corymbosa* L.

14.1. ***Oldenlandia biflora*** L. Sp. Pl.119.1753; Naskar Aq. & Semi-Aq. Plants 143.1990. *(Vern. Name: Gandha Bhaduli).* Slender annual herb.

14.2. ***Oldenlandia corymbosa*** L. Sp. Pl. 119.1753; Naskar Aq. & Semi-Aq. Plants 143.1990. *(Vern. Name: Khet Papra).* Glabrous prostrate herb.

14.3. ***Oldenlandia diffusa*** (Willd.) Roxb. Fl. Ind. 1: 444.1820; Naskar Aq. & Semi-Aq. Plants 143.1990. Diffuse glabrous annual herb.

14.4. ***Oldenlandia herbacea*** Roxb. Fl. Ind. 1: 424.1820; Naskar Aq. & Semi-Aq. Plants 145.1990. Erect glarbous annual or biannual herb.

14.5. ***Oldenlandia pumila*** (Linn.f.) DC. Prodr. 4: 425.1830; Naskar Aq. Semi-Aq. Plants 145.1990. *(Vern. Name: Panki).* Annual herb.

All these species of the *Oldenlandia* L. were collected by Naskar and described (l.c.) in respect of their taxonomy, key notes and utilities. These collections were made from the lower Ganga delta regions.

15. *PAEDERIA* L.

15.1. ***Paederia foetida*** L. Mant. 1: 5.1767; Hook.f. Fl. Brit. India 3: 195.1880; Watt Dict. Econ. Prod. 6(1): 2.1892; Prain Beng. Plants 1: 423.1903 (Rep. ed. 1963). *(Vern. Name: Ghanda Bhadulia).*

Slender twining shrub with foetid smell. Flowers in panicles. Styles 2, twisted, papilose. Fruit dorsally compressed. *Fls. & Frts.:* July-October.

Hedges and thickets in general.

Field Collection: Namkhana, *K.N.-1920.*

Distributions: Throughout India. Native of Indo-Malaysia.

Economic Importances: Leaves posses tonic and astringent properties; used in soups and other food preparations for invalids and convalescents, particularly those suffering from bowel complaints.

16. *PAVETTA* L.

16.1. ***Pavetta indica*** L. Sp. Pl. 110.1753; Hook.f. Fl. Brit. India 3: 150.1880; Watt Dict. Econ. Prod. 6(1): 114.1892; Prain Beng. Plants 1: 417. 1903 (Rep. ed. 1963); Wlh. India 7: 282.1966. *(Vern. Name: Kukur-Chura* or *Jui.).*

Stout bushy shrub, 0.5-1.5 m high; bark thin, smooth, yellowish; young branches terete, glabrous. Leaves membranous, glabrous beneath, variable in shape and size, elliptic oblong or elliptic lanceolate. Flowers white odorous, in terminal sessile corymbose pubescent cymes. Fruit globose, black, smooth. *Fls. & Frts.:* March-October.

Occasionally on the village thickets and shrubberies.

Field Collection: Canning, *K.N.-172.*

Distributions: Throughout India. Sri Lanka, Malaya Archipelago, South China and N. Australia.

Economic Importances: Ripe fruits sweetish and is eaten either raw or pickled; flowers also edible: roots are said to possess purgative, aperient, diuretic and tonic properties and are prescribed in viceral obstructions, jaundice, headache, urinary diseases and dropsical affections.

17. *XEROMPHIS* Rafin.

Key to the Species

1a Spines axillary; leaves less than 8 cm long; calyx lobes persistent ... 1. *X. spinosa* Keay.